Manual of Water Supply Practices

M80

Biological Drinking Water Treatment

American Water Works Association

Manual of Water Supply Practices—M80

Biological Drinking Water Treatment

Sr. Editorial Manager, Books: Suzanne Snyder
Manager – Manuals: Molly Beach
Manager – Publishing Operations: Gillian Wink
Specialist – Copyright and Permissions: Peggy Tyler
Director of Publishing: John Fedor
Technical Editor: Wiley
Cover Composition: Michael Labruyere
Cover Image: Galkin Grigory/Shutterstock.com
Production: Innodata

Library of Congress Cataloging-in-Publication Data

Names: American Water Works Association, editor.
Title: M80 - biological drinking water treatment / [edited] by American Water Works Association.
Description: First edition. | Denver, CO : American Water Works Association, [2025] | Includes bibliographical references and index. |
Identifiers: LCCN 2024030791 (print) | LCCN 2024030792 (ebook) | ISBN 9781647171957 (paperback) | ISBN 9781613007204 (pdf)
Subjects: LCSH: Water--Purification--Biological treatment.
Classification: LCC TD475 .M28 2025 (print) | LCC TD475 (ebook) | DDC 628.1/62--dc23/eng/20241023
LC record available at https://lccn.loc.gov/2024030791
LC ebook record available at https://lccn.loc.gov/2024030792

ISBN 978-1-64717-195-7 ISBN, electronic 978-1-61300-720-4

http://doi.org/10.12999/AWWA.M80ed1

American Water Works Association

American Water Works Association
6666 West Quincy Avenue
Denver, CO 80235-3098
awwa.org

Contents

Figures

This page intentionally blank.

Tables

Preface

Biological treatment processes are a sustainable method for removing an array of both inorganic and organic contaminants from water. In biological treatment processes designed for water treatment facilities, microorganisms found in natural water systems grow on engineered surfaces or media (i.e., form a biofilm) and biologically transform or degrade compounds in the process influent. In many cases, conventional filters designed for particle removal are being operated in a way that achieves both particle removal and biological treatment (referred to as biological filtration or biofiltration).

Biological treatment of drinking water has been used since the early 1800s through operation of carbon filters and slow sand filtration without an oxidant residual. As the industry gained a greater understanding of the role of biological instability in managing distribution system water quality, significant work began in the 1980s to optimize biological treatment design and operation. Since that time, the application of biological treatment has expanded with the promulgation of new regulations, such as the Stage 1 and 2 Disinfectants and Disinfection Byproducts Rule in the United States, greater industry acceptance of biological treatment in drinking water with the publication of additional guidance and case studies, and the need for more robust treatment processes to treat both emerging contaminants and alternative water supplies.

Biological treatment processes may occur under either aerobic or anoxic conditions. In aerobic biological treatment, the microorganisms require oxygen to create energy and oxidize contaminants, such as natural organic carbon, dissolved iron and manganese, and ammonia. In contrast, anoxic biological treatment involves microorganisms that mediate electron transfer from an electron donor (e.g., acetic acid) to an electron acceptor (e.g., nitrate) and use the energy generated during the process for growth. Anoxic treatment is used to treat contaminants, such as nitrate and perchlorate.

In the last couple of decades, many project reports, guidebooks, and manuals have been published on individual applications of biological treatment, such as biofiltration. This first edition of AWWA Manual M80, *Biological Drinking Water Treatment,* is the first guidance document covering all common biological treatment processes in practice today, as shown in P-1. This manual begins with an introductory chapter presenting an overview of biological treatment applications, processes, history, and key terminology. The subsequent chapters each focus on one of the four common applications of biological treatment (P-1).

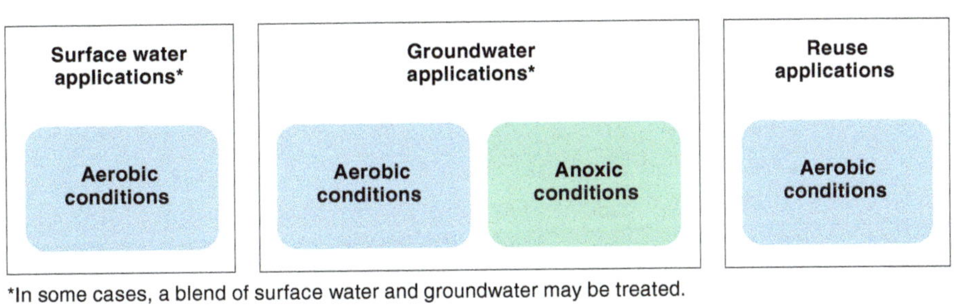

P-1 Four common applications of biological treatment

This page intentionally blank.

Acknowledgments

The AWWA Technical & Educational Council, the Water Quality & Technology Division (WQTD), and the Biological Drinking Water Treatment Committee gratefully acknowledge the contributions made by those volunteers who drafted, edited, and provided the significant and critical commentary essential to creating M80. The M80 Steering Committee members, chapter authors, and peer reviewers dedicated many hours in the preparation for this first edition to ensure the overall technical quality, consistency, and accuracy of the manual.

Chairs

Ashley Kent, Arcadis, Austin, Tex.
Lynn Stephens, Brown and Caldwell, Seattle, Wash.

Steering Committee Review Members

Ashley Kent, Arcadis, Austin, Tex.
Josh Fortenbaugh, AWWA, Denver, Colo.
Chance Lauderdale, HDR, Tampa, Fla.
Mark LeChevallier, Dr. Water Consulting, Morrison, Colo.
Eva Nieminski, Utah Division of Drinking Water, Salt Lake City, Utah
Lynn Stephens, Brown and Caldwell, Seattle, Wash.

Chapter Authors

Christina Alito, HDR, Vienna, Va.
Jess Brown, Carollo Engineers, Costa Mesa, Calif.
Amlan Ghosh, Corona Environmental, Flower Mound, Tex.
Jennifer Hooper, CDM Smith, Bellevue, Wash.
Asher Keithley, U.S. Environmental Protection Agency, Cincinnati, Ohio
Ashley Kent, Arcadis, Austin, Tex.
Darren Lytle, U.S. Environmental Protection Agency, Cincinnati, Ohio
Brian Noma, Minnesota Department of Health, Saint Paul, Minn.
Jennifer Nyfennegger, Carollo Engineers, Sarasota, Fla.
Bruce Rittman, Biodesign Swette Center for Environmental Biotechnology, Tempe, Ariz.
Lynn Stephens, Brown and Caldwell, Seattle, Wash.
Leigh Terry, University of Alabama, Tuscaloosa, Ala.
Ramola Vaidya, HDR Inc., Vienna, Va.
Giridhar Upadhyaya, Brown and Caldwell, Irvine, Calif.

Chapter Peer Reviewers

Jess Brown, Carollo Engineers, Costa Mesa, Calif.
Jimmy Chen, US Environmental Protection Agency, Washington, D.C.
Denise Funk, Brown and Caldwell, Snellville, Ga.
Daniel Gerrity, Southern Nevada Water Authority, Las Vegas, Nev.
Mary Jo Kirisits, University of Texas at Austin, Austin, Tex.
Eberhard Morgenroth, Institute of Environmental Engineering, Zurich, Switzerland

Lutgarde Raskin, University of Michigan, Ann Arbor, Mich.
Bruce Rittman, Biodesign Swette Center for Environmental Biotechnology, Tempe, Ariz.
JoAnn Silverstein, University of Colorado, Boulder, Colo.
Joel Thompson, Stantec, Leesburg, Va.

Case Study Authors

Robert C. Andrews, University of Toronto, Toronto, Ont.
Alma Beciragic, Arcadis, Charlotte, N.C.
Derek R. Belanger, Tighe & Bond, Worchester, Mass.
Charles Bott, Hampton Roads Sanitation District, Virginia Beach, Va.
Christopher E. Clark, Alaska Native Tribal Health Consortium, Anchorage, Alaska
Toby Dixon, Eugene Water and Electric Board, Springfield, Ore.
Ali Dorri, Microvi Biotechnologies, Hayward, Calif.
Roderick Dunn, City of Columbus Water Division, Columbus, Ohio
David Friese, Applied Process Technology, Orinda, Calif.
Daniel Gerrity, Southern Nevada Water Authority, Las Vegas, Nev.
Jared Giacomelli, Eugene Water and Electric Board, Pleasant Hill, Ore.
Nissim Gore-Datar, Arcadis, Austin, Tex.
Martin A. Lawrence, Chart Industries, Tampa, Fla.
Ray Leipold, Eugene Water and Electric Board, Springfield, Ore.
Kevin Linder, Aurora Water Department, Aurora, Colo.
Michael J. McKie, C3 Water, Breslau, Ont.
Keren Nof, Aqwise, Herzliya, Israel
Aleksey Pisarenko, Trussell Technologies, Inc., San Diego, Calif.
Ameen Razavi, Microvi Biotech, Hayward, Calif.
Germano Salazar-Benites, Hampton Roads Sanitation District, Virginia Beach, Va.
Lonny Sayles, Eugene Water and Electric Board, Springfield, Ore.

Art Coordinators

Nicole Allward, Dalhousie University, Halifax, N.S.
Brandy Martinez, Brown and Caldwell, Garland, Tex.
Meric Selbes, Hazen and Sawyer, Fairfax, Va.

Participating Utilities

Aurora Water Department, Aurora, Colo.
City of Columbus, Ohio
City of Denton, Tex.
City of Gilbert, Iowa
City of Taylorville, Ill.
Crescenta Valley Water District, La Crescenta, Calif.
Greene County, Xenia, Ohio
Eugene Water and Electric Board, Springfield, Ore.
Hampton Roads Sanitation District, Virginia Beach, Va.
San Diego Water Authority, Calif.
Sdot-Yam MBDEN Drinking Water Plant, Aqwise, Kibbutz Sdot-Yam, Israel
Sunny Slope Water Company, Pasadena, Calif.
Town of Putnam, Putnam, Conn.

Chapter 1

Biological Treatment Overview

Ashley Kent and Lynn Stephens

This Manual of Practice provides guidance on planning, designing, and operating biological treatment processes for treating surface water, groundwater, and reuse. This Manual of Practice synthesizes research findings from studies supported by the Water Research Foundation (WRF), AWWA, US Environmental Protection Agency (USEPA), and others, as well as full-scale experience gained over the last few decades. The purpose of this Manual of Practice is to provide knowledge on common applications of biological treatment and summarize best practices for planning, designing, operating, and monitoring biological treatment processes.

This introduction chapter provides an overview of biological drinking water treatment and defines key concepts that should be understood before reading any of the subsequent chapters. Each subsequent chapter of this Manual of Practice focuses on one of the following biological treatment applications described in Section 1.2.

- Chapter 2: Aerobic Surface Water Biological Filtration
- Chapter 3: Aerobic Biotreatment of Groundwater
- Chapter 4: Anoxic Biological Treatment
- Chapter 5: Aerobic Biofiltration for Reuse

The subsequent chapters all follow a similar organization targeting the key knowledge areas outlined in Figure 1-1.

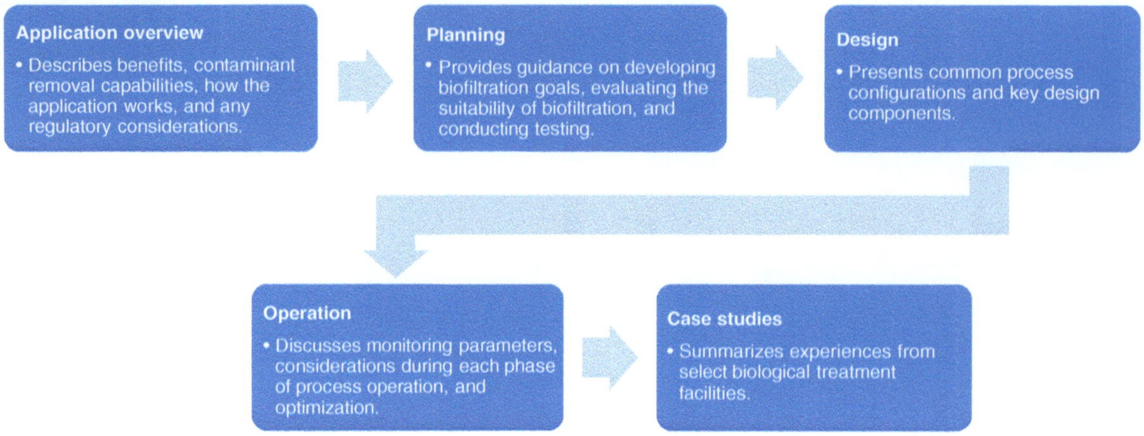

Figure 1-1 Organization of subsequent application Chapters 2–5

1.1 BIOLOGICAL TREATMENT DEFINITION AND BENEFITS

In biological treatment processes, microorganisms found in natural water systems grow on engineered surfaces or media (i.e., form a biofilm) and biologically transform or degrade compounds in the process influent. Biological treatment processes transform an array of compounds including the following:

- **organic carbon compounds,** including biologically available dissolved organic carbon (DOC) and disinfection by-product (DBP) precursors;
- **inorganic compounds,** such as iron, manganese, nitrate, ammonia, arsenic, perchlorate, selenium, and chromium; and
- **trace organic compounds,** such as synthetic and volatile organic compounds (VOCs), taste and odor (T&O) compounds (e.g., 2-methylisoborneol [MIB] and geosmin), nitrogen-containing DBPs (such as N-nitrosodimethylamine), cyanotoxins (often referred to as algal toxins), pharmaceuticals and personal care products (PPCPs), and potential endocrine disrupting chemicals (EDCs).

Through the removal of bulk organic carbon compounds, biological treatment can also improve distribution system water biostability and thus limit biological regrowth (LeChevallier et al. 2015, Evans et al. 2010, Rittmann and Snoeyink 1984), corrosion, and nitrification. Additionally, because contaminant removal is achieved through natural microbial processes, biological treatment minimizes energy, chemical, and residual management requirements. Thus, biological treatment is typically a cost-effective and sustainable (i.e., "green") treatment process.

1.2 BIOLOGICAL TREATMENT REMOVAL MECHANISMS

In biological treatment processes, microorganisms oxidize or reduce contaminants. In other words, microorganisms mediate the transfer of electrons between acceptors and donors (i.e., through redox reactions). The end products of biological treatment are typically either easier to remove, compared with the original contaminants (e.g., transformation of dissolved to particulate iron and manganese), or have fewer known health effects (e.g., carbon dioxide and nitrogen). Figure 1-2 provides some common examples of these transformations for both aerobic and anoxic treatment conditions; electron donors are shown in purple, and electron acceptors are shown in green.

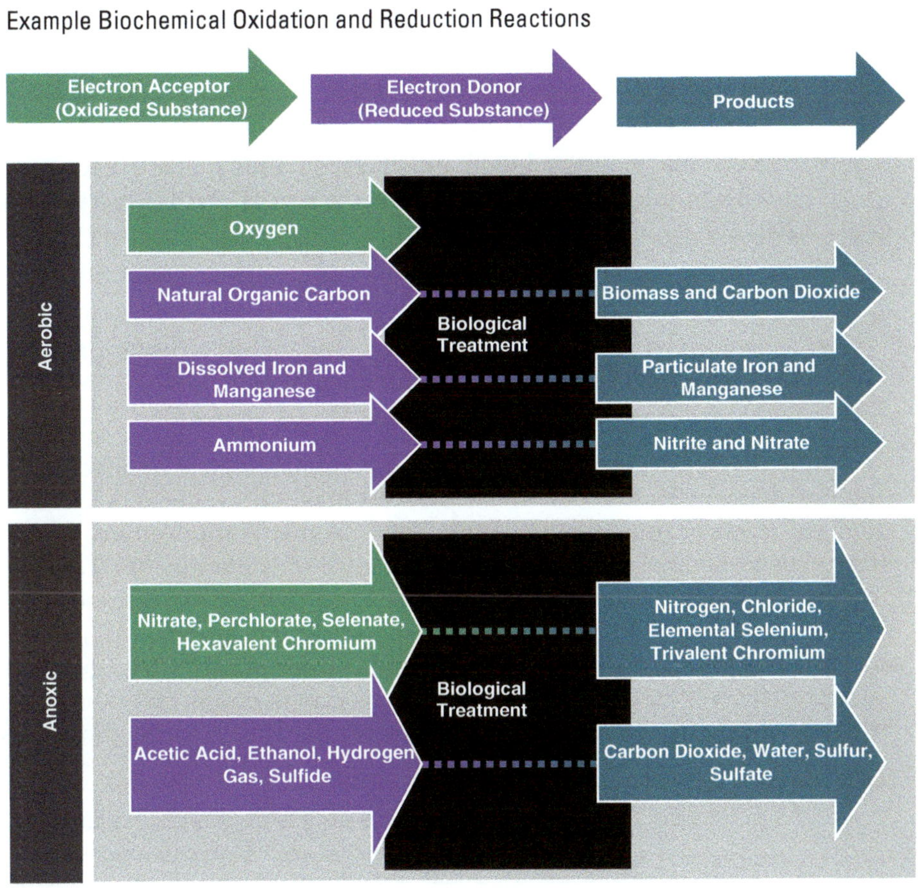

Figure 1-2 Example biochemical oxidation and reduction reactions

Contaminant removal mediated by microorganisms that require oxygen to create energy is called aerobic biological treatment. Aerobic treatment is used to oxidize contaminants that are in a reduced form, such as natural organic carbon, dissolved iron and manganese, and ammonia. For example, electron transfer from biodegradable organic carbon (i.e., the electron donor) to oxygen (i.e., the electron acceptor) produces carbon dioxide (CO_2), water, and energy; this reaction—aerobic respiration—supports microbial growth and activity. If sufficient oxygen is not present in the process influent (i.e., for groundwaters), it is added before treatment. The carbon source for these applications is either natural organic carbon in the source water or carbon dioxide.

In contrast, anoxic biological treatment involves microorganisms that use an alternative electron acceptor, such as nitrate, instead of molecular oxygen.

Anoxic treatment is used to transform contaminants that are oxidized (e.g., nitrate and perchlorate) by reducing them (e.g., to nitrogen and chloride). For example, electron transfer from organic carbon (i.e., the electron donor) to perchlorate (i.e., the electron acceptor) results in the transformation of perchlorate to chloride. Sufficient electron donor is typically not present in the process influent; thus, an electron donor (e.g., acetic acid or hydrogen gas) is typically added before treatment to promote anoxic microbial activity and reduction reactions.

Aerobic and anoxic treatment involve complex biological systems in which many different types of reactions can occur simultaneously. In addition to primary reactions that generate energy, secondary microbial reactions, which do not directly generate energy for microorganisms, also occur. These secondary reactions may also result in contaminant transformation, such as the following:

- biological oxidation or degradation due to extracellular enzymes (Tobiason et al. 2016, Mouchet 1992, Adams and Ghiorse 1987);
- bometabolic biodegradation, in which contaminants are biodegraded in reactions involving microbial enzymes (Greenstein et al. 2018, Alvarez-Cohen and Speitel 2001);
- use of a secondary substrate available only in limited quantities to support cellular growth (Rittmann et al. 1995);
- consumption as a macronutrient (e.g., phosphorus consumption to build biomass) (Keithley and Kirisits 2019) or a micronutrient (Gottschalk 1986); and
- absorption into the biofilm matrix (Fang et al. 2009).

In many applications, there may also be nonbiological removal mechanisms occurring simultaneously. Microbial reactions can make the water quality more favorable for chemical removal processes (Nealson 2006). In many applications, biological treatment may occur in parallel with the physical removal of contaminants (e.g., a biofilter that biochemically oxidizes dissolved manganese to particulate manganese and also physically removes particles). Additionally, if granular activated carbon (GAC) media is used, there may be adsorption occurring. Though for some contaminants (e.g., geosmin, MIB) biological degradation of the adsorbed materials can occur over time; for other contaminants (e.g., organic carbon), this mechanism will decrease over time if the media are not replaced.

Refer to the subsequent chapters for a more detailed explanation of the removal mechanisms for each application type.

1.3 COMMON BIOLOGICAL TREATMENT APPLICATIONS

The most common applications of biological treatment for drinking water are summarized subsequently. There are additional biological treatment processes, such as slow sand filtration or soil-aquifer treatment, which are not covered in the first version of this Manual of Practice.

1.3.1 Aerobic Surface Water Biological Filtration (Chapter 2)

Aerobic surface water biofiltration is the practice of operating conventional gravity filters (i.e., rapid-rate filters operated above 2 gpm/ft^2 [AWWA/ASCE 2005]) with low or absent oxidant residuals to encourage microorganisms in the source water to form a biofilm on granular filter media (Evans et al. 2010). Thus, these filters are designed for particle removal as the primary treatment goal. Additional biological treatment goals commonly include oxidation of DOC, inorganic contaminants, such as dissolved manganese, and trace organic compounds, such as T&O compounds. The target contaminants for removal are the electron donors, oxygen is the electron acceptor, and DOC from the source water provides a carbon source.

Surface water biofilters typically include upstream coagulation and flocculation, and in many applications, clarification. They often also include upstream preoxidation, such as ozonation. Upstream oxidation transforms organic carbon into compounds that are more amenable to biodegradation (Westerhoff et al. 2005). Biofiltration can then remove DOC compounds before the distribution of water. Because oxygen (i.e., the electron acceptor) is available naturally in the source water (and may be increased through upstream ozonation), oxygen supplementation is not required. However, other chemicals may be added to optimize performance and promote biological activity (e.g., nutrients or pH adjustment). Downstream treatment includes disinfection.

1.3.2 Aerobic Biotreatment of Groundwater (Chapter 3)

Aerobic groundwater biological treatment is the practice of developing a biofilm on manufactured media with upstream oxygenation. In some cases, these processes may be designed both for biological oxidation and subsequent removal of oxidized particulates (i.e., biofiltration). Biological treatment goals commonly include oxidation of ammonia, dissolved iron and manganese, and arsenite, As(III) (As^{3+}). The target contaminants for removal are the electron donors, oxygen is the electron acceptor, and carbon dioxide provides a carbon source. Treatment is conducted as either a single stage or multiple stages (e.g., removal of iron followed by removal of manganese in a second step).

The primary pretreatment step is aeration, which is needed because of low source water dissolved oxygen concentrations; there are multiple aeration/oxygenation delivery systems used, including both open configurations (e.g., cascading tower) and injection systems (e.g., bubble diffusers). Additionally, in some cases, performance has been reported to improve with the addition of a nutrient (e.g., phosphate) and/or with pH adjustment (Lytle et al. 2020, 2013). Last, postfilter disinfection, if not already practiced, may be required to inactivate microbes in the biofilter effluent.

1.3.3 Anoxic Biological Treatment (Chapter 4)

Anoxic groundwater biological treatment is the practice of adding substrate (i.e., carbon source or inorganic compound) to promote microorganisms to form a biofilm on media where oxygen is not present for the treatment of a groundwater source. There are multiple configurations for anoxic groundwater biological treatment. In fixed-bed bioreactors, biofilm develops on granular media in pressure vessels or open basins. Fluidized bed bioreactors are upflow reactors where biofilm develops on fluidized granular media. Biofilm can also be developed on one side of a submerged membrane or grown in a continuously stirred tank reactor (CSTR). Biological treatment goals commonly include the reduction of inorganic (e.g., nitrate, perchlorate, chromium) and organic contaminants (e.g., VOCs).

These contaminants serve as the electron acceptors, and an electron donor (e.g., acetic acid or H_2) is added for facilitating the biologically mediated contaminant removal. Additionally, in some cases, nutrients (e.g., phosphate) may also be added to promote biological activity. If required, biological oxidation of organic carbon and/or turbidity removal is accomplished by further downstream treatment. Additionally, downstream treatment often includes reoxygenation and disinfection.

1.3.4 Aerobic Biofiltration for Reuse (Chapter 5)

Aerobic biofiltration may also be implemented in a reuse application in combination with preoxidation and/or advanced oxidation processes (AOPs). In this application, rapid-rate GAC biofilters are designed both for biological oxidation as well as particulate and pathogen removal and are thus similar in many aspects to aerobic biofiltration for surface water. However, differences in the influent water quality (i.e., from wastewater or advanced treated wastewater) must be considered. Biological treatment goals for reuse applications commonly include the oxidation of DOC and trace contaminants, such as PPCPs, pesticides, and EDCs, which may be present in wastewater effluent. The target contaminants for removal are the electron donors, oxygen is the electron acceptor, and DOC from the source water (i.e., wastewater effluent or advanced treated wastewater) provides a carbon source.

Biofiltration in combination with ozonation is used in indirect potable reuse applications, in which an environmental buffer such as a lake or reservoir exists between the advanced treated water discharge and drinking water treatment facility intake, and direct potable reuse (DPR) applications, in which advanced treated water is directly used for potable-water production (e.g., "pipe-to-pipe"). Biofiltration in combination with preoxidation and/or AOPs can provide a more economical alternative treatment scheme relative to reverse osmosis (RO) and ultraviolet advanced oxidation processes, when treatment goals allow for it, and especially for inland facilities where the cost of disposing RO concentrate is challenging (Noibi et al. 2020). However, biofiltration may also be implemented with ozone to increase contaminant removal and reduce membrane fouling upstream of other advanced treatment processes. Thus, upstream (and downstream) treatment varies from one application to the next. As ozonation is placed upstream, which adds oxygen to the biofilter influent, no further oxygen addition is required. However, nutrient addition (e.g., phosphate) and/or pH adjustment may be required in some cases to promote biofilm formation.

1.4 HISTORY OF BIOLOGICAL DRINKING WATER TREATMENT

Biological treatment of drinking water has been used since the early 1800s, as shown by the progression of biological treatment practices in Figure 1-3. However, these early applications were largely unmanaged (i.e., also known as incidental, de facto, or unintentional). Biological processes were treated as a "black box" in drinking water treatment for many. Full-scale case studies, guidance documents on best practices, and proven strategies for optimizing biofilter operation and design were limited until recently. However, significant collaboration across the industry over the last few decades has expanded the application of biological treatment processes through increased research and knowledge sharing.

AWWA has played a large role in facilitating knowledge sharing. AWWA founded the Biological Drinking Water Treatment Committee (BDWTC) in 2008 to increase the industry's understanding of biological treatment. The BDWTC has hosted annual workshops, conference sessions, and symposia on biological treatment, in addition to supporting research and development opportunities. Additionally, international symposia on biological drinking water treatment were hosted in 2013, 2016, 2018, and 2020 (Evans et al. 2019a). This is the first AWWA guidance document on biological treatment and the first national biological treatment guidance document covering best practices for surface water, aerobic and anoxic groundwater, and reuse applications.

The following sections summarize the evolution of biological treatment for surface water, groundwater, and reuse applications.

1.4.1 Increasing Implementation of Surface Water Biofiltration

Incidental biofiltration (i.e., operation of carbon filters without an oxidant residual) in surface water treatment traces back to 4000 BCE (USEPA 2000). Further, biofiltration in the form of slow sand filtration (i.e., 0.02–2 gpm/ft^2) has been observed since the early 1800s in Europe (USEPA 2000) and the United States (Baker 1934). However, the first rapid-rate biofilters (above 2 gpm/ft^2 [AWWA/ASCE 2005]) were not observed until the 1970s in Europe and 1980s in the United States (Uhl 2008).

Many water systems in the United States began using chlorination in the early 1900s to reduce waterborne disease outbreaks (USEPA 2000). Although the first rapid-rate GAC filters (i.e., incidental biofilters) were installed in the United States in the 1930s (Baker 1934), the first publications on biological treatment for drinking water did not emerge until the 1980s (e.g., Bouwer and Crowe 1988, Rittman and Snoeyink 1984), when the industry was gaining a greater understanding of the effect of biological instability in distribution

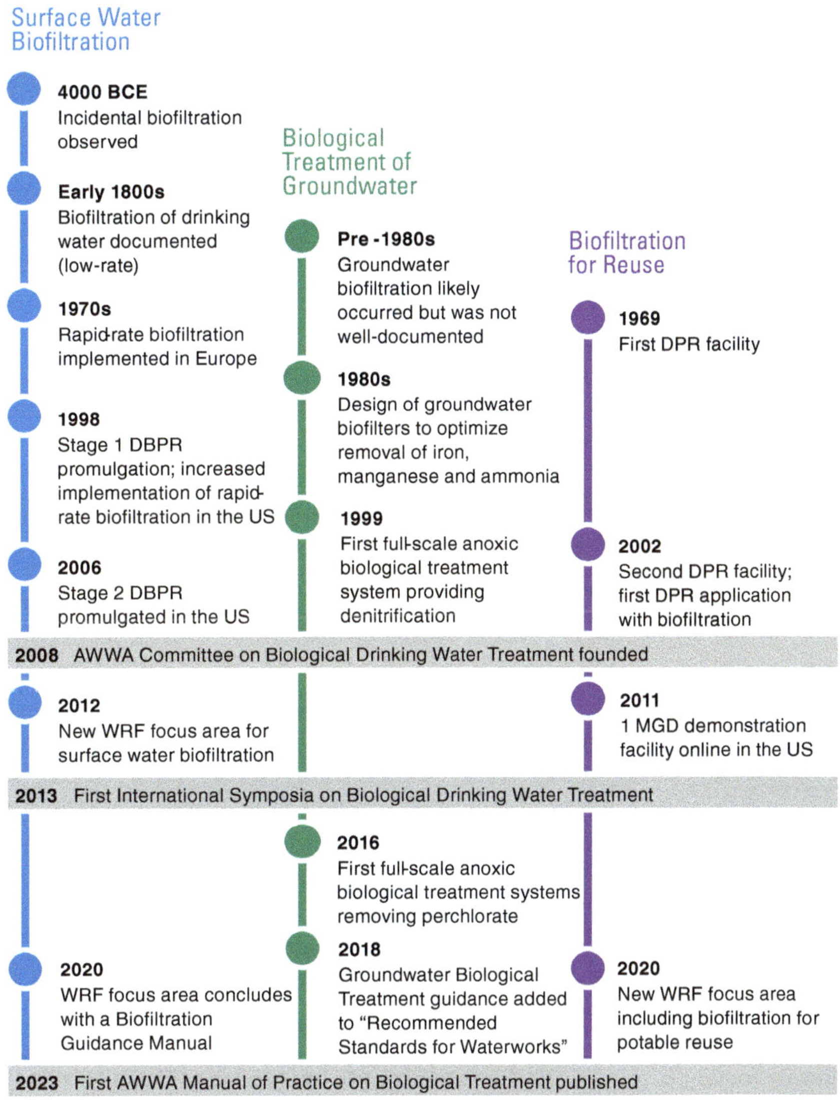

DBPR—Disinfectants and Disinfection Byproducts Rule, DPR—direct potable reuse, WRF—Water Research Foundation

Figure 1-3 Implementation of biological drinking water treatment

systems and the health effects of DBPs. Biofiltration implementation in the United States grew greatly in occurrence with the passing of the Stage 1 Disinfectants and Disinfection Byproducts Rule (DBPR), which was promulgated in 1998 and required compliance by 2002 (USEPA 2001). The DBPR rule required tighter control of DBPs through more stringent DBP maximum contaminant levels, limiting disinfection residuals, and requiring organic carbon removal. To meet the requirements of these rules, many utilities converted from chlorinated to nonchlorinated rapid-rate filters, thereby operating the first incidental rapid-rate biofilters for surface water treatment plants. According to a WRF survey, the largest increase in managed biofiltration facilities in a given year for those surveyed occurred in 2004 (Figure 1-4). Those facilities noted organic carbon removal and/or DBP reduction as goals for biofiltration (Evans et al. 2019b). Broader implementation of biofiltration in the United States continued following the Stage 2 DBPR in 2006.

In 2007, the WRF funded a study on "Biological Drinking Water Treatment Perceptions and Actual Experiences in North America" (Evans et al. 2010). The study concluded

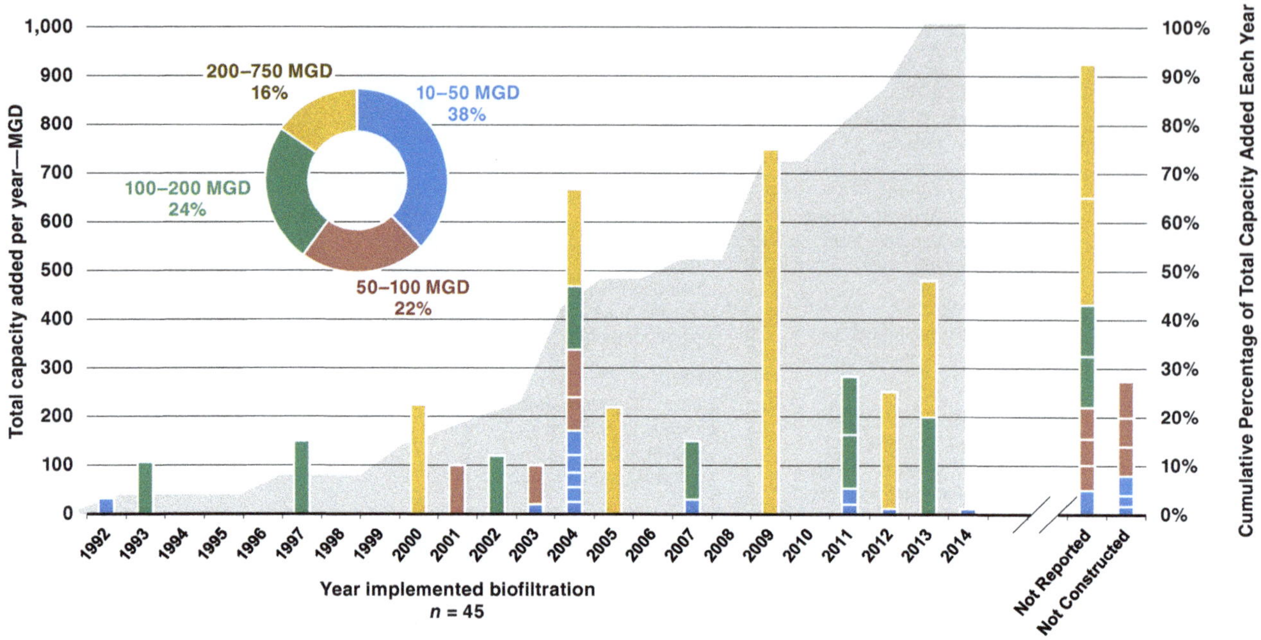

Source: Reprinted with permission from Evans et al. (2019b)

Figure 1-4 Increasing implementation of surface water biofiltration in the United States

that lack of full-scale experience and public perception of intentional biological growth in drinking water treatment were limiting greater implementation of biological treatment. Thus, in 2012, the WRF initiated a new focus area: "Biofiltration: Defining Benefits and Developing Utility Guidance." Over 4 million dollars in funding was provided to develop guidance documents and communicate the benefits of biofiltration (WRF 2019). Research studies documented case studies and current practices, assessed monitoring tools, and evaluated a range of operational and design strategies to optimize performance. As regulatory focus on unregulated contaminants evolved, studies also concluded that aerobic surface water biofiltration is effective for removing inorganic contaminants, such as manganese), and trace organic compounds, such as algal toxins, pharmaceuticals, and T&O compounds (Lauderdale et al. 2018, Zhu et al. 2010). The WRF focus area on surface water biofiltration closed with the publication of "Biofiltration Guidance Manual for Drinking Water Facilities" (2020), which compiles the recommendations developed across the focus area.

1.4.2 Optimizing Design of Groundwater Biological Treatment Systems

Biological treatment of groundwater likely occurred for many years before it was recognized; treatment was observed under conditions that were not favorable for chemical iron oxidation (Mouchet 1992). Europe began designing aerobic groundwater biofiltration systems for iron and manganese removal from groundwater in the 1980s (Mouchet 1992, Hettler 1982). Subsequently, implementation of managed aerobic biofiltration for groundwater treatment also increased in the United States over the last 20 years (Brown et al. 2017, 2013; Mouchet 1992) because of the increased awareness and acceptance of managed biological drinking water treatment, as well as the increased distribution of system water quality concerns about nitrification, DBP, and manganese.

Additionally, implementation of biological treatment has expanded to include anoxic removal of oxidized contaminants, such as nitrate and perchlorate, from groundwater (Brown et al. 2017). Europe began designing groundwater biofiltration systems for denitrification in the 1980s (Richard 1989). However, implementation of anoxic ground water treatment in the United States has been limited until recently. In 1999, the first full-scale anoxic biological treatment system targeting nitrate removal came on-line in the United States in rural Oklahoma (Oklahoma Water Resources Board 1999). Then, in 2016, the first full-scale anoxic biological treatment systems came on-line (Brown et al. 2018, Webster and Litchfield 2017) in Rialto, Calif. (fluidized bed bioreactor treating perchlorate), and Delano, Calif. (two-stage, fixed-bed bioreactor system treating nitrate). Most recently, the application of anoxic biological treatment has expanded to include the treatment of VOCs, hexavalent chromium, and uranium (Upadhyaya et al. 2019, 2012; Brown et al. 2017).

Increasing full-scale applications of both aerobic and anoxic biological treatment are anticipated in the United States because of the need for more water sources, the sustainability and cost efficiency of the process, and the ability of the processes to treat a wide range of contaminants in an evolving regulatory landscape. For example, in 2018, the first policy statements were published on aerobic and anoxic biological treatment in "Recommended Standards for Waterworks," which is also known as "Ten States Standards" (Great Lakes 2018).

1.4.3 Water Reuse, an Emerging Application

Utilities have been considering potable reuse for decades in response to greater water demands and increased water scarcity. In 1969, the first DPR facility began operating in Windhoek, Namibia. The facility was expanded from 1.1 to 2.0 million gallons per day (mgd) in the 1990s (du Pisani 2006). Although both indirect and direct water reuse plants have been designed and built in the past, recent implementation of biological filtration into water reuse accelerated the growth of water reuse plants. Utilities are beginning to consider biofiltration, in combination with preoxidation and/or AOPs, to treat wastewater effluent more economically and sustainably (Gerrity et al. 2014, Schimoler and Kealy 2013). The New Goreangab Water Reclamation Plant, constructed in 2002 in Namibia, was the expansion of the first DPR facility and included biofiltration with GAC media downstream of ozonation (du Pisani and Menge 2013). Several studies are either underway or recently completing evaluation of ozone and biofiltration for potable reuse applications in the United States. For example, the Pure Water 1-MGD demonstration facility in San Diego, Calif., has been on-line since 2011, the El Paso Water pilot study in Texas was completed in 2016, and the F. Wayne Hill Water Resources Center pilot study in Gwinnett County, Ga., was completed in 2018 (WRF 2018).

1.5 PLANNING FOR BIOLOGICAL TREATMENT

Establishing goals for biological treatment is a key first step in either evaluating or implementing biological treatment. Treatment goals should consider historical water quality, current or planned treatment characteristics, and applicable regulations, and should be specific and quantitative. Contaminant removal goals should define a specific effluent concentration (or percent removal) for each target contaminant. Operational goals should define specific full-scale target values for each operational parameter, such as filtration rates, run times, clean bed head loss and head loss accumulation, and filter productivity (i.e., unit filter run volumes).

On the basis of defined goals, an applicability evaluation should be conducted. Applicability evaluations are generally desktop studies, which may include literature reviews, the use of models, and a review of historical data. Applicability evaluations should assess the potential effectiveness of the process being considered to meet the established goals and should establish a business case for conducting additional testing.

Last, when warranted, testing should be conducted before full-scale implementation of biological treatment or optimization strategies. Testing objectives include the treatment goals defined for full-scale implementation and may include additional objectives related to assessing design alternatives (e.g., media profile changes or chemical addition) and conducting testing under a range of conditions anticipated full-scale. Multiple testing scales are often recommended on the basis of the testing objectives established because each testing scale has specific advantages and disadvantages. A brief description of each testing scale is provided, as follows:

- **Bench:** Bench-scale testing is generally conducted to test multiple design or operational parameters in a more controlled environment. Thus, bench-scale treatment units (e.g., columns) are often very small (i.e., less than 2-in. diameter) and have a lower cost per unit to construct and operate (e.g., require less water). However, because of these limitations, bench-scale experiments generally do not match full-scale filter production or hydraulic performance and do not allow for testing backwashing procedures.
- **Pilot:** Pilot-scale testing is generally conducted once the design and operational variables under consideration have been filtered to four to six options. Pilot-scale treatment units and operating conditions are generally relevant to anticipated full-scale conditions (e.g., media sizes, depths, and flow), but may not allow for evaluating all variables, such as backwashing procedures.
- **Demonstration:** Demonstration-scale testing design and conditions match full-scale (i.e., could be a single full-scale treatment unit), and thus, testing is generally conducted to verify that the selected design and operating conditions meet the established treatment goals. Demonstration-scale testing may also be used on existing treatment units for testing optimization strategies by changing a variable on an isolated treatment unit.

The planning section of each subsequent chapter outlines the process for determining whether implementation of each biological treatment application will meet the facility's treatment and operational goals.

1.6 DESIGN OF BIOLOGICAL TREATMENT SYSTEMS

Although design parameters for biological treatment processes vary for each application type, there are a few key design concepts for biological systems that are important to understand. Biological treatment systems rely on microorganisms in biofilms transforming contaminants.

- **Media selection:** The characteristics of the media, including the surface area available for biofilm development and the properties of that surface, affect contaminant removal. Although many biological treatment systems use inert media (e.g., anthracite, sand, plastic media or membranes), adsorptive media (i.e., GAC) may be advantageous depending on the contaminant removal goals and influent water quality or if contaminant toxicity is a concern.
- **Contact time:** The rate at which contaminants are biologically transformed varies for different contaminants and is influenced by several factors, such as the media selected and influent water quality. Thus, the time that water to be treated is in

contact with the microbial community is a critical design parameter. Additional contact time may be provided by slowing the rate of water or providing additional surface area for biofilm development (e.g., a larger media volume).

- **Influent water quality amendments:** The rate and extent of contaminant transformation in biological systems is dependent upon the influent water quality, and some contaminants can be biologically transformed across a wider range of water quality than others. Thus, in some cases, chemical supplementation may be required to meet treatment goals. Chemical supplementation may include providing an electron donor/acceptor or nutrients to support the growth of the microbial community or altering chemical conditions to make the transformation of a specific contaminant more favorable (e.g., changing the pH of the influent).

Other design recommendations discussed in the subsequent chapters include pre- and posttreatment requirements, treatment configuration alternatives, filtration system and bioreactor design criteria, and residuals handling. Additionally, for biological treatment processes designed for physical removal, additional design criteria specific to those mechanisms (e.g., conventional filtration design criteria for biofilters) may also apply.

The design section of each subsequent chapter describes common process configurations, including upstream and downstream treatment considerations, for each application and key design components, such as media profiles, underdrains, chemical feed, backwashing, and biomass control strategies.

1.7 OPERATION AND MONITORING OF BIOLOGICAL TREATMENT SYSTEMS

There are three phases of operation for biological systems that have different performance expectations and operational requirements. Figure 1-5 illustrates one example of how biological contaminant removal performance may vary for each phase. However, the specific duration and performance achieved are site-specific and based on many factors, such as the influent water quality and treatment application. Additionally, if nonbiological removal mechanisms are present, overall performance may be higher (e.g., with GAC media, adsorption of contaminants may be observed before or in addition to biological transformation).

- **Initial startup/acclimation:** The initial startup or acclimation of biological processes is the time required for the biofilm to establish on media and adapt to the

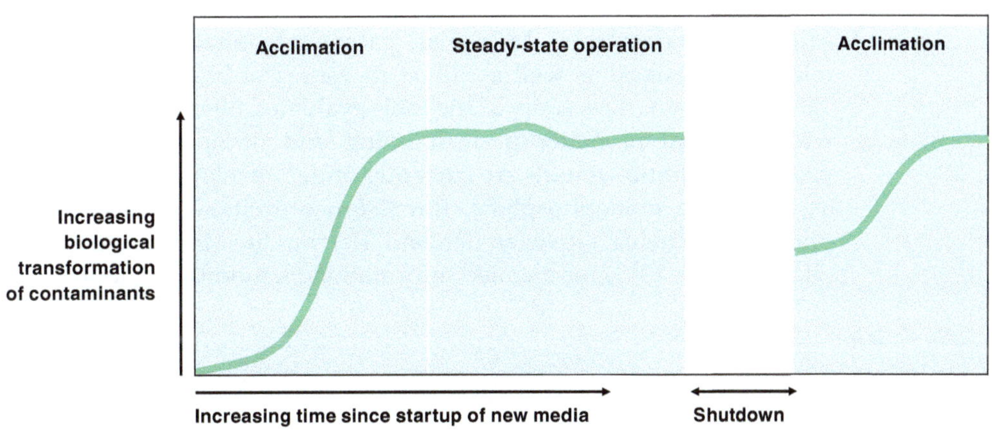

Figure 1-5 Example schematic of operational phases for biological treatment

environment before achieving steady-state removal of targeted contaminant(s). For processes with new GAC media, a decrease in contaminant removal may be observed during this phase because of the gradual exhaustion of the GAC media until steady-state biological treatment is achieved.

- **Steady-state operation:** Steady-state contaminant removal is observed once a mature and active biofilm (i.e., biomass or microbial community) is established on the media, and biodegradation or biotransformation is observed. Biological processes perform best under steady-state operational conditions. Some minor contaminant removal performance variations may be observed because of changes in the influent water quality; however, variations should be minor compared with those observed during initial startup.
- **Shutdown and restart:** Shutdown of biological processes may, depending on the shutdown conditions and length, result in a decrease in biomass and/or performance during restart. The absence of key nutrients, changes in water quality, or drying of the biofilm may affect the microbial community. Site-specific operating procedures for shutdowns and restart can be implemented to mitigate effects on performance.

The operational section of each subsequent chapter discusses key strategies for operating biological treatment systems, including monitoring, operational phases, optimization, and troubleshooting.

Monitoring is a critical component for the successful operation of a biological treatment system because microbial communities rapidly react and change in response to changes within their environment (e.g., flow and water quality). Key water quality parameters for monitoring will vary on the basis of the application type and treatment goals. However, all monitoring programs should include the following groups of parameters:

- **Operational parameters:** Operational monitoring should capture any changes affecting the facility's water production or hydraulic goals (e.g., flow and head loss accumulation rate).
- **Process integrity:** Process integrity monitoring should capture any changes to the design of the treatment system over time (e.g., changes in media characteristics).
- **Water quality parameters:** Water quality monitoring should include general water quality parameters (e.g., temperature and pH) in addition to all parameters with treatment goals.
- **Microbial parameters:** Microbial parameters should include indicators of microbial changes, as well as direct measures of biomass or microbial activity. Table 1-1 summarizes some of the tools evaluated over the past couple of decades with the goal of better understanding and optimizing biological treatment processes; additional tools are currently under development and evaluation, such as adenosine monophosphate, fluorescence excitation–emission matrixes, and photoelectrochemical oxygen demand. Refer to the Monitoring Guidance Manual (Hooper et al. 2019) for detailed information on monitoring tools.

Table 1-1 Summary of biological process monitoring tools

<table>
<tr><th>Monitoring Tool</th><th>Medium</th><th>Description</th><th>Benefits/Limitations</th></tr>
<tr><td>Change in DOC</td><td>Water</td><td>The difference in the DOC concentration in the biofilter influent and effluent</td><td>• Measure of bioactivity as well as contaminant removal performance (if organic carbon removal is targeted)
• Well-established and widely used method available through most commercial labs
• If a reliable correlation can be developed, the change in DOC can be used for some systems.</td></tr>
<tr><td>Carboxylic acids</td><td>Water</td><td>The quantity of acids generated from preoxidation, including acetate, oxalate, and formate</td><td rowspan="2">• Indicator of biostability of the treated water
• Most valuable for utilities with upstream ozonation
• Higher cost and longer turn-around time (three weeks) compared with DOC
• Typically reserved for optimization studies and initial testing of treatment effectiveness
• Carboxylic acids have been reported to have more reproducible results than the AOC or BDOC assays.</td></tr>
<tr><td>AOC/BDOC</td><td>Water</td><td>Organic carbon that is degradable (BDOC) or consumed to produce new biomass (AOC)</td></tr>
<tr><td>Biofilter effluent aqueous ATP</td><td>Water</td><td>The cellular energy present in the biofilter effluent</td><td>• Indication of microbial sloughing
• Onsite analysis using commercially available test kits
• Results may be highly variable
• Best correlates with the application of an oxidant</td></tr>
<tr><td>ATP on biofilter media</td><td>Media</td><td>The cellular energy present in the biofilm on the process media</td><td>• Measure of biomass
• Onsite analysis using commercially available test kits
• Not linked to performance changes following acclimation
• Widely used by many studies
• Most valuable during acclimation and troubleshooting</td></tr>
<tr><td>Biofilm formation rate</td><td>Coupon</td><td>The ATP in the biofilm that develops on a coupon installed in a pipe loop over a set duration of time; separate pipe loops receive biofilter influent and effluent flow</td><td>• Measure of biofilter bioactivity as well as finished water biostability
• Recommended biological monitoring tool (Hooper et al. 2019)
• Onsite analysis using commercially available test kits
• Requires installation of small pipe loops and coupon changeouts
• Not applicable if the influent has an oxidant residual</td></tr>
</table>

(continued)

Table 1-1 Summary of biological process monitoring tools (*Continued*)

Monitoring Tool	Medium	Description	Benefits/Limitations
DO consumption	Water	The difference in the DO concentration in the biofilter influent and effluent	• Indicator of aerobic microbial activity • Online, continuous monitoring • Not applicable for biofilters with upstream ozonation
EPS	Media	Materials produced by bacteria that support biofilms, such as proteins and polysaccharides	• In some studies, has been correlated with increased head loss accumulation • Recommended for troubleshooting, but not routine operational monitoring
Enzyme activity (e.g., phosphatase or esterase activity)	Media	The relative quantity of specific enzyme, produced by microorganisms	• Simple test kits are available for use with luminometer or fluorimeter. • May only target a fraction of the microbial community • May help identify if a nutrient deficiency exists (Keithley and Kirisits 2019) • In some studies, correlations have been found between esterase activity and DOC or THM-FP removal, and between phosphatase activity and phosphate-P removal (McKie et al. 2020) or head loss accumulation (Keithley and Kirisits 2019) • Currently, not widely used at full-scale
Microbial community characterization (e.g., 16S rRNA sequencing or metagenomics)	Media	The phylum and genus of microorganisms identified in a sample based on DNA (i.e., the types of microorganisms)	• May help evaluate whether a water quality or operational change shifts the microbial community • Typically reserved for research studies • More expensive per sample and requires an external laboratory • Results may vary on the basis of the laboratory • Require interpretation of health significance of results • Requires significant further research on correlations with contaminant degradation

(continued)

Table 1-1 Summary of biological process monitoring tools (*Continued*)

Monitoring Tool	Medium	Description	Benefits/Limitations
qRT-PCR	Media	The quantity of a specific microbial gene present	• Allows for evaluation of relative quantities of genes related to specific functions, such as ammonia oxidation • Typically reserved for research studies • More expensive per sample and requires an external laboratory • Requires significant further research on correlations with contaminant degradation
Direct measurement of bacteria on media	Media	The number of bacteria grown on culture medium under set conditions (e.g., HPC or coliforms)	• Direct measure of bacteria present • Only captures a fraction of the overall microbial community • Long turn-around time (greater than a day) to get results

Source: Hooper et al. 2019, Evans et al. 2013

AOC—assimilable organic carbon, ATP—adenosine triphosphate, BDOC— biodegradable organic carbon, DO—dissolved oxygen, DOC—dissolved organic carbon, EPS—extracellular polymeric substance, HPC—heterotrophic plate count, qRT-PCR—quantitative (real-time) polymerase chain reaction, rRNA—ribosomal RNA, THM-FP—trihalomethane formation potential

1.8 REFERENCES

Adams, L.F. and W.C. Ghiorse. 1987. "Characterization of Extracellular Mn2+-Oxidizing Activity and Isolation of an Mn2+-Oxidizing Protein from Leptothrix discophora SS-1." *Journal of Bacteriology*. 169(3):1279-1285.

Alvarez-Cohen, L. and G.E. Speitel Jr. 2001. "Kinetics of Aerobic Cometabolism of Chlorinated Solvents." *Biodegradation*. 12(2):105-126. doi: 10.1023/a:1012075322466.

AWWA/ASCE (American Water Works Association/American Society of Civil Engineers). 2005 (4th ed.). *Water Treatment Plant Design*. Denver: AWWA.

Baker, M.N. 1934. "Sketch of the History of Water Treatment." *Journal AWWA*. 26(7):902-938. https://doi.org/10.1002/j.1551-8833.1934.tb14361.x

Bouwer, E.J. and P.B. Crowe. 1988. "Biological Processes in Drinking Water Treatment." *Journal AWWA*. 80(9):89-93. https://doi.org/10. 1002/j.1551-8833.1988.tb03103.x

Brown, J., B. Murphy, and V. Hart. 2013. "Biofilter Acclimation & Operation at 10°C for the Removal of Ammonia, Iron, and Manganese." Proc. AWWA Annual Conf. and Exhibition, Denver, Colo.

Brown, J., J. Riley, C. Cleveland, P. Carlo, and G. Upadhyaya. 2018. "Two-Stage, Fixed-Bed Biological Treatment: From Piloting to Full-Scale Implementation." Proc. AWWA International Symposium on Biological Treatment, Austin, Tex.

Brown, J., R.S. Summers, E. Dickenson, and G. Upadhyaya. 2017. "Biological Removal of Hazardous Pollutants in Drinking Water Treatment." In *Hazardous Pollutants in Biological Treatment Systems: Fundamentals and a Guide to Experimental Research* (F. Cecen, U. Tezal, editors). Volume 1. London: IWA Publishing.

du Pisani, P. 2006. "Direct Reclamation of Potable Water at Windhoek's Goreangab Reclamation Plant." *Desalination*. 188(1-3):79-88. 10.1016/j.desal.2005.04.104.

du Pisani, P. and J. Menge. 2013. "Direct Potable Reclamation in Windhoek: A Critical Review of the Design Philosophy of New Goreangab Drinking Water Reclamation Plant." *Water Science & Technology: Water Supply*. 13(2):214-226. 10.2166/ws.2013.009.

Evans, A.N., J.T. Carter, J. Brown, and C. Lauderdale. 2019b. "North American Biofiltration Practices." *AWWA Water Science*. 1(4):e1152.

Evans, P.J., E. Opitz, P. Daniel, and C. Schulz. 2010. "Biological Drinking Water Treatment Perceptions and Actual Experiences in North America." Project #4129. Denver: WRF (Water Research Foundation) and Department of Defense.

Evans, P.J., J.L. Smith, M.W. LeChevallier, O.D. Schneider, L.A. Weinrich, and P.K. Jjemba. 2013. "Biological Filtration Monitoring and Control Toolbox: Guidance Manual." Project #4321. Denver: WRF.

Evans, A., L. Williams Stephens, and L. Terry. 2019a. "Harness Innovative Knowledge and Tools to Drive Biofilter Performance." *Journal AWWA*. 111(9):98-101.

Fang, W., J.Y. Hu, and S.L. Ong. 2009. "Influence of Phosphorus on Biofilm Formation in Model Drinking Water Distribution Systems." *Journal of Applied Microbiology*. 106(4):1328-1335.

Gerrity, D., E. Owens-Bennett, T. Venezia, B. Stanford, M. Plumlee, J. Debroux, R.S. Trussell. 2014. "Applicability of Ozone and Biological Activated Carbon for Potable Reuse." *Ozone: Science & Engineering*. 36(2):123-137.

Gottschalk, G. 1986. *Bacterial Metabolism*. New York: Springer-Verlag.

Great Lakes – Upper Mississippi River Board of State and Provincial Public Health and Environmental Managers. 2018. "Recommended Standards for Water Works." St. Paul, Minn.: Minnesota's Bookstore Communications Media Division.

Greenstein, K.E., J. Lew, E.R.V. Dickenson, and E.C. Wert. 2018. "Investigation of Biotransformation, Sorption, and Desorption of Multiple Chemical Contaminants in Pilot-Scale Drinking Water Biofilters." *Chemosphere*. 200:248-256.

Hettler, J.P. 1982. "Une Station de Deferrisation Biologique." *T.S.M. L'Eau*. 77(10):481.

Hooper, J., K. Vickstrom, P. Evans, C. Alito, S. Black, and C. Lauderdale. 2019. "Guidance Manual for Monitoring Biological Filtration of Drinking Water." Project #4620. Denver: WRF.

Keithley, S.E. and M.J. Kirisits. 2019. "Enzyme-Identified Phosphorus Limitation Linked to More Rapid Headloss Accumulation in Drinking Water Biofilters." *Environmental Science & Technology*. 53(4):2027-2035. doi: 10.1021/acs.est.8b04573.

Lauderdale, C., C. Alito, J. Hooper, K. Dowdell, and E. Wert. 2018. "Optimizing Biofiltration for Various Source Water Quality Conditions." Project #4555. Denver: WRF.

LeChevallier, M.W., O.D. Schneider, L.A. Weinrich, P.K. Jjemba, P.J. Evans, and J.L. Hooper. 2015. "An Operational Definition of Biostability for Drinking Water." Project #4312. Denver: WRF.

Lytle, D.A., C. White, D. Williams, L. Koch, and E. Nauman. 2013. "Innovative Biological Water Treatment for the Removal of Elevated Ammonia." *Journal AWWA*. 105(9):E524-E539.

Lytle, D.A., D. Williams, C. Muhlen, E. Riddick, and M. Pham. 2020. "The Removal of Ammonia, Arsenic, Iron and Manganese by Biological Treatment from a Small Iowa Drinking Water System." *Environmental Science: Water Research & Technology*. 6(11):3142-3156.

McKie, M.J., L. Taylor-Edmonds, S.A. Andrews, R.C. Andrews. 2020. "Effective Enzyme Activity: A Proposed Monitoring Methodology for Biofiltration Systems with or Without Ozone." *Water Research*. 183:116069. https://doi.org/10.1016/j.watres.2020.116069.

Mouchet, P. 1992. "From Conventional to Biological Removal of Iron and Manganese in France." *Journal AWWA*. 84(4):158-167.

Nealson, K.H. 2006 (3rd ed.). "The Manganese-Oxidizing Bacteria." In *The Prokaryotes*. Volume 5. New York: Springer-Verlag.

Noibi, M., J. Hooper, D. Funk, and K. Bell. 2020. "Direct Potable Reuse Using Full Advanced Treatment Compared with Non-reverse Osmosis Based Ozone and Biofiltration Treatment at Gwinnett County: A Cost Comparison." *AWWA Water Science*. 2(6):e1210.

Oklahoma Water Resources Board. 1999. "Towns Turn to Technology to Treat High Nitrates." *Oklahoma Water News Bimonthly Newsletter of the Oklahoma Water Resources Board*.

Richard, Y.R. 1989. "Operating Experiences of Full-Scale Biological and Ion-Exchange Denitrification Plants in France." *Water and Environment Journal*. 3(2):154-167.

Rittmann, B., C. Gantzer, and A. Montiel. 1995. "Biological Treatment to Control Taste-and-Odor Compounds in Drinking Water Treatment." In *Advances in Taste-and-Odor Treatment and Control*. Denver: WRF.

Rittmann, B.E. and V.L. Snoeyink. 1984. "Achieving Biologically Stable Drinking Water." *Journal AWWA*. 76(10):106-114.

Tobiason, J.E., A. Bazilio, J.E. Goodwill, X. Mai, and C. Nguyen. 2016. "Manganese Removal from Drinking Water Sources." *Current Pollution Reports*. 2(3):168-177.

Uhl, W. 2008. "Biofiltration Processes for Organic Matter Removal." In *Biotechnology: Environmental Processes* (H.J. Rehm, G. Reed, A. Puhler, P. Stadler, editors). Weinheim, Germany: John Wiley & Sons.

Upadhyaya, G., B. Dunagan, C. Mattingly, and J. Brown. 2019. "Resilient Biological Hexavalent Chromium Removal with a Two-Stage, Fixed-Bed Biotreatment System." *AWWA Water Science*. 1(4):e1151.

Upadhyaya, G., P. Ghosh, T.M. Clancy, Y. Bi, J. Carpenter, K.F. Hayes, and L. Raskin. 2012. "Drinking Water Production Using an Anaerobic Fixed-Bed Bioreactor from Water Sources Contaminated with Nitrate and Uranium." 2012 AWWA Water Quality Technology Conf. & Exposition, Toronto, Canada.

USEPA. 2001. "Stage 1 Disinfectants and Disinfection Byproducts Rule." EPA 816-F-01-014. Washington: Office of Water (4607).

USEPA. 2000. "The History of Drinking Water Treatment." EPA-816-F-00-006. Washington: Office of Water (4606).

Webster, T.S. and M.H. Litchfield. 2017. "Full-Scale Biological Treatment of Nitrate and Perchlorate for Potable Water Production." *Journal AWWA*. 109:5:30-40.

Westerhoff, P., R.S. Summers, Z. Chowdhury, and S. Kommineni. 2005. "Ozone-Enhanced Biofiltration for Geosmin and MIB Removal." Project #2775. Denver: WRF.

WRF. 2019. *Biofiltration: Defining Benefits and Developing Utility Guidance*. https://www.waterrf.org/news/biofiltration-defining-benefits-and-developing-utility-guidance). Accessed March 13, 2025.

WRF. 2018. "Demonstration of High Quality Drinking Water Production Using Multi-Stage Ozone Biological Filtration: A Comparison of Direct Potable Reuse (DPR) with Existing Indirect Potable Reuse (IPR)." Project #4777. https://www.waterrf.org/research/projects/demonstration-high-quality-drinking-water-production-using-multi-stage-ozone (accessed Mar. 7, 2023).

Zhu, I.X., T. Getting, and D. Bruce. 2010. "Review of Biologically Active Filters in Drinking Water Applications." *Journal AWWA*. 102(12):67-77.

Chapter 2

Aerobic Surface Water Biological Filtration

Lynn Stephens, Ashley Kent, and Jennifer Nyfenegger

Aerobic biological treatment for surface water is a sustainable treatment approach for removing an array of both organic and inorganic contaminants. In surface water applications, biological treatment occurs within conventional gravity filters by removing or limiting the oxidant residual, such as chlorine, across the filters, within which microbial activity develops naturally. This chapter addresses the biofiltration process located in surface water treatment plants downstream of pretreatment and upstream of primary disinfection. Other configurations, such as biological roughing filters at the head of the treatment train, are not covered in this chapter. This chapter discusses the target contaminants and regulatory considerations, biological removal mechanisms, planning and process evaluation best practices, design criteria for operating conventional filters as biologically active filters (BAF) or biofilters, and operational guidance. This chapter also includes case studies of three full-scale surface water biofilters.

2.1 APPLICATION OVERVIEW

Biological filtration or biofiltration is defined by the AWWA Biological Drinking Water Treatment Committee as "an operational practice of managing, maintaining, and promoting biological activity on granular media in the filter to enhance the removal of organic and inorganic constituents before treated water is introduced into the distribution system" (Brown et al. 2016). This chapter covers gravity filters with a surface loading rate of at least 2 gpm/ft^2 (AWWA/ASCE 2012). Growth of indigenous microorganisms is promoted by minimizing or eliminating an oxidant residual across filters. Additionally, in some cases, biofiltration may be paired with upstream oxidation (e.g., ozone) to transform organic carbon into a more biodegradable form, which may increase bioactivity in downstream biofilters and result in greater overall removal of organic carbon.

Biofiltration, in the form of slow sand filtration, has been used for drinking water treatment since the early 1900s. Besides slow sand filters, the intentional use of biofiltration in North America was limited until the 2000s. Until that time, incidental biofiltration would be realized primarily because of the absence of an oxidant residual in the filter influent. However, following the promulgation of the US Environmental Protection Agency's (USEPA) Stage 1 and Stage 2 Disinfectants and Disinfection By-products Rules (DBPRs) in the early 2000s, biofiltration has been intentionally targeted in North America (Evans et al. 2019). Over the last 20 years, industry focus on sustainable treatment (biofilters remove an array of contaminants with little or no additional energy or chemical requirements), emerging contaminants' removal, and distribution system biostability has resulted in extensive research and knowledge-sharing to optimize biofiltration performance for surface water treatment. Biofiltration has recently been reported as widely accepted across North America (Nieminski and Perry 2015). The Water Research Foundation (WRF) had a focus area on biological filtration that culminated in the "Biofiltration Guidance Manual for Drinking Water Facilities" (Brown et al. 2020). This chapter focuses on rapid-rate biological filtration and does not cover slow sand filtration.

The following section presents an overview of the benefits of surface water biofiltration, discusses removal mechanisms, and summarizes regulatory considerations.

2.1.1 Benefits

The primary function of a biological filter is to remove particles and pathogens, similar to a nonbiological filter. Additionally, biofiltration can simultaneously remove multiple organic contaminants, including natural organic matter (NOM), disinfection by-product (DBP) precursors, taste and odor (T&O) compounds, cyanotoxins and anthropogenic compounds, such as pharmaceuticals and personal care products (PPCPs), pesticides, flame retardants, and plasticizers. Biofiltration can also remove inorganic compounds, such as iron, manganese, and ammonia. It also increases the biostability of the drinking water distribution system by removing biodegradable substances, such as assimilable organic carbon (AOC) and biodegradable organic carbon (BDOC), that could be food for microorganisms in the distribution system.

One of the main benefits of biofilters is that there is no additional energy, and minimal, if any, additional chemical requirements, compared with direct or conventional filtration. Contaminants are typically reduced to innocuous end products that are less harmful. Further, no additional residual waste is produced during the removal of the contaminants. Thus, biofiltration provides a cost-effective and sustainable treatment approach.

2.1.2 Contaminant Removal

Table 2-1 summarizes biofiltration removal performance (surface-water specific), listed by contaminant group. Removal of each contaminant group depends on multiple site-specific factors, such as source water quality, pretreatment, filter media type, empty bed contact time (EBCT), and temperature. A good resource for a more detailed investigation is WRF Project #4559, "Simultaneous Removal of Multiple Contaminants Using Biofiltration" (Dickenson et al. 2018).

2.1.2.1 Particles/pathogens. Biofilters are often the sole filtration process in a water treatment plant; therefore, biofilters must consistently meet turbidity requirements (Brown et al. 2016). Regulatory agencies in the United States require water systems to filter and disinfect surface water sources to remove pathogens in drinking water, including ***Giardia lamblia*** cysts and ***Cryptosporidium*** oocysts. Utilities that choose biofiltration aim

Table 2-1 Contaminant drivers and removal potential by biofiltration* (surface-water specific)

Typical Target Contaminants	Driver	Biofiltration Removal Potential: Without Pre-ozonation	Biofiltration Removal Potential: With Pre-ozonation	References
Turbidity	Regulatory compliance	High	High	Brown et al. (2020), Goldgrabe et al. (1993), LeChevallier et al. (1992)
NOM (TOC, DOC, AOC, BDOC, humic and fluvic acids, carboxylic acids, aldehydes, ketones, organic nitrogen)	Distribution system biostability, reduced chlorine demand, DBP precursors	Moderate	Moderate to high	Lauderdale et al. (2018), Brown et al. (2017), LeChevallier et al. (2015), Kohl and Dixon (2012), Simpson (2008), Wert et al. (2008)
T&O compounds (MIB, geosmin)	Aesthetics	Low to moderate	High	Dickenson et al. (2018), Hooper et al. (2019), Lauderdale et al. (2018), McKie et al. (2015), Evans et al. (2013), Persson et al. (2007), Emelko et al. (2006), Bouwer and Crowe (1988)
Cyanotoxins†	Health effects	Low	High	USEPA (2016)
Emerging contaminants (EDCs, PPCPs, pesticides)	High-quality water, long-term health effects unknown	None to moderate	None to high	Dickenson et al. (2018), Zearley and Summers (2012), Snyder et al. (2007)
Inorganics (iron, manganese, ammonia)	Secondary drinking water standards, reduced chlorine demand	Moderate to high	Moderate to high	Lauderdale et al. (2018), Brown et al. (2017), LeChevallier et al. (2015), Kohl and Dixon (2012), Simpson (2008), Wert et al. (2008)

AOC—assimilable organic carbon, BDOC—biodegradable organic carbon, DBP—disinfection by-product, DOC—dissolved organic carbon, EDCs—endocrine disrupting compounds, MIB—2-methylisoborneol, NOM—natural organic matter, PPCP— pharmaceutical and personal care product, T&O—taste and odor, TOC—total organic carbon

*Removal may depend on multiple site-specific factors, such as source water quality, pretreatment, filter media type, EBCT, and temperature.

†Good removal is generally expected for microcystins, anatoxin-a, and cylindrospermopsin. However, removal of saxitoxin may be limited.

to optimize costs by combining particle/pathogen filtration with biological treatment. Most utilities implement single-stage biofiltration, in which a single biofilter is used. Many utilities are also compliant with the Partnership for Safe Water (Brown et al. 2016), producing filtered water with turbidities less than 0.1 NTU. Disinfection credits for pathogen inactivation are attained before and/or after biofiltration. Design and operation of gravity media filters for particle/pathogen removal is further discussed in AWWA's Manual of Practice 37: Coagulation and Filtration.

2.1.2.2 NOM, DBP precursors, and DBPs. Biofiltration is mostly used to remove NOM, which is present in all water bodies. A survey of biofiltration facilities across North America found that biofilters typically removed between 10 and 20 percent of total organic carbon (TOC), although a wide range of removal was reported (from less than 5 to over 70 percent) (Terry and Summers 2018, Brown et al. 2016). Organic matter is the most ubiquitous precursor for DBP. Therefore, the preferred method of DBP formation control is removing the precursor organic matter before the disinfection step. Biofilters reduce the concentration of DBP precursors present in the water and/or formed in the treatment process. Depending on the DBP precursors and operational conditions, regulated trihalomethanes (THMs) and haloacetic acid (HAA5) formation can be reduced in the range of 10–70 percent (Cuthberson et al. 2020, Selbes et al. 2016). The removal of TOC and consequently DBP precursors is highly dependent on how biofilters are operated, including the temperature and extent of upstream oxidation. For instance, an upstream application of ozone can enhance the biodegradability of organic matter and thus improve their removal through a biofilter.

The removal of preformed DBPs can also be achieved through biofiltration, but the amount of removal depends on the biodegradability of DBPs and the operational conditions of biofilters. Some DBPs, such as THMs, are less biodegradable, and therefore limited removal is observed in biofilters. On the other hand, HAAs and many other unregulated DBPs (e.g., nitrogenous DBP precursors) are readily biodegradable, and higher removal can be achieved with biofiltration (Cuthberson et al. 2020, Stanford et al. 2019, Liu et al. 2017, Xie et al. 2004).

In addition to mitigating the formation of regulated DBPs, biofilters can also remove some dissolved organic nitrogen (Farré et al. 2011), which can be a precursor of nitrogenous DBPs. These emerging nitrogenous DBPs include, but are not limited to, nitrosamines, haloacetonitriles, and halonitromethanes. While removal of preformed nitrosamines is well-recognized with biofilters, the removal of nitrosamine precursors is not well understood. In some cases, biofiltration can increase the concentration of nitrosamine precursors (Liu et al. 2017), which can lead to a higher nitrosamine formation during downstream chloramination (Krasner et al. 2012). Source water characteristics, media type, and backwash procedures may affect the formation of nitrosamine precursors (Evans et al. 2019).

A fraction of the TOC is present as AOC, which is naturally present and can also be formed in greater quantity as a by-product of oxidation processes, such as ozone. Oxidation can increase BDOC and AOC, such as aldehydes, ketones, and carboxylic acids. While these constituents are commonly associated with preoxidation (e.g., ozone treatment), they can be detected as naturally occurring in the source water, as well. Biofilters can remove a nontrivial fraction of BDOC (i.e., 25–100 percent) (Brown et al. 2016). Biofiltration benefits the distribution system stability and water quality by improving BDOC removal, which, if not removed, leads to issues such as DBP formation, biofilm formation, bacteriological regrowth, loss of disinfectant residuals, and increased corrosion within drinking water distribution systems (LeChavellier et al. 2015).

2.1.2.3 T&O compounds. Geosmin and 2-methylisoborneol (MIB) are common T&O compounds that are nontoxic but cause aesthetic concerns and can be a cause of customer complaints. T&O compounds can be detected by people at low concentrations, as low as 4 and 9 ng/L for geosmin and MIB, respectively (McGuire et al. 1981). MIB and geosmin removal can range between 15 and 50 percent (Dickenson et al. 2018) with greater removal observed for geosmin than MIB (Dickenson et al. 2018). T&O compound removals down to nondetect levels have been shown with ozonation and biofiltration (Williams et al. 2012). Biofilters have been shown to control T&O over the long term and that the routine replacement of granular activated carbon (GAC) filter media is not required (Schneider and LeChavallier 2017).

2.1.2.4 Cyanotoxins and emerging contaminants. Blue-green algae, also known as cyanobacteria, are frequently found in surface water supplies. Exposure to drinking water contaminated with elevated concentrations of cyanotoxins, specifically microcystin and cylindrospermopsin, can cause liver and kidney damage. More utilities are discovering cyanobacteria in their source waters and this trend is anticipated to continue with climate change (Jeffrey et al. 2015). Algal cells can be physically removed through conventional filtration or an oxidation approach in which the cells that can lyse the toxins can be used

Table 2-2 Removal ranges of emerging contaminants with biofilters

	Moderate		
High >85%	50–85%	15–50%	Low 0–15%
Benzo[*a*]pyrene	Acetaminophen	Caffeine	Atenolol
DDT	Chlorobenzene	DEET	Carbamazepine
Fluoxetine	Clofibric acid	Erythromycin	Meprobamate
Formaldehyde	Dimethoate	Diclofenac	Primidone
Ibuprofen	Phenol	Gemfibrozil	Sucralose
Molinate	Triclosan	Trimethoprim	Sulfamethoxazole
p-toluenesulfonic acid	Microcystin-LA	Saxitoxin C1	TCEP
Saxitoxin C2	Microcystin-LR		Trichloromethane

Source: Modified from Dickenson et al. 2018
DDT—Dichloro-diphenyl-trichloroethane , DEET—N,N – diethyl-meta-toluamide, TCEP—Tris(2-carboxyetheyl)phosphine

as long as the dose and contact time are adequate for toxin removal. Although the removal of many cyanotoxins through water treatment is being researched, biofiltration has shown promise for the removal of cyanotoxins. Approximately 50 percent removal of both microcystin-LA and microcystin-LR have been observed in studies with biofiltration based on bench-scale testing of sand filters with an EBCT of 15 min with initial concentrations ranging from 3 to 20 μg/L (Ho et al. 2007). While saxitoxin C1 and C2 can be removed in biofilters, saxitoxins GTX2 and GTX3 can be formed. For low DOC source waters, anatoxin-a concentrations can be reduced to near zero with ozone doses as low as 1 mg/L (Newcombe and Nicholson 2004). There have been several studies on the biofiltration removal of some cyanotoxins (microcystins and saxitoxins), but other cyanotoxins have been less studied (nodularin, anatoxin-a, and cylindrospermopsin) and more case studies are needed (Dickenson et al. 2018).

Biofiltration can reduce emerging contaminants (anthropogenic compounds produced from human activity and detected at trace concentrations), such as PPCPs and pesticides. However, the extent or rate at which these constituents are degraded varies (Table 2-2), because some contaminants are more susceptible to biodegradation (e.g., acetaminophen, ibuprofen, triclosan) than others (e.g., atrazine, carbamazepine, meprobamate, sucralose) (Dickenson et al. 2018, Zearley and Summers 2012). Studies have shown varying removal rates of trace chemical constituents using biofiltration, as presented in Table 2-2. Furthermore, removal of these constituents is highly site-specific and dependent on how biofilters are designed and operated. EBCT, temperature, and ozone demand/dose have all been shown to affect contaminant removal (Halle et al. 2015, Lee et al. 2012, Zearley and Summers 2012). Long-term health effects from exposure to these compounds at low concentrations are unknown.

2.1.2.5 Inorganic compounds. Manganese is naturally occurring in most surface water supplies and can also be introduced into the treatment process from treatment chemicals. Biofilters have been shown to remove inorganics, including both total and dissolved iron and manganese. The primary removal mechanisms of particulate inorganics (e.g., oxidized upstream) occur because of physical filtration, whereas soluble iron and manganese are primarily removed by iron- and manganese-oxidizing bacteria present on the filter media (Kohl and Dixon 2012). Typically, increased removal of dissolved manganese is observed at higher pH (greater than 7.5) and dissolved oxygen (DO) greater than 5 mg/L. While DO is important, sufficient DO is generally present in the influent of all surface water biofilters.

If ammonia is present in the source water, biofilters may also oxidize ammonia to nitrite via ammonia-oxidizing bacteria and/or ammonia-oxidizing archaea. Nitrite can further be oxidized to nitrate by nitrite-oxidizing bacteria. However, ammonia is typically not a contaminant of concern in surface water supplies because of the low levels typically found. Biological oxidation of manganese, iron, and ammonia from groundwater sources is further discussed in Chapter 3 of this manual.

2.1.3 Removal Mechanisms

This section describes the two primary removal mechanisms in a biofilter: physical separation of particle and pathogen and biodegradation/biotransformation. Surface water biofilters are typically conventional gravity filters operated to encourage biological activity in a biofilm. Because the primary function of most surface water biofilters is particle and pathogen removal, physical separation (i.e., filtering) of particles is a key removal mechanism. Particles may be physically strained if they are larger than the media void space, can settle onto filter media, or can collide with the filter media via interception or diffusion (i.e., Brownian motion). Upstream charge neutralization (coagulation) is important for removing particulate contaminants. These mechanisms are described further in AWWA's M37.

In addition to particle removal and adsorption, another removal mechanism observed in surface water biofilters is biodegradation/biotransformation. When minimal or no oxidant residual is present across conventional gravity drinking water filters, a biofilm of microorganisms (from the source water) attaches to the granular filtration media, with greater biomass observed at the top of filters where nutrients and oxygen are more abundant. Drinking water biofilms generally contain 10–90 percent by weight extracellular polymeric substances (EPS) (Flemming and Wingender 2010). EPS are primarily proteins and carbohydrates excreted by microorganisms that provide the biofilm structure and support the stability and adhesion of biofilms (Wimpenny and Colasanti 1997, Stoodley et al. 1999).

Over time, biomass increases to a steady-state quantity, and the microorganisms acclimate (refer to Chapter 1) to the site-specific water quality and conditions. The presence, abundance, diversity, and activity of microorganisms vary between biofilters on the basis of the unique microbiota in the water source and site-specific conditions, such as upstream treatment processes, water quality, media type, and operational practices (de Vera et al. 2018, Moll et al. 1998, Pinto et al. 2012). The microbial community may also vary with filter depth (Moll et al. 1998). Additionally, not all microorganisms present in the influent proliferate in a biofilter, and some microorganisms will detach more frequently than others (i.e., be more present in the filter effluent) (Pinto et al. 2012); detached microorganisms should be subsequently inactivated in the primary and secondary disinfection processes downstream. Once fully acclimated, the microbiome in a specific biofilter is generally stable over time (Pinto et al. 2012). Depending on the contaminant targeted and site-specific conditions, contaminant removal via biodegradation/biotransformation may happen in a filter after a few weeks of operation to several months (Evans et al. 2019, Dickenson et al. 2018, Brown et al. 2016, Zearley and Summers 2012).

Biological removal mechanisms include redox reactions and secondary microbial reactions, which are described in more detail in Chapter 1. Through redox reactions, microorganisms transfer electrons between electron acceptors (i.e., oxygen for this application) and electron donors (i.e., dissolved organic or inorganic contaminants). DO concentrations in surface waters are generally sufficient to support microbial oxidation of organic and inorganic contaminants; if ozonation is upstream of biofiltration, the biofilter influent is typically supersaturated. Additionally, the BDOC concentration is generally sufficient in surface water biofilters to support cellular processes. However, greater organic carbon

Table 2-3 Example pathways for biological contaminant removal

Electron Acceptor	+	Electron Donor	→	End Products
Example 1. Removal of organic carbon (aerobic respiration)				
Oxygen	+	BDOC	→	Carbon dioxide + water + energy
Example 2. Transformation of ammonia to nitrite and nitrate contaminants				
Oxygen	+	Ammonia	→	Nitrite + hydrogen ion + water + energy
Oxygen	+	nitrite	→	nitrate + energy

BDOC—biodegradable organic carbon

removal and microbial activity are often observed if upstream oxidation transforms DOC into a more readily biodegradable form (Brown et al. 2020, Snyder et al. 2007).

Table 2-3 summarizes two examples of contaminant transformations or redox reactions that are mediated by microorganisms in surface water biofilters. The first reaction, aerobic respiration, supports microbial growth and energy production (i.e., primary cellular functions); energy is stored in cells as adenosine triphosphate (ATP) molecules (Bonora et al. 2012). In addition to supporting cellular processes, and thus, a healthy microbial community, the aerobic respiration process also removes organic carbon, the most targeted contaminant for drinking water biofilters (Brown et al. 2016).

The extent and rate of biological transformation of contaminants varies on the basis of several environmental factors, including overall water quality (e.g., temperature, pH, alkalinity), contaminant concentrations, the microbial community, and the time available for biotransformation. Some contaminants are more readily biotransformed (e.g., NOM), whereas other contaminants may require a greater contact time (e.g., manganese and atrazine) (Evans et al. 2019, Dickenson et al. 2018, Greenstein et al. 2018, Zearley and Summers 2012). Certain contaminants, in some cases, may also interfere with the removal of other contaminants (e.g., presence of ammonia may interfere with manganese oxidation) (Mouchet 1992, Vandenabeele et al. 1995, Kohl and Dixon 2012); however, this is not as common in surface water as groundwater because of the low concentrations of contaminants typically in surface water (Evans et al. 2019). Thus, adjusting the contact time in the biofilter or environmental conditions in the biofilter influent (e.g., pH adjustment or upstream oxidation) may promote greater contaminant removal (Lauderdale et al. 2018).

Additionally, if GAC is used as filter media, adsorption of contaminants may also occur. Adsorption of contaminants decreases over time. Adsorption may occur for months or up to many years depending on the contaminant targeted and site-specific conditions (Brown et al. 2020). Over that time, contaminants may also desorb from GAC, or, in the special case of biofilters, biodegradation of the adsorbed contaminant may regenerate the adsorption site allowing for additional adsorption (Brown et al. 2020).

2.1.4 Regulatory Considerations

Biofiltration has been reported as a widely accepted operational practice and does not have any regulated design or monitoring requirements beyond those in place for granular media filtration (Nieminski and Perry 2015). When granular media filters are operated biologically, the regulatory requirements (e.g., related to particle and pathogen removal) that apply for other granular media filters apply to biofilters, as well. Like other granular media filters, biofilters also need to meet the requirements of the Safe Drinking Water Act (SDWA) and Surface Water Treatment Rule, including the Interim Enhanced Surface Water Treatment Rule, Filter Backwash Recycling Rule, and Long Term 1 and Long Term 2 Enhanced Surface Water Treatment Rules.

Biofilters typically achieve higher removal of DBP precursors than conventional filters, which may also help utilities achieve the Stage 2 DBPR. Biofilters also improve distribution system biostability, which reduces regrowth and aids with compliance of the Total Coliform Rule.

2.2 PLANNING

Biofiltration for surface water has been well proven at full scale (e.g., Evans et al. 2019). However, performance varies on the basis of site-specific design and operating conditions. Thus, the following sections outline best practices for evaluating biofiltration before full-scale implementation through establishing treatment goals, conducting an applicability evaluation, and developing a robust testing program.

2.2.1 Treatment Goals

Establishing treatment goals is an important first step in evaluating biofiltration. A range of water quality and performance data should be reviewed for multiple years to allow for identifying seasonal and temporal trends (e.g., less organic carbon removal might be required in winter compared with summer). In addition to historical water quality, treatment goals should consider regulatory requirements and anticipated treatment by upstream and downstream processes. When data are not available (e.g., new facilities), models or estimations of water quality may be needed. Example biofiltration goals that are site-specific, quantifiable, and consider seasonality are provided, as follows:

- achieve an effluent turbidity less than X at least 95 percent of the time;
- achieve a biofilter effluent concentration of X mg/L for contaminant A;
- achieve X percent removal of contaminant B in summer and Y percent removal of contaminant B in winter;
- maintain biofilter run times of at least X hours in summer and Y hours in winter to achieve unit filter run volumes (UFRVs) of X gal/ft^2 and Y gal/ft^2 in summer and winter, respectively; and
- maintain terminal biofilter head loss of no more than X feet.

2.2.2 Applicability Evaluation

Once treatment goals are established, a desktop evaluation should be conducted to evaluate the suitability of biofiltration on the basis of biofilter influent water quality and treatment goals and develop a business case for proceeding with biofiltration testing. This evaluation should also identify testing objectives and testing conditions (i.e., optimization strategies) likely to meet the treatment goals (i.e., inform testing design).

A desktop evaluation may be performed to benchmark historical biofilter influent water quality and treatment performance (i.e., upstream, downstream, and across filtration). At least a couple of years of data should be analyzed to understand the range of water quality and treatment characteristics that should be considered using statistics such as 5th and 95th percentiles or through calculation of outliers. Analysis should also consider seasonal trends and temporal changes. If historical data are not available (e.g., new facilities or because of upstream treatment changes) and sampling to collect data is not possible, water quality and treatment may need to be estimated analyzing some relevant case studies and/or using a model (e.g., using the USEPA Water Treatment Plant [WTP] Model, USEPA Drinking Water Treatability Database, AWWA Water Quality and Treatment: A Handbook on Drinking Water, or other models) for the biofilter influent water quality before considering potential biofilter performance.

A desktop evaluation of biofiltration can assess the potential for achieving the established treatment on the basis of peer facilities and published data. This assessment should also identify potential designs and operational and optimization strategies that should be tested. In addition to conducting literature reviews (e.g., journals and WRF reports) or modeling biomass development and/or biodegradation of contaminants (e.g., BIOFILT Model developed by Hozalski and Bouwer [2001, 2000]), two tools specifically for surface water biofiltration have been developed to aid utilities in developing a business case for investing in testing:

- **Biofiltration Knowledge Base (www.biofiltrationknowledgebase.org):** The Biofiltration Knowledge Base (Brown et al. 2016) is a compendium of full-scale biofiltration experience from over 45 utilities across North America, including data from planning, design, implementation, and operation of biofiltration. The website, which includes case studies, a library, and reports summarizing data submitted by utilities, is a good starting place for finding peer facilities with similar characteristics as the facility under consideration.
- **Biofilter Conversion Assessment Tool (CAT):** The CAT is a downloadable Microsoft Excel tool (https://www.waterrf.org/resource/biofilter-conversion-assessment-tool) that assesses the site-specific suitability of biofiltration and provides recommended optimization strategies using responses to 24 multiple choices or "select all that apply" questions. The CAT also compares select responses with data from the Biofiltration Knowledge Base. Although the CAT does not cover every situation, it provides a high-level assessment regarding the most common decision criteria for proceeding with biofiltration testing.

Last, an applicability evaluation should acknowledge the limitations with a desktop study. Data from other facilities may not match actual site-specific conditions. Additionally, desktop studies are often not able to accurately predict the effect of changing water quality or operating conditions. Thus, if the applicability evaluation is favorable, a key recommendation should include site-specific testing.

2.2.3 Testing

Robust biofiltration testing programs optimize design criteria, improve capital and operating cost estimate accuracy, and improve full-scale performance. Although surface water biofiltration is well proven, performance may vary depending on site-specific water quality and treatment conditions. Best practices are summarized in the following section for establishing testing objectives, determining testing scale(s), designing the test, and evaluating testing outcomes.

2.2.3.1 Testing objectives. The first step in designing a testing program is to define the objectives of testing. Biofiltration testing is generally conducted to confirm that treatment goals can be achieved and to optimize biofilter design and performance. Testing objectives should include verification that full-scale treatment goals can be achieved, but may also include additional objectives specific to defining successful testing. Although biofiltration testing may also be conducted for research studies with objectives relating to developing a mechanistic understanding (i.e., how do microorganisms remove contaminants), research studies are not further discussed in this Manual of Practice. Common testing objectives include the following:

- **optimize biofilter design**, such as media profiles, backwash and filter-to-waste protocols and chemical additions;
- **optimize biofilter operation**, such as filter hydraulics, water production, and EBCT;

Table 2-4 Summary of benefits, limitations, and common objectives of each testing scale

Key Benefits	Key Limitations	Common Testing Objectives
	Bench	
• Conditions are strategically controlled to allow for testing of specific variable(s). • Cost-effective testing of many design and operational strategies	• Does not match full-scale filter production, hydraulic performance, turbidity removal, or backwash protocols • May not capture the effect of seasonal water quality changes because of controlling variables • Hydraulic performance and turbidity removal are not scalable.	• Contaminant biodegradation or biotransformation • Media type and EBCT • Optimization strategies (for biodegradation or biotransformation)
	Pilot	
• Tests seasonality (i.e., testing conducted over a longer duration reflecting changes in biofilter influent water quality) • Tests water production and compliance (e.g., effluent turbidity) • Conditions represent full-scale process design • Establishes expectations for full-scale performance	• Generally limited to two to six columns (i.e., parameters tested in parallel) • Greater cost per filter column and greater water, energy, and space requirements than bench-scale • Requires a control column because of changing influent water quality; confounding factors could cause ambiguous results	Confirmation of bench testing objectives at a more representative scale, plus the following: • Particulate removal • Hydraulic performance • Screening-level investigations of backwashing protocols, such as the effect of oxidant addition
	Demonstration (e.g., partial conversion)	
• Confirms results of testing at bench and pilot with conditions matching full-scale • Tests or matches full-scale backwashing protocols • Could be most cost-effective for testing a single optimization strategy if a full-scale filter can be isolated for testing • Provides operator training on biofilter operation or optimization strategies	• Generally limited to two filters (i.e., parameters tested in parallel) with one being a control • Greater cost per filter column and greater water, energy, and space requirements than pilot-scale if side-stream demonstration filters are procured • Requires a control column because of the changing influent water quality; confounding factors could cause ambiguous results	Confirmation of pilot testing objectives at a more representative scale, plus the following: • Complete investigation of backwashing protocols

EBCT—empty bed contact time

- **verify that full-scale treatment goals (Section 2.2.1) are achieved**, including particle filtration and biodegradation of organic, inorganic, and emerging contaminants;
- **test performance under challenging conditions,** representing extreme historical water quality and operating conditions and/or predicted future worst-case scenarios (e.g., through contaminant spiking);
- **develop a business case for full-scale implementation,** by obtaining sufficient data, based on duration or volume of data, to increase confidence in biofiltration performance; and
- **provide training opportunities for operators** to learn how to operate and monitor biofilters.

2.2.3.2 Testing scales. Next, the testing scale(s) should be determined on the basis of the established testing objectives. Testing at multiple scales is recommended proceeding from scales that allow testing of a greater number of parameters and in more controlled conditions to scales more representative of anticipated full-scale design and operating conditions (i.e., conditions that will scale up more accurately). Each testing scale was defined in Section 1.5. Table 2-4 summarizes key considerations for surface water biofiltration testing at each scale.

2.2.3.3 Testing design. Testing programs are most successful when carefully planned and designed. A detailed biofiltration testing plan should begin with a summary of the testing objectives (Section 2.2.3.1) and the outcomes of the applicability evaluation (Section 2.2.2) including historical water quality and performance and a summary of the current or anticipated full-scale design and operational parameters. In addition, the testing plan should capture the schedule, key design criteria of the filter design to be evaluated, testing conditions, and sampling regime to assess performance.

2.2.3.3.1 Testing schedule. The testing schedule should account for the time spent developing a testing plan, procuring equipment and materials, conducting testing, and reporting testing results. Testing durations often range from three months to over a year. Shorter studies may be sufficient for targeted bench studies and/or a demonstration test using a full-scale filter (e.g., to test a single optimization strategy during a specific time of year). However, most testing programs target capturing performance across at least a couple of seasons with the final scale of testing before implementation capturing performance across all seasons. The duration of testing should allow for acclimation (defined in Section 1.7) and steady-state operation (described in Section 2.4.3) under anticipated typical operating conditions before changing any operating conditions (i.e., baseline testing). Additional time should be allowed for the system to reach a steady state

Table 2-5 Description of key filter design factors for testing

Design Factor	Description
Seasonal water quality variations	Generally, site-specific (e.g., settled water) fed continuously; some bench systems may use batches of water or a lab water.
Number of filters	Selected on the basis of testing objectives to testing each variable; a control filter is typically included.
Empty Bed Contact Time (EBCT)	Designed to match anticipated average and minimum full-scale EBCTs.
Flow rate, as defined in Section 1.7	Generally, in the gpm (L/min) range and based on anticipated full-scale maximum and minimum loading rates for pilot and full-scale systems; bench-scale systems are typically in the mL/min range and selected to achieve the target EBCT.
Media L/d ratio	For filters targeting particulate removal, a L (depth of media) to d (media effective size) of >1,000 is generally recommended.
Filter column diameter	Selected on the basis of scale/economy: • ≤2-in. diameter or the filter column diameter to media diameter ratio is >15:1 for bench scale. • 4- to 8-in. diameter for pilot scale. • In the range of 50 ft^2 or as large as an existing full-scale filter for demonstration systems.
Media type	Selected to match the current or anticipated full-scale design.
Media depth	Selected to match the current or anticipated full-scale design; bench system media depth is generally selected on the basis of the column size and targeted EBCT.
Media effective size	Selected to match the current or anticipated full-scale design; depending on the bench column diameter, bench media may be sized smaller to avoid wall effects.

EBCT—empty bed contact time

following each planned or unplanned change in operating conditions (e.g., a shutdown or influent water quality).

2.2.3.3.2 Test filter design. The test filter design should match either the current or anticipated full-scale design but will vary on the basis of the testing scale. Key design criteria are described in Table 2-5. In many cases, multiple test filters are designed (e.g., with different media profiles) to allow for the testing of design alternatives under consideration.

New media (i.e., virgin) for test columns may be acquired from a vendor, or spent media may be sourced from existing full-scale biofilters or another biofiltration facility. The use of virgin media ensures the design media's effective size and alleviates any interactions due to the biofilm and precipitates that may have accumulated on the media over time. However, it may require weeks or months to establish a biofilm on virgin media before beginning testing, and washing is required to remove fines before installation. Additionally, if GAC is used, separating biological degradation from physical adsorption of contaminants may be challenging. Thus, the main advantages of using spent media are the shortened acclimation time (i.e., a biofilm is already established) and, if GAC is used, the reduced significance of adsorption as a removal mechanism. However, with spent media, acclimation to a different water quality may be required, accumulated precipitates must be considered, and sieve analysis should be performed to validate that the media represent the desired design effective size.

Bench and pilot system backwashing protocols are not representative of full scale because of sidewall effects and are often selected on the basis of what is available for the testing system. Although backwashing protocols can be optimized (e.g., with media expansion goals) to ensure media are cleaned, the design of full-scale backwashing protocols should be based on demonstration testing and/or knowledge from other full-scale facilities. Test filters are, however, beneficial at any scale for testing the effect of backwashing with an oxidant on the microbial community, if that strategy is being considered full scale.

2.2.3.3.3 Testing conditions. In addition to testing ambient or typical conditions, testing should also incorporate imposed challenge conditions representative of the range of water quality and operational conditions anticipated full scale. Ambient conditions may include the use of a filter aid polymer, if anticipated full scale, as well as pretreatment, such as the addition of upstream ozonation, or optimization through filter influent chemical addition. Chemicals commonly used for optimization include oxidants (i.e., for head loss management), nutrients, and pH adjustment (e.g., for improved manganese removal). Common challenge conditions include flow (i.e., lowest anticipated EBCT), water temperature (i.e., testing during both cold and warm seasons), and contaminant spiking (i.e., to mimic the highest anticipated future concentration of contaminants). Depending on testing objectives, these conditions may be tested over the study duration or in different combinations for short durations (e.g., a contaminant spiked for two weeks). The test plan should outline a schedule for testing conditions, including anticipated doses for all chemicals.

2.2.3.4 Testing outcomes. Understanding expected outcomes is the final component of a well-designed testing plan. Conclusions should acknowledge limitations of the testing scale (Section 2.2.3.2) and compare actual outcomes with expected outcomes. For example, if a testing duration of less than a year was conducted, all water quality and operational conditions were likely not tested. Recommendations should include additional testing, if warranted at either the current scale or by proceeding to a testing scale more representative of full-scale conditions (refer to Section 2.2.3.2). The completion of an initial testing plan is generally not the end of a well-designed testing program. Ending testing too early often results in more expensive designs and suboptimal full-scale performance. More information on how to design a biofiltration testing program along with decision matrixes for evaluating testing outcomes and next steps is available in the Biofiltration Guidance Manual for Drinking Water Facilities (Brown et al. 2020).

2.3 DESIGN

This section describes design considerations for implementing biological filtration for the treatment of surface water.

2.3.1 Process Configurations

Biofiltration for surface water treatment relies on granular media. There may or may not be an oxidant added upstream, such as ozonation. There are several biological direct filtration systems that do not implement preoxidation. Pretreatment can vary from upstream softening to processes such as ballasted sedimentation. Figure 2-1 shows some example process flow diagrams for the most common applications. As mentioned in Section 2.2, an understanding of treatment goals, applicability, and testing should drive other process selections.

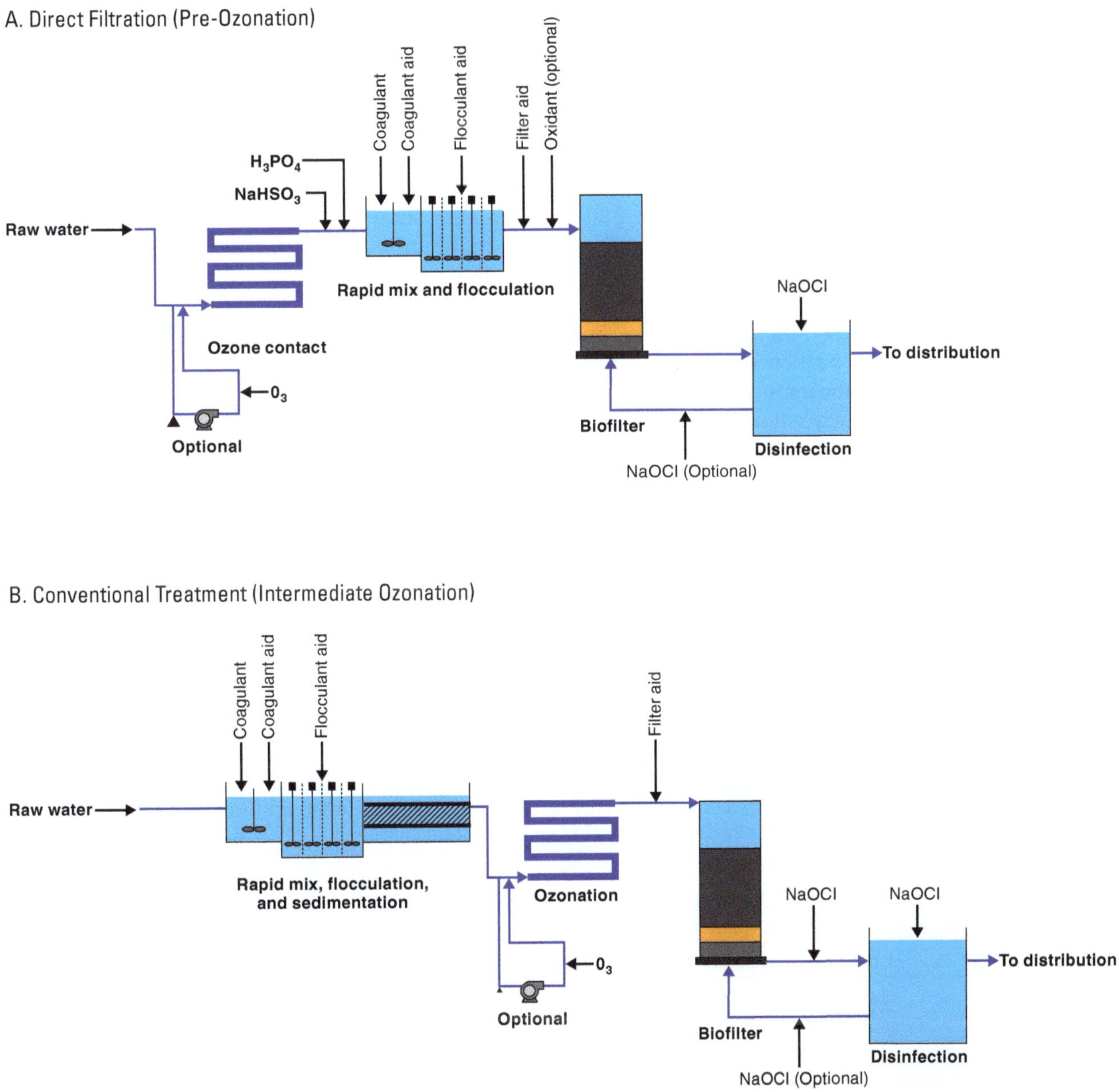

Figure 2-1 Example process flow diagrams for most common intentional biofiltration processes: (A) direct filtration (pre-ozonation) and (B) conventional treatment (intermediate ozonation)

2.3.2 Design Components

Numerous references are available that provide rigorous filter design guidance (AWWA's M37, Crittenden et al. 2005). This section focuses on additional design elements important to biological treatment performance, including pretreatment, filter underdrains, filter trough height, media type and configuration, EBCT, chemical feeds, backwash protocols, chlorine/oxidant addition, and residuals handling. An additional resource for the biological filtration design for surface water treatment applications is the WRF's Guidance Manual for Surface Water Treatment (Brown et al. 2020), which is publicly available.

2.3.2.1 Pretreatment. Application of ozone or other strong oxidants results in breaking down the NOM present in the source water to AOC. The AOC formed during preoxidation further enhances biofiltration. Because ozone has a high oxidation potential, it is typically used together with biofilters. As a strong oxidant, ozone not only converts NOM into AOC/BDOC but also oxidizes metals and transforms trace organic constituents. Furthermore, the addition of ozone can also be used for additional disinfection credits within the treatment process. The ozone dose needed for disinfection credit is often higher than the dose needed for biodegradation and filter productivity.

Upstream processes or pretreatment can be designed to promote biofiltration in the filters and/or to target holistic treatment across the water treatment plant. Optimization of upstream processes may include adjusting coagulant and preoxidant dosages and types. Table 2-6 summarizes chemicals added to the treatment process for filters with biofiltration and typical dose ranges. Filter feed quality should be assessed (in addition to raw water quality) to understand constituents that biofiltration will need to remove and the level of nutrients available for biological activity (coagulation, flocculation, and sedimentation will reduce raw water nutrient concentrations).

2.3.2.2 Media type, size, and configuration. Filter media serves two purposes in biofilters: particle removal (i.e., filtration) and physical support for the growth of microbial communities. Optimizing filter media design criteria, including media type, depth, and effective size, should be taken into consideration to achieve performance objectives. The effective size of the media affects filter hydraulics, particle retention, and biomass levels. The most common media types of surface water biofilters are GAC, anthracite, and sand. The majority of biofilters are comprised of dual media, with GAC or anthracite over sand, although several biological plants have operated successfully using mono-media

Table 2-6 Common chemicals added in biofiltration applications (WRF Project #4719)

Chemical	Use
Filter aid polymer	Added upstream to reduce filter ripening periods with respect to turbidity removal. Consideration for dosage due to head loss accumulation with biological activity.
Ozone	Added to raw water (pre-ozonation) or settled water (intermediate ozonation) to break down organic compounds into more readily degradable compounds and increase DO levels.
Hydrogen peroxide	May be added to feedwater or backwash water to help maintain uniform hydraulic conditions in the bed and minimize underdrain clogging.
Nitrogen and phosphorus	Depending on filter feedwater quality, nutrients may be added to improve constituent removals.
Quenching agent	Can be used to quench an oxidant residual before filtration, such as calcium thiosulfate.
pH adjustment	A base, such as soda ash, can be added if pH is too acidic.
Chlorine	May not be added at all to promote biofiltration or added intermittently to control biofiltration growth and head loss levels. Chlorine can be used as preoxidant if residual levels are quenched or very low.

Source: Adapted from Brown et al. 2020
DO—dissolved oxygen

configurations (Upadhyaya et al. 2017). When 43 full-scale facilities were surveyed as a part of the WRF's Knowledge Base project (Brown et al. 2016), 74 percent of utilities had dual media. Table 2-7 compares different media types.

2.3.2.3 EBCT. EBCT is a design parameter that describes the length of time water is in contact with the filter media. It is calculated by dividing the total volume of the media bed (including pore space) by the flow rate across the bed. EBCT is expressed in units of minutes. EBCT is an important parameter because it is strongly correlated with the rate of biological removal. In general, typical EBCTs range from 5 to 15 min (Dickenson et al. 2018), and longer EBCTs typically yield higher removal. Short EBCTs (less than 2 min) do not provide sufficient contact time for the removal of influent contaminants, and longer EBCTs (more than 30 min) provide diminishing returns on the removal of influent contaminants for the capital cost to implement such a long EBCT. Lower EBCTs (around 5 min) are generally acceptable for organic carbon removal. But many trace contaminants or compounds of emerging concern, and even inorganic compounds, such as manganese, can require longer EBCTs, especially when conditions are more challenging, such as low temperature.

To determine the optimal EBCT, the cost of the filter design and contaminant removal goals should be balanced. Methods to evaluate the effect of EBCT on biofilter performance are discussed in Section 2.2.3.

2.3.2.4 Filter underdrains. The purpose of the filter underdrain is to support the filter media and to maintain hydraulic flow during production and backwashing. Filter underdrain and backwash system selection criteria are discussed in AWWA's M37 and Brown et al. 2020. Specific to biofiltration, excess biomass accumulation may lead to fouling of filter underdrain equipment. There are three types of underdrains used for biofiltration that provide good results: gravel, block, and nozzle. Gravel underdrains are the least susceptible to biofouling, but they provide less effective production and backwash distribution compared with block and nozzle underdrains. Gravel underdrains require a higher profile and increase the overall cost of the filter box.

Fouling of filter underdrains can limit production rates/water efficiency, and in rare cases, lead to underdrain failures (Lauderdale et al. 2011). Underdrains can foul because of biofouling or mineral scaling if they are not effectively cleaned during backwash cycles. Additional monitoring can mitigate this issue by measuring backwash pressure or backwash underdrain head loss, ideally at the filter, to assess if filter underdrain fouling is an issue. A trend of increasing clean-bed head loss over time can also be an indicator of filter underdrain fouling.

Table 2-7 Media selection considerations

GAC	Anthracite
• Provides adsorption capacity, although this is temporary without regular media changeout, which can be beneficial as biofilter acclimates. • More conducive to maintaining higher biomass levels; this can be advantageous in situations where biofilters experience one or more environmental stressors (e.g., cold temperatures, regular presence of chlorine). • Holds contaminants for longer on the media through adsorption, resulting in the slow biodegradation of contaminants, thus releasing sites for further adsorption. • Catalytically reacts with oxidants and head loss control strategies can be more limited. • GAC has demonstrated greater contaminant removals for some contaminants, such as manganese.	• May limit biological activity in biofilters that experience one or more environmental stressors. • Less expensive (approximately five times less). • Less susceptible to abrasion.

Source: Adapted from Lauderdale et al. 2018
GAC—granular activated carbon

2.3.2.5 Backwash system. Effective backwashing is critical to breaking up and controlling biomass and extracellular substances on the media, which can increase head loss accumulation, especially during warmer months. Most facilities use air scour and a simultaneous air scour/backwash. Of 43 full-scale facilities surveyed, 70 percent used air scour (Brown et al. 2016). Backwash steps typically include air scour, simultaneous air scour/backwash, low-rate backwash, high-rate backwash, low-rate backwash, settling time, and filter- or rinse-to-waste. Backwashing frequency typically is between 24 and 96 hours but can be as high as 200 hours (Brown et al. 2016).

Air scour is recommended for the backwashing regime, especially if deep-bed filters are used. According to WRF Project #4459 research, approximately 70 percent of utilities with biofiltration have air scouring for the removal of particle accumulation (Brown et al. 2016). A surface wash system alone (without air scour) may not be effective in meeting backwash cleaning goals in most biofilter applications (Brown et al. 2016).

Low- and high-rate backwashes should be included in the design, as well as filter-to-waste. Filter-to-waste or rinse-to-waste is also recommended to reduce the effect of particle breakthrough following a backwash. Filter-to-waste discharges filtered water to the backwash washwater handling system until the filtered water achieves the utility's turbidity water quality goal.

Most full-scale facilities practicing biofiltration (76 percent of 43 facilities surveyed) typically operate with a nonchlorinated backwash (Brown et al. 2016). Operating with an unchlorinated backwash typically requires a separate storage tank or channel separate from the clearwell. Less biological activity has been found on anthracite media filters with a chlorinated backwash (Hooper 2015). Filters with GAC media see fewer effects from a chlorinated or chloraminated backwash because of the catalytic reaction of chlorine or chloramines with the GAC media. The desired design configuration is site-specific and depends on water quality goals. Introduction of chlorine periodically in the backwash can control biomass and mitigate biomass release during filter runs, and the ability to have this capability is advised.

As the microorganisms proliferate, the biofilm may restrict the pore space within the granular media. This induces additional head loss in addition to the captured solids/particulates that are present in the feedwater. Therefore, hydraulics of the treatment process should be considered because higher head loss is typically associated with biofilters when compared with conventional filters. To mitigate higher head loss conditions, a more frequent backwash strategy may be necessary with biofilters. If implemented, this can result in a higher amount of filter backwash waste.

2.3.2.6 Head loss mitigation strategies. In addition to larger media size (Section 2.3.2.2) and limiting filter aid coagulants/polymers, head loss accumulation due to biomass growth can be managed with periodic backwashing with an oxidant residual and/or addition of a low oxidant dose to the filter influent. Best practice for biofilter design is to include injection ports for the flexibility to dose an oxidant to the backwash as well as the filter influent. Typical oxidants used for this purpose include chlorine and hydrogen peroxide. Of the reporting utilities in the Biofiltration Knowledge Base, 24 % use a chlorinated backwash regularly or when biofouling or algal growth occurs in the filter box.

2.3.2.7 Posttreatment requirements. This section discusses posttreatment requirements following biofiltration. Existing regulations should be protective of public health for biofiltration. However, there are some factors to consider for the design or modification of an existing surface water treatment plant:

- In many cases, chlorine demand will be lower and more stable because of the additional organics removal through biofiltration.
- There can be periodic biological sloughing and higher heterotrophic plate count (HPC) levels in the filtrate, but typical final disinfection should be adequate to control HPC bacteria.

- An unchlorinated backwash can promote increased biological activity on a filter compared with a chlorinated backwash. Most full-scale facilities with anthracite media practicing biofiltration operate with a nonchlorinated backwash (Brown et al. 2016). Bifurcation of the clearwell or an additional means to achieve an unchlorinated backwash should be considered, along with the additional cost. A means to dose chlorine into the backwash should be included in case clean-bed head loss increases over time.

2.3.2.8 Residuals handling. Residuals handling is treated similarly for biofilters and conventional filters. For more details on residuals handling in granular media filters, refer to Cornwell and Roth (2011) or Chapter 18 of the 5th edition of ***Water Treatment Plant Design*** (AWWA/ASCE 2012).

Biofiltration produces less waste when compared with other treatment processes, such as membrane filtration, because contaminants are being broken down to innocuous end products rather than sequestered in the waste stream. Biofiltration may also reduce the amount of primary coagulant or powdered activated carbon (PAC) needed to achieve a similar organic carbon reduction across the treatment facility, which may decrease overall solids handling requirements. However, biofiltration can increase head loss accumulation rates and therefore increase backwash frequency, which should be accounted for in solids handling design.

If using biofiltration for the removal of dissolved inorganic compounds, care must be taken to holistically manage all sources, including the biofilters themselves. For example, some of the manganese solids that are filtered out in the biofiltration process and removed during backwashing can be released back into the dissolved state during solids handling and anoxic conditions. If recycling, it is recommended to assess the return stream's contribution of manganese.

2.4 OPERATIONS

Biofiltration monitoring is an important aspect of operations to understand biofiltration performance. Biofilter operation may include the following phases or modes of operation: startup and acclimation; steady-state operation; and shutdown, idling, and restart. Although there are no regulated monitoring requirements specific to biofiltration beyond those in place for conventional filtration (Nieminski and Perry 2015, AWWA's M37), best practices for biofiltration incorporate monitoring and evaluation of biological, water quality, filter hydraulics, and operational parameters, in addition to any specific contaminants of interest for which the process is used. Biofiltration treatment objectives (e.g., removal of target contaminants, hydraulic performance) should be used to guide performance monitoring and optimize performance. An operations and maintenance plan should also be developed that outlines expectations and protocols for the various phases of operation and troubleshooting.

2.4.1 Monitoring

A variety of tools are available for biofiltration monitoring (Brown et al. 2020, Hooper et al. 2019). Utilities should develop a monitoring strategy and data management plan, including a comparison of biofiltration performance data against site-specific filter treatment objectives. The plan should include monitoring tools and methods, sampling location (e.g., filter influent/effluent), sampling quality assurance and quality control, and frequency of analysis specific to each phase of operation (i.e., startup/acclimation, steady state, shutdown/idling/restart). The plan should also detail data management, analysis, and reporting procedures. For routine operation, the monitoring strategy should characterize the process and develop a baseline for troubleshooting or optimization.

This subsection describes the monitoring tools that are recommended for plants operating biofilters to inform operators on the health of their biofilters, upstream and downstream operations, and anticipate potential operational or water quality performance changes (e.g., decrease in contaminant removal or hydraulic issues). A well-designed monitoring program should do the following:

- account for physical characteristics to confirm media and underdrains are within original and operational specifications;
- assess the biological community;
- alert operators to potential operational changes, such as filter head loss accumulation and underdrain fouling; and
- include chemical and water quality parameters related to treatment goals that can be key indicators of the biological health of the filters.

Additional monitoring or operational actions may be required for troubleshooting specific issues. For example, a facility should monitor clean-bed head loss over time to identify hydraulic issues. If filter hydraulics have been compromised, then a more rigorous or chlorinated backwash can be used. If filter effluent water quality has been compromised, then media ATP can be used to evaluate whether loss in biomass has occurred. Media ATP may prove better than the biofilm formation rate for troubleshooting because analysis can be conducted quickly, whereas the biofilm formation rate requires two weeks of incubation. Additional tools may prove helpful in determining the cause of upset conditions, such as carboxylic acid analysis, as well as operational tools, like oxidant demand and the ratio of upstream oxidant dose to TOC.

2.4.1.1 Operational parameters. Table 2-8 lists typical hydraulic and operating parameters.

2.4.1.2 Water quality parameters. Table 2-9 summarizes recommended monitoring for water quality parameters. General water quality parameters that are typically monitored on biofilters include turbidity, temperature, and pH. Turbidity is required for regulatory compliance, and utility goals for filtered water quality can be based on Partnership for Safe Water or SDWA requirements with desired safety factor. Biofilters can achieve turbidity removal similar to that of conventional filters. However, biofilter filtrate can have slightly higher particle counts associated with biofilm sloughing from the media. Monitoring of temperature may provide causation for a change in a constituent's removal related to biofiltration, as biological removal rates are strongly affected by water temperature (e.g., approximately double for every 10°C increase in water temperature). Biofiltration performance can also be affected by pH, so influent pH is commonly monitored and controlled within a site-specific range for optimal removal of turbidity, manganese, and other target contaminants. Continuous monitoring of these parameters is commonly accomplished through the use of supervisory control and data acquisition systems. These parameters should also be verified manually via grab samples for quality control.

Additional water quality monitoring will depend on desired constituent removal and water quality goals. For example, if a goal of biofiltration is to mitigate DBP formation, then DBP precursor removal (i.e., TOC, DOC, and/or UV254) would likely be monitored on a weekly or biweekly basis. If iron and manganese removal is required, monitoring for influent and effluent particulate and dissolved iron and manganese is important to assess performance. Target compounds associated with episodic events, such as T&O associated with algal blooms (e.g., MIB, geosmin), may be monitored frequently (e.g., daily) during the event and less often at other times.

2.4.1.3 Microbial parameters. Monitoring the biofiltration process for biological parameters is important for decoupling the biological component from chemical and physical processes occurring within the filter. Understanding "biofilter health" can provide

Table 2-8 Recommended monitoring for operational parameters

Parameter	Sampling Location	Sampling Method	Sampling Frequency	Performance Indication
Flow/hydraulic loading rate	Biofilter effluent	Inline	Continuous	Hydraulic loading rate (calculated as flow divided by filter surface area) affects mass transfer of contaminants and EBCT through the filter media. Hydraulic loading rate is typically 0.5–10 gpm/ft^2.
EBCT	N/A	Calculated	Continuous	EBCT (typically 5–15 min but can range from 2 to >20 min in practice) directly relates to biological contaminant removal. When filter media depth is set, EBCT varies with flow rate/ hydraulic loading rate.
Clean-bed head loss	Across the filter	Inline	Pressure differential at beginning of filter run	Indicated hydraulic performance. Clean-bed head loss (at the beginning of filter run) varies with filter design (typically 1–4 ft) and hydraulic loading rate. During acclimation, increasing clean-bed head loss can be an indicator of biological activity increasing over time. During routine operation, increasing clean-bed head loss may indicate insufficient cleaning of the underdrain/media and mitigation may be required to prevent excess growth/biofouling.
Terminal head loss	Across the filter	Inline	Pressure differential at end of filter run	Terminal head loss (at end of filter run) varies with filter design (typically 8–12 ft, although older filter designs may have lower ranges such as 4–6 ft).
HLAR	Across the filter	Inline for head loss; HLAR is calculated	Continuous/per filter run	HLAR is the rate at which head loss increases over time during the filter run. Indicates particle loading/removal and biological growth/fouling.
Filter run time	N/A	Calculated	per filter run	Time between backwashes, typically resulting from time to reach terminal head loss, effluent turbidity limit, or user-defined maximum filter run time. Typically, 18–96 h (depends on filtration rate). Can indicate filter performance, particularly when loading rate does not change.
UFRV	N/A	Calculated	per filter run	Universal calculation that accounts for filtration rate to assess filtration performance. Typically, 8,000–16,000 gal/ft^2. Changes over time may indicate change in filter performance, efficacy of backwash, or biofouling.
Backwash pressure	Backwash piping at filter	Inline	Continuous during backwashing	Evaluate for increases over time during high-rate backwash rate, which may indicate fouling and require mitigation to prevent mechanical failure of the underdrain.
Water production efficiency	N/A	Calculated	Continuous	Indicates net amount of water produced (accounts for water used for backwashing).

Source: Adapted from WRF Project #4719 (Brown et al. 2020)
EBCT—empty bed contact time, HLAR—head loss accumulation rate, UFRV—unit filter run volume

Table 2-9 Recommended monitoring for water quality parameters

Parameter	Sampling Location	Sampling Method	Sampling Frequency	Performance Indication
Turbidity	Biofilter effluent (required); biofilter influent or effluent of upstream treatment process	Online	Continuous	Indicates filtration performance. Effluent monitoring used for regulatory compliance. Calculate solids loading rate from influent turbidity and flow rate to help understand changes in hydraulic performance.
pH	Biofilter influent and effluent	Grab or online	Continuous	Affects biological growth and activity. Influences effectiveness of nutrient supplementation, control of Mn removal/release, and particle removal. pH for biofiltration is typically between 6 and 9 S.U. Optimal pH may be site-specific.
Temperature	Raw, biofilter influent or effluent	Grab or online	Continuous	Affects biological growth and activity.
DOC	Biofilter influent and effluent	Grab	Regular (e.g., two times per week)	Indicates organic carbon removal.
UV254	Biofilter influent and effluent	Grab or online	Regular (e.g., two times per week)	Indicates aromaticity of organic carbon compounds; used to calculate SUVA.
AOC	Biofilter effluent	Grab	During special studies, research, or optimization (as applicable)	Indicates regrowth potential in the distribution system.
BDOC	Biofilter effluent	Grab	During special studies, research, or optimization (as applicable)	Indicates regrowth potential in the distribution system.
Carboxylic acids	Biofilter influent and effluent	Grab	Regular (e.g., weekly) during optimization; during special studies and research (as applicable)	Oxidation by-product from ozonation of NOM used as a surrogate to estimate AOC. Use if oxidant is applied upstream of biofilter.
Phosphorus (orthophosphate)	Biofilter influent and effluent	Grab	Occasional (e.g., every month); consider more frequent sampling if supplementing phosphorus to optimize performance	Affects biological growth and activity. Monitoring influent phosphorus using analytical method for low-level orthophosphate is critical for tailoring nutrient dose.
Nitrogen (ammonium, nitrate, nitrite)	Biofilter influent and effluent	Grab	Occasional (e.g., every month); more frequent sampling if supplementing ammonia to optimize performance	Affects biological growth and activity. Monitoring influent nitrogen is critical for tailoring nutrient dose if influent concentration is limiting (e.g., carbon-to-nitrogen ratio >10:1).

(continued)

Table 2-9 Recommended monitoring for water quality parameters *(Continued)*

Parameter	Sampling Location	Sampling Method	Sampling Frequency	Performance Indication
Manganese	Biofilter influent and effluent	Grab	Regular (e.g., two times per week) if target compound; otherwise, occasional (e.g., every month)	Indicates removal efficiency and/or release of legacy Mn present on the filter media, which may occur under certain conditions.
ORP	Biofilter influent	Grab or online	Continuous or regular monitoring (e.g., two times per week) if Mn is a target compound or removed through biofilters; if not, occasional (e.g., monthly)	ORP and pH relate to iron and manganese removal as they control the speciation as soluble or particulate forms.
Target compounds	Biofilter influent and effluent	Grab	Regular (e.g., two times per week)	Removal rate of target compounds indicates biofilter performance.
Oxidant residual	Biofilter influent	Grab or online	Continuous or regular (e.g., daily)	Used for process control to avoid overdosing and harm to biological activity.
DBPs (THMs/ HAAs)	Biofilter effluent or finished water	Grab	Regular	Indicates treatment performance for removal of DBP precursors. DBP formation potential is measured in biofilter effluent and DBPs are measured in finished water.
DO	Biofilter influent and effluent	Online	Continuous	Most beneficial for non-ozone applications. Can be an indicator of biological performance upset.

AOC—assimilable organic carbon, BDOC—biodegradable organic carbon, DBP—disinfection by-product, DO—dissolved oxygen, DOC—dissolved organic carbon, HAA—haloacetic acid, NOM—natural organic matter, ORP—oxidation-reduction potential, SUVA—specific ultraviolet absorbance, THMs—trihalomethane

a reference point for performance and aid in determining whether further enhancement of bioactivity is warranted or possible. The biological process monitoring tools presented in Table 2-10 are readily available for immediate use and have been shown to provide reliable data, be cost-effective, and operator-friendly (Hooper et al. 2019). These tools include biofilm formation rate, DO consumption, and ATP. Other biological monitoring tools (e.g., EPS, microbial community characterization) are not recommended for regular monitoring, but may be useful during startup (e.g., acclimation/reacclimation), optimization, troubleshooting, and/or research studies. Utilities should consider collecting baseline samples for parameters recommended during troubleshooting to have a basis of comparison against results collected during normal operation. Refer to WRF Project #4620 (Hooper et al. 2019) for more details on biological monitoring tools.

2.4.1.4 Process integrity. Table 2-11 summarizes recommended monitoring for process integrity. Careful monitoring of filter media and underdrain performance is crucial to biofilter operation and water efficiency. At a minimum, biofiltration requires the same monitoring as conventional media filtration (e.g., sieve analysis, visual checks for cracks/mudballs, track media expansion and check for boils during backwash, solids

Table 2-10 Recommended monitoring for biological parameters

Parameter	Sampling Location	Sampling Method	Sampling Frequency	Performance Indication
Biofilm formation rate	Biofilter influent and effluent	Grab from coupon rack	Every two weeks	Indicates biological activity by measuring ATP of biofilm growth on a coupon after an incubation period (e.g., two weeks). (Note: Not recommended when biofilter influent contains residual chlorine l.)
DO consumption	Biofilter influent and effluent	Inline using LDO probes	Continuous	Indicates biological activity. This method is only recommended *for biofilters without upstream ozone.*
ATP	Filter media	Grab	Occasional during acclimation/reacclimation and troubleshooting; during special studies, research, or optimization (as applicable)	Indicates biological activity. ATP monitoring is recommended during acclimation/reacclimation and troubleshooting.
EPS	Filter media	Grab	During troubleshooting; during special studies, research, or optimization (as applicable)	Indicates hydraulic performance. Hydraulic challenges are common when excess EPS is produced.
Microbial community structure	Filter media	Grab	When sustained hydraulic issues are experienced; during special studies, research, or optimization (as applicable)	Provides information on "who" is present in the system. Requires the use of molecular microbial methods; typically performed at specialty labs and universities.
Enzyme activity	Filter media	Grab	During filter optimization or when sustained hydraulic issues are experienced	Indicates effective activity of the filter biology. It can be used to indicate a nutrient (phosphorus) deficiency (Keithley and Kirisits 2019).

ATP—adenosine triphosphate, DO—dissolved oxygen, EPS—extracellular polymeric substance, LDO—luminescent dissolved oxygen

Table 2-11 Recommended monitoring for process integrity

Parameter	Sampling Location	Sampling Method	Sampling Frequency	Performance Indication
Media characteristics	Filter media	Grab	Regular (e.g., annual)	Allows assessment of particle size distribution through sieve analysis (e.g., effective size, uniformity coefficient) and evaluation of media depth
Chemical levels and feed flow/dose (e.g., oxidant dose)	Upstream of biofilters; backwash supply	Grab	Continuous	Oxidant (e.g., ozone, chlorine, hydrogen peroxide) dose and nutrient supplementation affect biological growth and activity, and contaminant removal. Continuous monitoring of chemical feed is critical for ensuring consistent treatment performance
Inline analyzer maintenance	As applicable	NA	Regular (e.g., every week)	Allows effective and consistent system performance

retention analysis). If an oxidant (e.g., ozone, chlorine, hydrogen peroxide) is fed upstream of the biofilters, its dosing rate should be closely monitored. In addition, the residual concentration should be monitored in the biofilter feedwater (as indicated in Table 2-11) and the process controlled accordingly to avoid overdosing and die-off of biological activity.

For more information, consult the following references:

- Integrated Design and Operation of Water Treatment Facilities by Susumu Kawamura (2000)
- Filter Evaluation Procedures for Granular Media by Daniel Nix and John Taylor (2018)
- Filter Maintenance and Operations Guidance Manual (WRF 2511) by Gary Logsdon et al. (2002)

2.4.2 Initial Startup/Acclimation

A startup of biofiltration is initiated when no or minimal oxidant (including disinfectant) has been ensured in the influent of filter. Acclimation is the time required to establish a mature and active biomass in the biofilter after startup. Acclimation may also be needed if a filter is placed offline for an extended duration (greater than two weeks), if a filter is drained/dried, or if significant changes in water quality occur. Acclimation first includes colonization of the biofilter media followed by adaptation of the microbial community. Whereas the colonization may occur within weeks, establishing a mature and active biomass capable of degrading the contaminants of interest may take several months. The length of time required to reach steady-state removal is dependent on the influent water characteristics, the target contaminant(s), and operating conditions, such as the following:

- filter loading rate,
- backwashing procedures,
- water quality (e.g., pH, temperature, nutrients, and disinfectant residuals),
- upstream process operation (ozone, coagulation, and prechlorination), and
- filter media characteristics (e.g., media type).

A fully acclimated biofilter will demonstrate steady-state contaminant removal of target contaminant(s) and relatively stable hydraulic performance (e.g., UFRV). Steady-state performance changes seasonally and it may be difficult to confirm if temperature and influent water quality are changing.

During acclimation, the biofilm formation rate or biomass, measured as ATP on filter media, can be sampled to confirm biomass has been established on the filter media, and samples should be collected until steady-state conditions exist. The reduction in biofilm formation rate from filter influent to effluent is recommended as a routine monitoring parameter and is preferred over ATP. As bioactivity acclimates, the difference in the biofilm formation rate from the influent to effluent will increase and then stabilize once steady-state conditions have been attained.

2.4.3 Steady-State Operation

Once a mature and active biomass is continuously removing contaminants of interest, steady-state operation is achieved. Subsection 2.3.1 discusses hydraulic, water quality, and biological monitoring tools to develop a monitoring strategy and a filter management plan.

2.4.4 Shutdown and Restart

Operational disruptions may have an adverse effect on the biofilter's ability to remove contaminants. Shutdown strategies are dependent on the timing and duration of the shutdown. A different approach should be taken if the shutdown lasts for the short term (24 hours to 2 weeks) or long term (more than two weeks). It is recommended that routine monitoring parameters (e.g., turbidity, head loss) measurements are recorded for comparison with the restart performance; then, the biofilter should be stored wet. Upon restart, a backwash is recommended to remove any biomass that has died off during the shutdown, as well as potential manganese (depending on conditions), algal growth, etc.

Once a biofilter is put back into service, an acclimation period may be required depending on the duration and timing of the shutdown (i.e., seasonal temperature effects). Depending on treatment objectives, there are multiple strategies for reducing acclimation period including the following:

- filling the filters with raw or settled water (undisinfected water) to supply food, and seed more biomass for difficult-to-remove constituents, which may increase the rate of acclimation, but also require extended filter-to-waste durations, which could affect water production capabilities;
- an extended subfluidization (low-rate) backwash;
- low initial filter loading rates;
- biofilter shutdown procedures (e.g., maintaining aerobic conditions, keeping filters submerged);
- timing (e.g., bringing biofilters on-line during warmer temperatures).

2.4.5 Optimization

When thinking about optimization of biofilters, a holistic approach considering both upstream and downstream treatment is recommended. Given the multiple benefits and capabilities of biofiltration, there may be opportunities to optimize upstream treatment to achieve enhanced organics removal through the biofiltration process, thereby reducing overall operating costs and decreasing the volume of plant waste streams. When optimizing the biofiltration process, it is important to consider the effect on finished water biostability to avoid the potential for downstream biological growth. Biofilter optimization strategies include the following:

- EBCT or flow adjustment,
- media selection,
- backwash protocol,
- nutrient augmentation,
- pH adjustment,
- preoxidant addition, and
- pretreatment approach.

These strategies can be tested at multiple scales. Pilot testing evaluates optimization strategies once steady state has been achieved. There are benefits and limitations of pilot-scale testing. The main benefit is that one can evaluate approaches without affecting finished water quality and can learn from mistreatment approaches. The important limitation of pilot-scale testing is that the backwash rates are not typically representative of full-scale because of sidewall effects. Filter underdrains are also not typical of full-scale.

Full-scale optimization can verify that approaches that were successful at pilot scale work at full scale as well. Both pilot and full-scale optimization should take place after

Table 2-12 Common challenges and potential solutions

Issue	Potential Solution
High turbidity in filter effluent	Check for excessive head loss. Breakthrough may occur at high head loss. Look for flow fluctuation. Widely varying flows will cause breakthrough. Verify correct filter aid dosages. Examine the backwash cycle for complete wash. Inspect for damaged bed due to backwashing.
Rapid buildup of head loss	Check the influent water-suspended solids. Verify correct filter aid dosages. Verify correct applied water flow rate. Examine the backwash cycle for complete wash. Inspect the head loss differential pressure sensor for air in one side, because this could give a false reading.
Insignificant buildup of head loss	Check applied water-suspended solids. Verify the applied water flow rate. Verify correct filter aid dosages. Check head loss differential pressure sensor for air in one side, because this could give a false reading. Examine filter effluent for suspended solids going out (filter breakthrough). Inspect for damaged bed due to backwashing. Backwash and check for complete cycle.
Rapid loss of filter media	Look for washout during backwash cycle. Examine for media in effluent indicating a damaged filter underdrain. Check for excessive scouring during backwash cycle time. Excessive scouring will grind media into fines.
High head loss through clean filter	Inspect differential pressure sensor for air in one side. Check for incomplete backwash cycle. Look for mudballs in filter media. Take a plug sample.
Flow indicated when effluent valve is closed	Inspect differential pressure sensor for air in one side. Check instrumentation loop for calibration. Examine valve for proper position.
Backwash stops before completing cycle	Look for sticking valve. Inspect for electrical relay hang-up. Check for timer out of sequence. Examine backwash water supply. Inspect electrical control (pump lockout).
Growth of higher-order organisms	Optimize backwashing procedures. Perform backwashing procedures at regular intervals. Reduce filter run time. Monitor influent water quality and pretreatment efficiency. Reduce organic and pathogenic content of the filter influent.

biological acclimation has occurred (i.e., relatively steady system performance). Long-term operation at full scale may also reveal optimization strategies that were not observed or anticipated during pilot testing (e.g., effects of electron donor carryover, chemical feed pump maintenance).

2.4.6 Troubleshooting

Operational challenges most often observed: variable hydraulic performance, such as shortened filter runs, increased head loss, and decreased UFRVs associated with biological growth; contaminant breakthrough; poor biological growth and activity; increased biofouling (e.g., biogrowth in sample lines); and biomass sloughing. For additional information about troubleshooting common challenges, refer to the Biofiltration Guidance Manual (Brown et al. 2020) and Biofilter Conversion Guidance Manual (Upadhyaya et al. 2017).

Common challenges that may arise during biofilter operation and mitigation measures are presented in Table 2-12 (Lauderdale et al. 2012, 2011; Kerri 1995).

2.5 FEATURED CASE STUDIES

2.5.1 Reducing DBPs Through Ozone-Biofiltration

Dublin Road Water Plant, Columbus, Ohio
Author: Rod Dunn

The City of Columbus sits in a water-rich region, downstream of three primarily agricultural (corn/soybean) watersheds that contain a handful of small Midwestern towns and villages. Water from these three watersheds fill four in-stream reservoirs, an upground reservoir, and a wellfield. This abundant source supplies two surface water treatment plants and a groundwater treatment plant, serving a common distribution system with over 3,500 miles of distribution mains. The system produces an average of 135 million gallons per day (mgd) to serve a population of 1.2 million people in central Ohio.

The addition of ozone and the conversion of conventional anthracite/sand/gravel filters to biologically active GAC/sand filters were part of a large treatment upgrade to the Dublin Road Water Plant in 2014 through 2015. The primary objective of the upgrade was DBP precursor removal. Additional objectives included atrazine removal, T&O control, and preparation for the future removal of emerging contaminants of concern. All the objectives would reduce the amount of money spent on PAC, which before 2015 totaled over a $1 million dollars a year.

2.5.1.1 Facility description. The Dublin Road Water Plant is a surface water treatment facility that obtains its source water from the Scioto River, downstream of Griggs (1.4 billion gallons [BG]) and O'Shaughnessy Reservoirs (4.8 BG). In addition, the John R. Doutt Upground Reservoir (9 BG) provides source water during times of drought. The approved treatment capacity for Dublin Road is 80 mgd, with an average annual production of 50 mgd.

Source water quality is highly variable through the year (Table 2-13). Turbidity varies from less than 1 NTU in the winter under the ice, to more than 500 NTU following spring and early summer thunderstorms. After farmers plant their crops in late April through early June, thunderstorms bring runoff laden with herbicides and nutrients. Atrazine (pre-emergent herbicide used on corn) can spike over 20 μg/L, well above the maximum contaminant levels (MCL) of 3 μg/L. The biggest nitrate spikes of the year are seen during early summer. As part of the upgrade, ion exchange treatment was installed to reduce these nitrate spikes (that may go above 15 mg/L as NO_3-N in the source water and are treated down to less than 80 percent of the MCL of 10 mg/L as NO_3-N in the treated water).

Figure 2-2 shows a process flow diagram for the Dublin Road Water Plant. Source water from the Scioto River is treated initially through first-stage sedimentation (alum coagulation/rapid mix/flocculation/sedimentation) for turbidity removal. The first-stage effluent is further treated through second-stage softening and sedimentation (lime addition/rapid mix/flocculation/sedimentation). After recarbonation to reduce the pH, the softened water is ozonated and sent through 18 GAC/sand biofilters. Fluoride, sodium hypochlorite for disinfection, and zinc orthophosphate for corrosion control (1:5 zinc to phosphate ratio) are added to the filter effluent before entering a series of clear wells for

Table 2-13 Water quality

Parameter	Location	Unit	Concentration (Average±SD)
TOC	Raw	mg/L	6.2±1.3
DOC	Raw	mg/L	5.6±1.2
UV254	Raw	cm^{-1}	0.21±0.05
Temperature	Raw	C	14.8±7.5
Total hardness	Raw	mg/L as $CaCO_3$	228±33.9
Total alkalinity	Raw	mg/L as $CaCO_3$	147±17.6
TOC	Biofilter influent	mg/L	2.7±0.4
DOC	Biofilter influent	mg/L	2.7±0.5
UV254	Biofilter influent	cm^{-1}	0.03±0.01
Total hardness	Biofilter influent	mg/L as $CaCO_3$	122±1.6
Total alkalinity	Biofilter influent	mg/L as $CaCO_3$	54±5.4

DOC—dissolved organic carbon, TOC—total organic carbon, SD—standard deviation

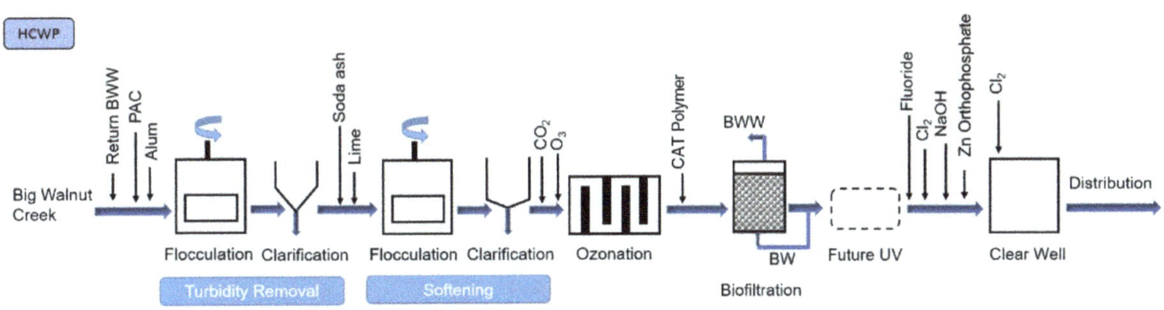

Source: WRF Project #4731 (Upadhyaya et al. 2021)

BW—backwash, BWW—backwash water, CAT—calcium thiosulfate, PAC—powdered activated carbon, UV—ultraviolet

Figure 2-2 Process flow diagram for the Dublin Road Water Plant

disinfection contact time. Side-stream ion exchange treatment to remove nitrate is available to treat a portion of the combined filter effluent before entering the clear well, if needed (approximately one week per year). UV disinfection before the clear well is currently under construction.

Additional chemical feed options are available to maximize our treatment options. This includes potassium permanganate, which can be added to the raw water before rapid mix, to assist with algae removal. PAC can be added to the rapid mix units for T&O control, atrazine removal, and for DBP precursor removal. A flocculent aid polymer can also be added to the rapid mix units. Sodium hydroxide and soda ash can be added for softening, peroxide addition with the ozone to aid in atrazine removal, calcium thiosulfate to quench any ozone leaving the contact basin, and a filter aid polymer.

The biological treatment process comprises 18 dual media (GAC and sand) BAF. Each filter contains 36 in. of GAC over 6 in. of sand. During the summer, a low dose of prechlorine (residual of less than 0.3 mg/L) is fed ahead of the biofilters to prevent mayflies from hatching inside the filter building. Typically, the filter loading rate is about 1.3 gpm/ft^2. The filters are generally backwashed on the basis of runtime (144 hours); however, backwash limits have been established for head loss (4 ft) and filter effluent turbidity (0.3 NTU) but are rarely reached. Biofilter backwash water is recycled to the head of the plant. Filter-to-

Figure 2-3 Photo of converted GAC/sand filters

waste occurs from filter startup until the turbidity drops below 0.1 NTU, which ranges from 30 to 60 min. A photo of the filters is shown in Figure 2-3. Filter design criteria are summarized in Table 2-14.

2.5.1.2 Facility biological treatment goals. The City of Columbus has the following goals for biological treatment:

- **Turbidity:** Meet regulatory requirements and Partnership for Safe Water, less than 0.1 NTU all of the time.
- **UFRV:** UFRV greater than 5,000 gal/ft^2.
- **THM/HAA5:** Reduce THM and HAA5 levels to less than 80 percent of MCL.
- **Atrazine:** Reduce atrazine levels to less than 80 percent of MCL.
- **T&O:** Reduce Geosmin and MIB to levels less than human threshold of detection.
- **AOC:** Reduce AOC sufficiently to prevent regrowth in the distribution system.

2.5.1.3 Operation and performance. The Dublin Road Water Plant (DRWP) worked with the WRF on a tailored collaboration project (WRF Project #4731 [Upadhyaya et al. 2021]) to demonstrate full-scale biofiltration optimization and integration of the bio-

Table 2-14 Biological treatment design criteria

Parameter	Unit	Design Criteria or Typical Operating Value
Flow		
Design	mgd	80
Filter loading rate		
Design	gpm/ft^2	3.5
Typical	gpm/ft^2	1.3
Media		
GAC		
Depth	in.	36
Eff size	mm	0.8–1.0
Sand		
Depth	in.	6
Eff size	mm	0.45–0.55
EBCT (design/typical)	min	10/15
Prechlorination		During summer to control mayflies
Residual Cl_2	mg/L	<0.3 mg/L
Backwash		
Trigger		
Run time	h	144
Turbidity	NTU	0.3
Head loss	ft	4
Backwash flow rate		
High	gpm/ft^2	13
Low	gpm/ft^2	5
Air scour	$scfm/ft^2$	2.5
Preoxidant (Cl_2) dose	mg/L	0–0.3
Filter-to-waste		Yes
Flow rate	mgd	Filter rate
Duration		Until NTU drops to <0.1 (ranges from 30 to 60 min)

EBCT—empty bed contact time, Eff—effluent, GAC—granular activated carbon, mgd—million gallons per day, NTU—nephelometric turbidity unit

filter into the overall treatment process. Some of the optimization and integration concepts being explored include phosphorus supplementation, prechlorination effects, filter shutdown and maintenance strategies, coagulation dose optimization, and ozone:TOC ratio optimization. Both water quality and hydraulic parameters are being used to track filter performance. In addition, an engineering and economic analysis will be undertaken to provide a holistic optimization of biofiltration and the upstream processes without compromising finished water quality or hydraulic performance.

Figure 2-4 shows typical TOC removal. Through ozonation and biofiltration, an additional 15 percent of TOC is removed.

2.5.1.4 Key success factors. While the City is still optimizing the ozone/BAF process, there are a few lessons learned:

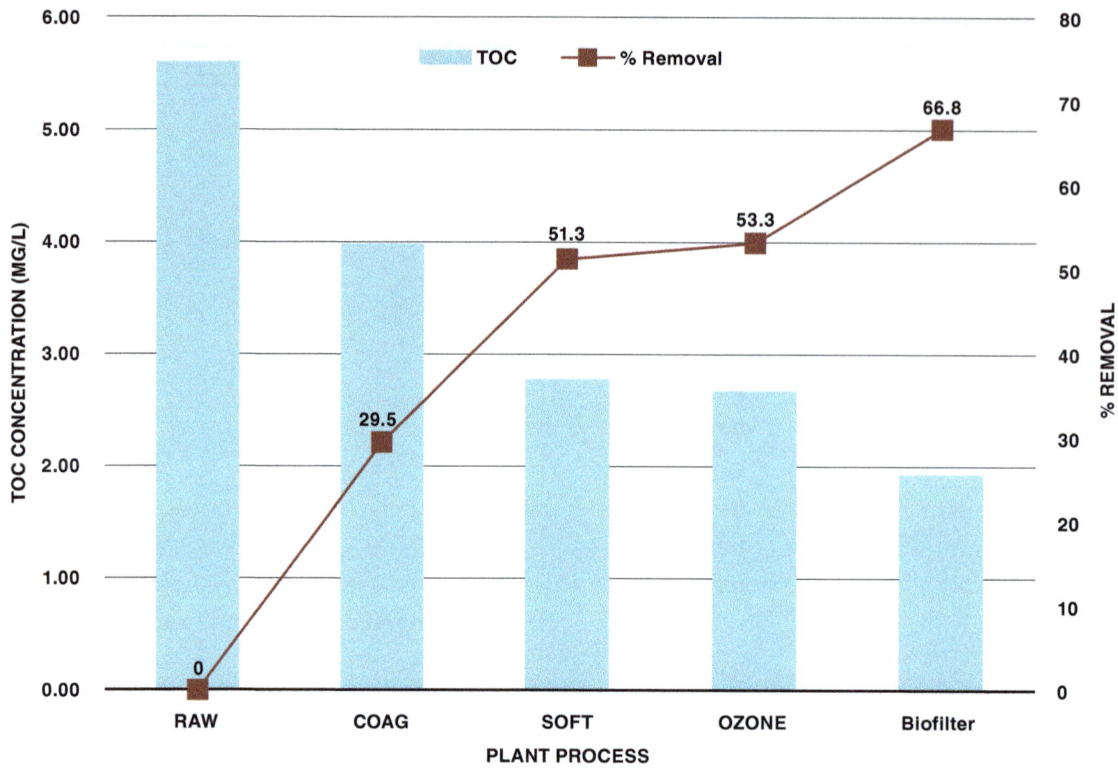

COAG—coagulation, TOC—total organic carbon

Figure 2-4 DRWP TOC removal

- The system is easy to operate and does not require additional operators.
- Higher-order organisms like spiders and mayflies were found in the filter building. Adding a small chlorine dose (less than 0.3 mg/L) after ozone but ahead of the biofilters eliminates the flies and spiders and does not affect DBP precursor removal.
- It was difficult for some operators to fully accept the biofilter concept and the use of biology as a beneficial treatment process.
- The new GAC/sand filters performed the same as the older anthracite/sand/gravel filters for turbidity removal.
- A pH drop of typically 1.0 S.U. is experienced across the filters in the summer (from pH 8.2 to 7.2). The pH drop trends with increasing temperature and increased TOC/DOC removal. A sodium hydroxide feed was installed after the biofilters to raise the pH to our finished water pH goal of 7.8 S.U.
- Nutrients, specifically phosphate, were not found to be limited following two stages of coagulation/softening. Additional details are provided in WRF Project #4731 (Upadhyaya et al. 2021).
- Increased biological activity and better removal are observed in the summer, but biological activity still exists in the winter.
- Despite reduced biological activity in the winter, the ozone/biofiltration process achieves good TOC, DBP precursor, atrazine, and T&O removal year-round.

2.5.2 Conversion to Biological Filtration in a Low-Organics, Cold Water System

Hayden Bridge Filtration Plant, Springfield, Ore.

Authors: Michael J. McKie, Toby Dixon, Jared Giacomelli, Lonny Sayles, Ray Leipold, Robert C. Andrews

The Eugene Water and Electric Board (EWEB) is Oregon's largest customer-owned utility serving Eugene, East Springfield, and the McKenzie Valley. More than 200,000 people receive water from EWEB's only water treatment facility, the Hayden Bridge Filtration Plant. The plant has been in operation since 1950.

2.5.2.1 Facility description. The Hayden Bridge Filtration Plant is currently rated at 88 mgd but generally operates at less than 50 percent of the design capacity. The plant treats water from the McKenzie River. As shown in Table 2-15, the river has low organic carbon, low turbidity, and a water temperature range of 2.7–17.9°C.

Aluminum sulfate, caustic soda, and PAC are added to the raw water, as necessary. Then, that is followed by hydraulic flocculation (i.e., mixing resulting from turbulent flow), sedimentation, cationic filter aid polymer addition, biofiltration, and chlorine disinfection (Figure 2-5).

Details pertaining to the existing full-scale filters, which would be converted to biofiltration, are provided in Table 2-16.

Typically, only half of the available filters are used for production. Filters are operated in constant flow mode and backwash on the basis of head loss, turbidity, or runtime, with head loss being the primary backwash initiation trigger. After a filter run, filters are placed into standby mode following backwash until all filters have been cycled through.

Table 2-15 Water quality

Parameter	Location	Unit	Concentration (Average±SD)
Temperature	Raw	°C	9.9±3.7
Turbidity	Raw	NTU	
DOC	Raw	mg/L	1.0±03
pH	Raw	N/A	7.6±0.1
Alkalinity	Raw	mg/L as $CaCO_3$	22±2.3
DO	Raw	mg/L	11.2±0.9
DOC	Biofilter influent	mg/L	0.5±0.2
pH	Biofilter influent	N/A	7.5±0.1
DO	Biofilter influent	mg/L	11.2±0.9

DO—dissolved oxygen, DOC—dissolved organic carbon, SD—standard deviation

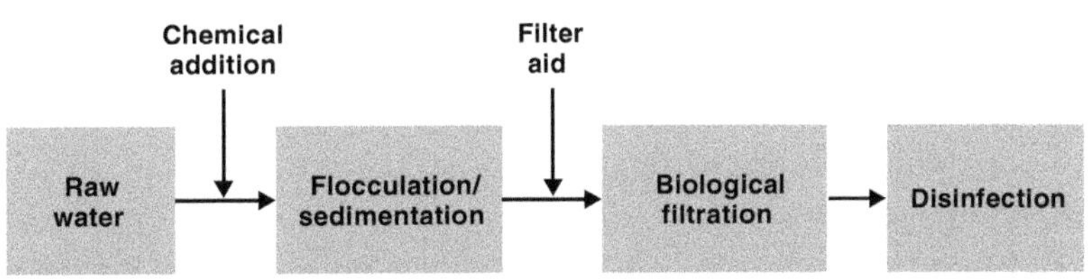

Figure 2-5 Process flow diagram for the Hayden Bridge Filtration Plant following conversion to biofiltration

Table 2-16 Biological treatment design criteria

Parameter	Unit	Design Criteria
Media type	N/A	Anthracite Sand
Effective size	Anthracite Sand	1.0 mm 0.5 mm
Anthracite depth	Shallow bed Deep bed	18″ 50″
Sand depth	Shallow bed Deep bed	10″ 15″
Total media depth	Shallow bed Deep bed	28″ 65″
Loading rate	gpm/ft^2 m/h	3 7.5
EBCT	Min	Design: 6–13
Backwash strategy	Min	Surface sweep: 2 (chlorinated) High rate: 4 High rate: 3
Trigger for terminating filter-to-waste	NTU	0.07

EBCT—empty bed contact time

During warm-water conditions, filters may be offline for a period of two to four days, depending on system demand; cold water conditions may increase the shutdown period to seven days or more.

2.5.2.2 Facility biological treatment goals. When considering the elimination of prechlorination and the potential conversion to biofiltration, EWEB had the following goals regarding operation of biological treatment:

- **Eliminate prechlorination:** Reduce the amount of sodium hypochlorite required such that a smaller and less expensive sodium hypochlorite production facility could be commissioned. The conversion to biofiltration was considered in parallel with changes in the disinfection process at the facility. Before 2019, chlorine gas was used for disinfection; however, purchasing chlorine gas resulted in cost uncertainty and increasing operational costs. To reduce potential cost variability and ensure adequate disinfectant supply, a sodium hypochlorite production facility was commissioned by EWEB in 2019. In conjunction with the implementation of sodium hypochlorite, elimination of prechlorination at the plant inlet was considered to reduce chlorine production requirements. It was anticipated that this process change would result in BAF, as the filters would no longer be exposed to an oxidant;
- **Improve filter production:** Maintain or increase UFRVs when compared with nonbiological operation, as biological acclimation was feared to increase head loss development;
- **Provide biological stability:** Reduce the need for flushing and swabbing (i.e., scouring the water main with foam "swabs" propelled by high velocity water) of the distribution system by removing biologically available organic carbon at the treatment plant; and

Figure 2-6 Hayden Bridge Filtration Plant pilot system

- **Achieve similar water quality:** Achieve the same finished water turbidity and organic carbon (e.g., DOC, THM, HAA) as before conversion to hypochlorite and biofiltration.

2.5.2.3 Operation and performance. Pilot studies were conducted to evaluate the effect of biofiltration on treated water quality (i.e., turbidity and organic carbon) and characterize the biological community (e.g., biomass development and enzyme activity). The pilot filters treated nonchlorinated water and were designed to match the shallow bed filter design (Table 2-15) used at the facility, as shown in Figure 2-6. One biological pilot filter was operated without exposure to chlorine, while the other filter was backwashed with chlorinated water as per the full-scale filters. Both filters were operated in constant flow mode, with backwash triggered automatically on the basis of loss of flow, turbidity breakthrough, or head loss triggers.

Results of the pilot testing indicated that nonchlorinated backwash resulted in significantly higher accumulation of biomass (as indicated by ATP on the filter media; Figure 2-7) and biological activity (as indicated by esterase and phosphatase enzyme activity; Figure 2-8). Biomass accumulation and activity were not affected by temperature, with stable or increasing trends during cold water periods (December 2018 to March 2019). Filtered water organic carbon concentrations and turbidity were like those observed in the full-scale conventional filters, thereby meeting the biofilter goals. Finally, significant improvements were observed with respect to UFRVs, as shown in Table 2-17.

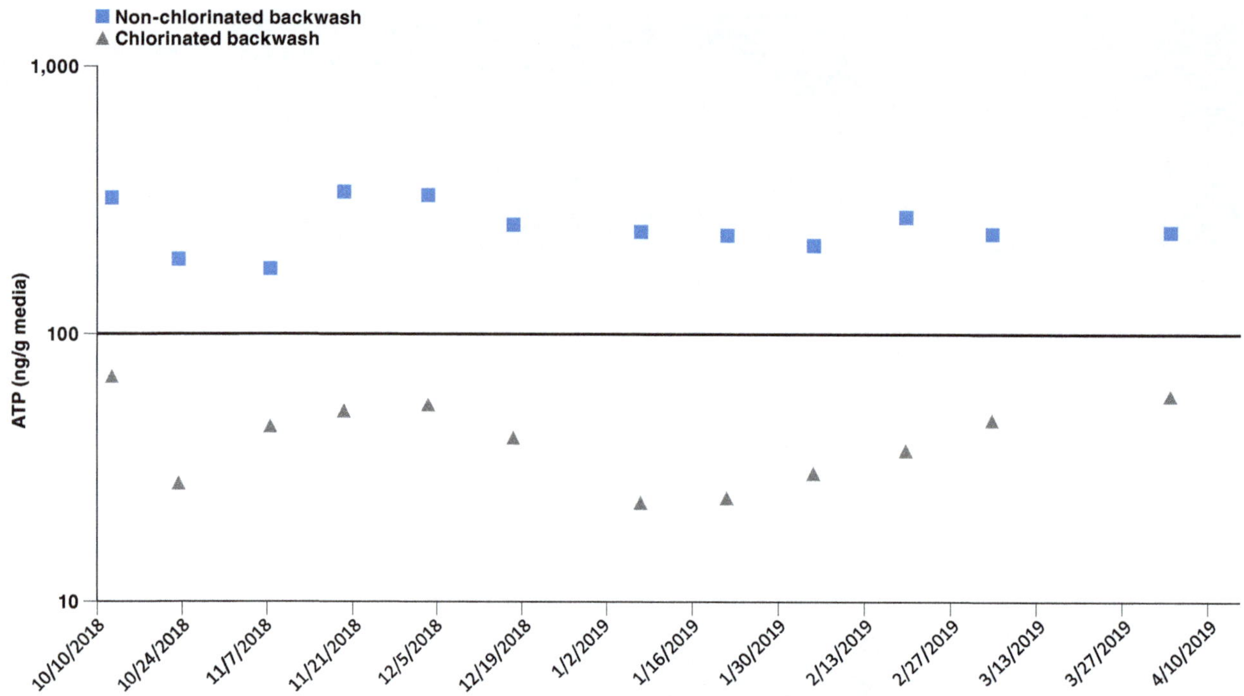

Figure 2-7 Biomass accumulation in filters backwashed with nonchlorinated and chlorinated water

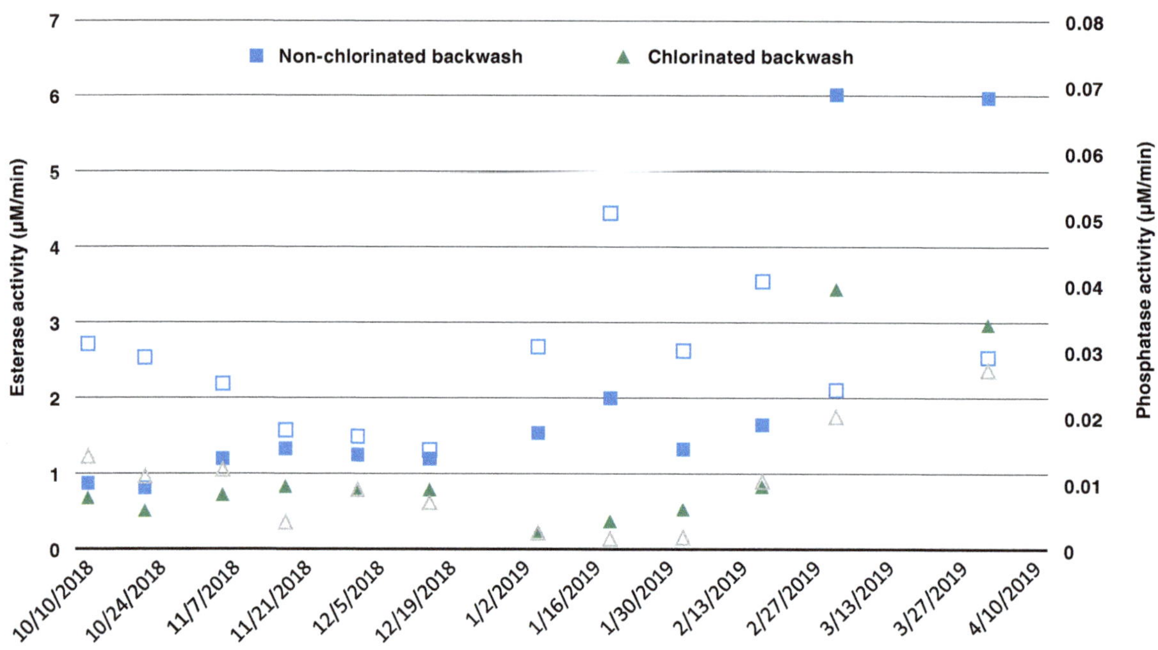

Note: Solid symbols = esterase activity; open symbols = phosphatase activity.

Figure 2-8 Enzyme activity associated with nonchlorinated and chlorinated backwash

Table 2-17 Pilot biofilters average performance

Parameter	Nonchlorinated Backwash	Chlorinated Backwash
ATP (ng/g media)	185	41
Esterase activity (μM/min)	3.5	1.75
Phosphatase activity (μM/min)	0.065	0.021
DOC removal (%; mg/L)	15; 0.08	10; 0.06
THM-FP removal (%; μg/L)	18; 8	14; 6
HAA-FP removal (%; μg/L)	24; 3.6	22; 3.2
UFRVs (gal/ft^2)	5,333	4,480

ATP—adenosine triphosphate, DOC—dissolved organic carbon, HAA-FP—haloacetic acid formation potential, THM-FP—trihalomethane formation potential, UFRV—unit filter run volume

Figure 2-9 Filters before (left) and after (right) maintenance cleaning

When considering the goals outlined for implementation, all four were achieved:

- Chlorine usage was reduced by more than 20 percent by eliminating prechlorination.
- Operating filters biologically increased UFRV by 19 percent and reduced backwash costs.
- Biological filtration resulted in 60 percent less flushing in the first year after conversion.
- Similar finished water turbidity and organic carbon were achieved with biological filters.

Despite the efforts put forth before implementation, unforeseen complications were experienced following full-scale implementation. Excess biofilm growth was observed in the filters, which are uncovered and exposed to the environment (Figure 2-9).

To prevent excess biofilm growth on the filter components (e.g., walls, backwash overflow troughs), the filters are taken out of service four times annually and manually cleaned with a chlorinated solution. Although there is not an anticipated effect on water quality, the cleanings are expected to provide improved long-term operation of the filters by allowing the operators to conduct regular structural and mechanical inspections of each filter to proactively identify issues before failure or whether emergency upgrades are required.

2.5.2.4 Key success factors and lessons learned. Pilot testing before full-scale implementation was critical for the success of biofiltration at the Hayden Bridge Filtration Plant in terms of implementation, operation, and treated water quality. Pilot testing allowed them to understand how removing prechlorination would affect filter operation (e.g., longer filter runs) and filter performance (e.g., lower DBP formation with equal turbidity removal). Understanding this process using pilot filters allowed the utility to be prepared for potential challenges upon full-scale implementation, including additional maintenance requirements as well as the effect of dosing PAC in response to high turbidity spikes and/or algal toxin presence. Their collaboration with the Drinking Water Research Group at the University of Toronto through the WRF Tailored Collaboration program was also vital to the success of the project because it allowed both organizations to leverage their knowledge and experience to ensure the best possible outcome. This project demonstrated the benefits associated with utility–academic partnerships and provides a framework for similar collaborations in the future.

2.5.3 Acclimating New Biofilters for Manganese Removal

Lake Lewisville WTP, Denton, Tex.
Authors: Ashley Evans, Alma Beciragic, and Nissim Gore-Datar

The City of Denton owns and operates two water treatment plants: the Lake Lewisville WTP (LLWTP), producing 30 mgd, and the Ray Roberts WTP (RRWTP), producing 20 mgd. Together, these plants provide water to over 125,000 customers in the North Texas area. In 2014, the City of Denton began exploring treatment alternatives for the LLWTP processes to ensure the continued delivery of high-quality drinking water. Water quality goals were established on the basis of source-to-tap water quality data, established goals and drivers, the existing regulatory environment and its horizon, and possible distribution system effects of blending treated water (from the LLWTP and RRWTP). An array of treatment process alternatives was examined for their ability to achieve water quality goals and position LLWTP for future water treatment demands. The City opted to upgrade the LLWTP with the addition of ozone and conversion to biofiltration. LLWTP upgrades were completed in April 2016.

The addition of ozone helped achieve water quality targets related to virus and *Giardia* inactivation, as well as T&O control. The conversion to biofiltration helped achieve the water quality target for turbidity (i.e., below 0.15 NTU for at least 95 percent of samples), improve iron and manganese removal, and provide a barrier for trace contaminants. Moreover, the biofiltration retrofit helped increase organic carbon removal (thereby improving distribution system biostability) while also decreasing DBP formation.

2.5.3.1 Facility description. The LLWTP treats water from Lewisville Lake, which is fed from the upstream Ray Roberts Lake and the Elm Fork of the Trinity River. Raw water quality data are summarized in Table 2-18 and biofilter influent water quality data are summarized in Table 2-19. When averaging data collected over the entire study period, the biofilter influent water turbidity at LLWTP averaged 1.8 NTUs. The biofilter influent TOC averaged 4.3 mg/L, and the biofilter influent total iron and manganese concentrations were 0.44 mg/L and 51 mg/L, respectively.

The LLWTP pairs elements of conventional drinking water treatment techniques (including coagulation, sedimentation, and flocculation) with advanced treatment including ozone and GAC biofiltration (Figure 2-10). Potassium permanganate is dosed into raw water to help oxidize dissolved iron, maintain the raw water pipeline, and

Table 2-18 Raw water quality (January 2014 through December 2014)

Parameter	Unit	Concentration (Average±SD)
pH	S.U.	8.1±0.4
Turbidity	NTU	10.9±7.3
Alkalinity	mg/L $CaCO^3$	93.0±11.7
TOC	mg/L	6.5±0.4
Total Iron	mg/L	0.31±0.15
Total Manganese	µg/L	2.9±1.1

TOC—total organic carbon, SD—standard deviation

Table 2-19 Biofilter influent water quality (November 2013 through December 2015)

Parameter	Unit	Entire Study Concentration* (Average±SD)	Summer Concentration† (Average±SD)	Winter Concentration‡ (Average±SD)
Temperature	°C	21.2±7.6	24.7±3.1	14.0±5.0
pH	S.U.	6.8±0.4	6.6±0.6	6.9±0.2
Turbidity	NTU	1.8±0.4	1.6±0.5	1.9±0.3
Alkalinity	mg/L $CaCO^3$	85.7±9.9	79.0±9.8	79.7±9.3
TOC	mg/L	4.3±0.5	4.3±0.7	4.1±0.1
Total iron	mg/L	0.44±0.14	0.43±0.15	0.58±0.06
Total manganese	µg/L	51±45	88±70	32±7.0

TOC—total organic carbon, SD—standard deviation
*Entire study duration: November 2013–Dec 2015
†Summer defined as May 2014–July 2014
‡Winter defined as November 2013–January 2014

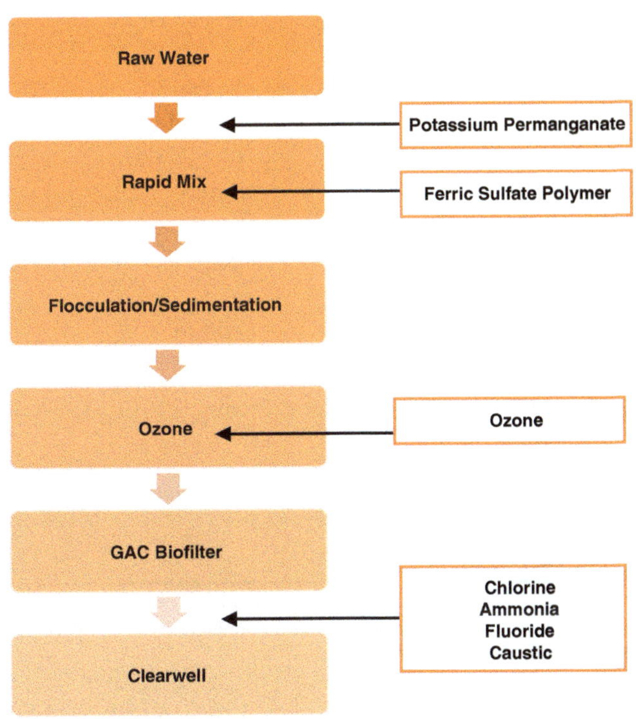

GAC—granular activated carbon, LLWTP—Lake Lewisville WTP

Figure 2-10 Process flow diagram for the LLWTP

Figure 2-11 Photo of the LLWTP biofilters

Table 2-20 Biological treatment design criteria

Parameter		Unit	Design Criteria
Plant capacity		mgd	30
Number of filters		N/A	16
Size of filters		ft^2 (ft × ft)	360 (21.8 × 16.5)
Media depth		in. (GAC) in. (Sand)	26 10
Media specifications	Effective size	mm (GAC) mm (Sand)	1.0–1.2 0.45–0.55
	Uniformity coefficient	GAC Sand	<1.5 <1.6
Underdrain		N/A	AWI Phoenix Underdrain System
Design maximum filter rate		gpm/ft^2	4.5
Design minimum EBCT		min	5.0
Backwash procedures*	Nonchlorinated with air scour	min	2
	High flow		≤15
	Low flow		3
	Settling		2
	Filter-to-waste		≥1

AWI—Anthratech Western Inc., EBCT—empty bed contact time, GAC—granular activated carbon, mgd—million gallons per day
*LLWTP has an option for chlorinated backwash, if needed. Backwash parameters may be operator adjusted. The extended backwash durations correlate with a lower maximum high backwash rate.

Table 2-21 Biofilter operational parameters during study

Parameter	Unit	Measured Values
Average filter rate during this study*	gpm/ft^2 (Filter 8)	2.6±0.3
	gpm/ft^2 (Filter 10)	2.1±0.6
Average EBCT during this study*	min (Filter 8)	8.8±1.8
	min (Filter 10)	9.3±2.4
Average run times during this study*	h (Filter 8)	24±15
	h (Filter 10)	25±16
Average UFRV	gal/ft^2 (Filter 8)	3,800±2,400
	gal/ft^2 (Filter 10)	3,400±2,300

EBCT—empty bed contact time, UFRV—unit filter run volume
*Data shown as average ± standard deviation.

prevent growth of zebra mussels. Ferric sulfate is applied as the coagulant. Once the water has passed through the GAC biofilters, it moves into a clearwell where pH is adjusted, and chlorine and ammonia are dosed for disinfection purposes.

There are 16 biofilters at LLWTP that contain GAC and sand media (Figure 2-11). The filters have an effective area of about 360 ft^2 and an average filter rate of about 2.1–2.6 gpm/ft^2. Additional design criteria (including typical operating ranges during acclimation) are summarized in Table 2-20. During this study, the filters, on average, were operated at about half their design capacity (Table 2-21). The GAC filters are typically backwashed with non-chlorinated water (albeit having the option for chlorinated backwash, if needed) and an air scour, followed by high/low flows and a settling period. Filter effluent is diverted to waste for at least 1 min before resuming normal operations.

2.5.3.2 Facility biological treatment goals. The LLWTP has the following goals biological treatment:

- **Turbidity:** Remove turbidity to below 0.15 NTU for at least 95 percent of samples.
- **Organic carbon:** Increase organic carbon removal to improve distribution system biostability and decrease DBP formation in coordination with implementation of upstream ozonation.
- **Iron and manganese:** Provide an additional barrier for iron and manganese (i.e., without chlorinated filtration).
- **Trace organic contaminants:** Provide an additional barrier for trace organic contaminants.

2.5.3.3 Operation and performance. The GAC biofilters at LLWTP were brought on-line in two phases. Filters 10 and 8 were observed for this study and brought on-line in November 2013 and May 2014, respectively, to characterize and compare seasonal trends in biofilter acclimation and performance. Filter 10 was monitored for almost two years (from February 2014 through December 2015), and filter 8 was monitored for eight months (from May through December 2014). Once acclimation was observed in Filter 8 (and approximately four months of data from steady-state operation were collected), monitoring was suspended.

Biofilter acclimation was determined using several filter performance indicators, namely, ATP concentrations on the filter media, and removal iron, manganese, turbidity, and TOC. Once these parameters reached steady-state levels (i.e., a consistency in overall data trends) acclimation was considered complete.

2.5.3.4 General acclimation considering TOC and ATP. TOC and ATP were analyzed to determine when biofilter acclimation had been achieved. Upon startup of the LLWTP biofilters, TOC removal was initially very high (i.e., roughly 40 percent); this is likely because of the adsorption of TOC to the surface of virgin GAC. As expected, TOC removal decreased over time (to approximately 21–23 percent) as adsorption sites became occupied and biodegradation became the more dominant removal mechanism (Brown et al. 2020).

Comparing TOC removal between the two filters studied helps illustrate the effect of seasonality on acclimation rate (i.e., the time needed to achieve "steady-state" operation). Filter 8, which was converted in the summer, acclimated in half the time as compared with the filter that was converted in the winter (Filter 10). Filter 8 reached steady state after approximately two months, whereas Filter 10 took four months (Figure 2-12).

ATP concentration (as an indicator of microbial activity on the filter) is another parameter that can help inform acclimation rate. ATP analysis reinforced that Filter 8 (summer startup) acclimated twice as fast as Filter 10 (winter startup), reaching steady-state 3.2×10^5 pg/g ATP in two-and-a-half months versus five months, respectively.

Considering both TOC removal and ATP, it can be observed that relatively steady-state TOC removal (21±3.7 percent and 22±12 percent for Filters 8 and 10, respectively) was observed, while ATP concentrations were continuing to increase; this suggests that biomass on the filter media was no longer limiting TOC removal.

2.5.3.5 Manganese acclimation. Manganese was another parameter monitored at LLWTP; Filter 8 reached steady-state manganese removal in two-and-a-half months, whereas Filter 10 required approximately seven months (Figure 2-13). Once again, summer

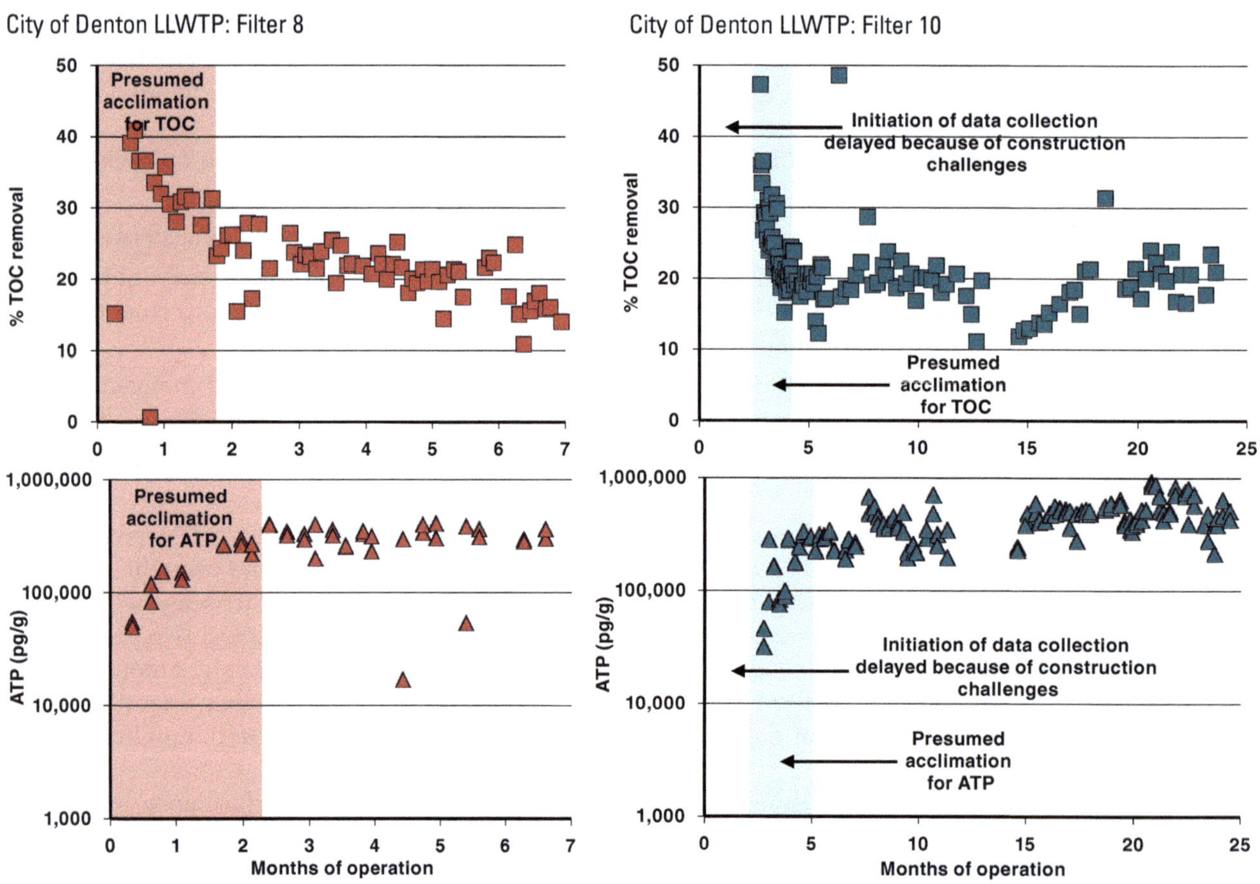

ATP—adenosine triphosphate, LLWTP—Lake Lewisville WTP, TOC—total organic carbon

Figure 2-12 Acclimation of ATP on filter media and TOC removal

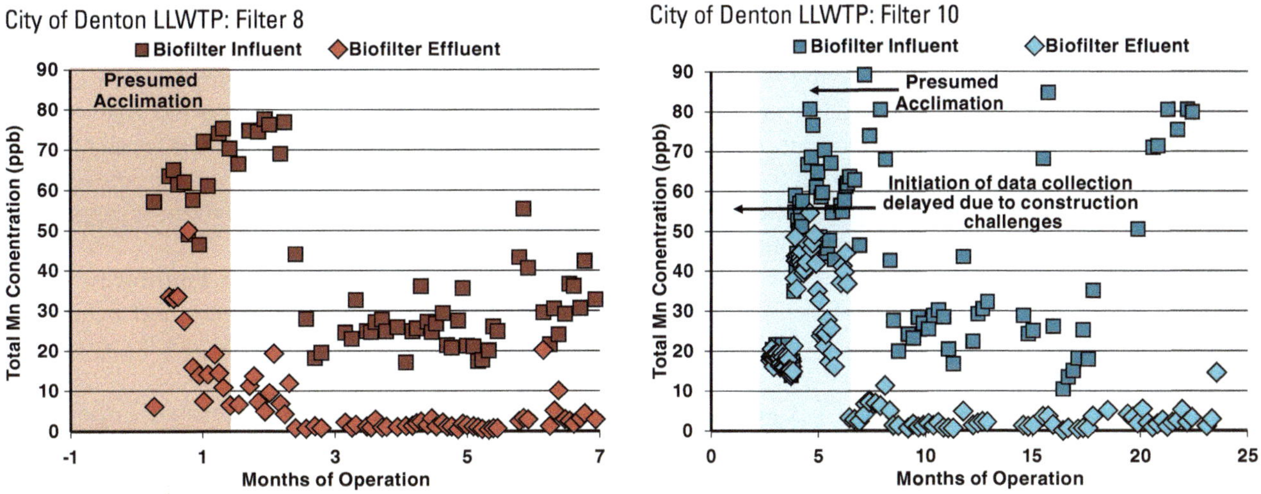

LLWTP—Lake Lewisville WTP

Figure 2-13 Acclimation of manganese removal

startup was associated with shorter acclimation times. Following acclimation, Filter 8 removed an average of 92 percent of total manganese, and Filter 10 achieved an average removal of 95%.

2.5.3.5.1 *Effect of seasonality on acclimation rates of various parameters.* Different water quality parameters achieve steady state at various rates; these rates can be influenced by the season in which the filter was brought on-line. When comparing ATP concentration with TOC and manganese removal, data suggest that TOC was the first of the three parameters monitored to acclimate, independent of filter startup season. Total manganese removal, however, was more sensitive to seasonality and stabilized last in the filters brought on-line in the winter. For filters brought on-line during the summer, both ATP and manganese removal stabilized at roughly the same rate.

2.5.3.5.2 *Additional water quality analyses.* No acclimation period was observed for turbidity or total iron removal; excellent particulate removal was exhibited by the filters upon startup (data not presented). Biofilter effluent turbidity was at or below 0.10 NTU for 95 percent of samples across the study, exceeding the LLWTP's goal of 0.15 NTU with average UFRVs greater than 3,000 gal/ft2 (Table 2-21). Biofilter effluent iron concentrations were 28±41 µg/L (data not presented), achieving an average of 92 and 94 percent total iron removal across Filters 8 and 10, respectively.

2.5.3.6 Key success factors and lessons learned. Unfortunately, because of challenges during construction and startup, the dissolved fraction of constituents was not measured. Monitoring of dissolved fractions is important to confidently differentiate physical particle removal from biological oxidation and subsequent removal across biofilters in which multiple removal mechanisms occur.

Not all water quality parameters will achieve steady-state removal at the same time. The acclimation rate may be affected by the season in which the filter was brought on-line, among other factors. The acclimation period was generally two to three times longer for filters brought on-line during the winter (i.e., at 9°C) compared with the summer (e.g., at 22°C). Construction sequencing for future projects should consider the effect of seasonality on biofilter acclimation.

High initial TOC removal (i.e., 40%) was observed and was likely due to the vast availability of adsorption sites on virgin GAC. The virgin GAC was able to capture TOC early on (before microbial establishment) and ensure that increased AOC produced by ozonation does not lead to a decrease in distribution system water quality stability. The capacity of virgin GAC to adsorb TOC while the biological community becomes established on the filter media allowed the City of Denton to continue to produce high-quality finished water.

When conducting ATP analysis, it is important to consistently sample and measure biofilter samples. Differences in the depth of media a sample was collected from or the duration into a filter run may result in different ATP concentrations on the media. Additionally, it is important to have the same person measure ATP each time and consistently perform each step. For example, if time reagents are allowed to sit at room temperature before analysis it may alter the results.

For additional details on this case study, as well as results from monitoring acclimation of a second WTP, which although in operation for over 10 years had been shut down for two months for winter maintenance, refer to the following *AWWA Water Science* publication:

Evans, A.N., J.T. Carter, B. Kuhnel, and V.E. Nystrom. 2021. "Seasonal impacts on Biofiltration Acclimation Dynamics and Performance." *AWWA Water Science*. 3(3):e1225. https://doi.org/10.1002/aws2.1225

2.6 REFERENCES

AWWA/ASCE (American Water Works Association/American Society of Civil Engineers). 2012 (5th ed.). *Water Treatment Plant Design*. Denver: AWWA.

AWWA. 2011 (3rd ed.). *m37 Operational Control of Coagulation and Filtration Processes*. Denver: AWWA

Bonora, M., S. Patergnani, A. Rimessi, E. De Marchi, J.M. Suski, A. Bononi, C. Giorgi, S. Marchi, S. Missiroli, F. Poletti, M.R. Wieckowski, and P. Pinton. 2012. "ATP Synthesis and Storage." *Purinergic Signalling*. 8(3):343-357.

Bouwer, E.J. and P.B. Crowe. 1988. "Biological Processes in Drinking Water Treatment." *Journal AWWA*. 80(9): 82-93.

Brown, J., G. Upadhyaya, J. Carter, T. Brown, and C. Lauderdale. 2016. "North American Biofiltration Knowledge Base." Project #4459. Denver: WRF.

Brown, J., G. Upadhyaya, J. Nyfennegger, G. Pope, A. Evans, J. Carter, S. Black, C. Alito, C. Lauderdale, J. Hooper, L. Stephens, E. Dickenson, E. Wert, and L. Weinrich. 2020. "Biofiltration Guidance Manual for Drinking Water Facilities." Project #4719. Denver: WRF (Water Research Foundation).

Cornwell, D.A. and D.K. Roth. 2011 (6th ed.). "Water Treatment Plant Residuals Management." In *Water Quality & Treatment: A Handbook on Drinking Water* (J.K. Edzwald, editor). New York: McGraw-Hill.

Crittenden, J.C., R.R. Trussell, D.W. Hand, K.J. Howe, and G. Tchobanoglous. 2012 (3rd ed.). *MWH's Water Treatment: Principles and Design*. Hoboken, N.J.: John Wiley & Sons.

Dickenson, E., E. Wert, E. Verdugo, K. Greenstein, J. Brown, G. Upadhyaya, G. Pope, O. Schneider, R. Marfil-Vega, and S. Summers. 2018. "Simultaneous Removal of Multiple Chemical Contaminants Using Biofiltration." Project #4559. Denver: WRF.

Emelko M.B., P.M. Huck, B.M. Coffey, and E.F. Smith. 2007. "Effects of Media, Backwash, and Temperature on Full-Scale Biological Filtration." *Journal AWWA*. 98(12): 61-73.

Evans, A., L.W. Stephens, and L. Terry. 2019. "Harness Innovative Knowledge and Tools to Drive Biofilter Performance." *Journal AWWA*. 111(9):98-101.

Farré, M.J., J. Reungoat, F.X. Argaud, M. Rattier, J. Keller, and W. Gernjak. 2011. "Fate of N-nitrosodimethylamine, Trihalomethane and Haloacetic Acid Precursors in Tertiary Treatment Including Biofiltration." *Water Research*. 45(17):5695-5704.

Goldgrabe J., R.S. Summers, and R.J. Miltner. 1993. "Particle Removal and Head Loss Development in Biological Filters." *Journal AWWA*. 85(12):94-106.

Hooper, J. 2015. "Monitoring Tools for Biological Filtration." ACE Workshop, Anaheim.

Hooper, J., K. Vickstrom, P. Evans, C. Alito, S. Black., and C. Lauderdale. 2019. "Guidance Manual for Monitoring Biological Filtration of Drinking Water." Project #4620. Denver: WRF.

Hozalski, R.M. and E.J. Bouwer. 2000. "Non-Steady State Simulation of BOM Removal in Drinking Water Biofilters: Model Development." *Water Research*. 35(1):198-208.

Hozalski, R.M. and E.J. Bouwer. 2001. "Non-Steady State Simulation of BOM Removal in Drinking Water Biofilters: Applications and Full-Scale Validation." *Water Research*. 35(1):211-223.

Jeffrey, L., K. Knight, and L. Williams. 2015. "Water Utility Planning Strategies to Mitigate Impacts of Climate Change in Central Ohio." Project #4585. Denver: WRF.

Kawamura. 2000 (2nd ed.). *"Integrated Design and Operation of Water Treatment Facilities."* New York: John Wiley & Sons.

Kohl P.M. and D.R. Dixon. 2012. "Occurrence, Impacts, and Removal of Manganese in Biofiltration Processes." Project #4012. Denver: WRF

Lauderdale C., P. Chadik, M.J. Kirisits, and J. Brown. 2012. "Engineered Biofiltration: Enhanced Biofilter Performance through Nutrient and Peroxide Addition." *Journal AWWA*. 104(5):298-309.

Lauderdale, C., C. Alito, J. Hooper, K. Dowdell, and E. Wert. 2018. "Optimizing Biofiltration for Various Source Water Quality Conditions." Project #4555. Denver: WRF.

LeChavellier, M., W.C. Becker, P. Schorr, and R.G. Lee. "Evaluating the Performance of Biologically Active Rapid Filters." *Journal AWWA*. 84(4):136-146.

LeChavellier, M., W. Schneider, L. Weinrich, P. Jjemba, P. Evans, J. Hooper, and R. Chappell. 2015. "An Operational Definition of Biostability in Drinking Water." Project #4312b. Denver: WRF.

Liu, C., C.I. Olivares, A.J. Pinto, C.V. Lauderdale, J. Brown, M. Selbes, and T. Karanfil. 2017. "The Control of Disinfection Byproducts and Their Precursors in Biologically Active Filtration Processes." *Water Research*. 124:630-653.

Logsdon, G. 2002. "Filter Maintenance and Operations Guidance Manual." Project #2511. Denver: Water Research Foundation.

McGuire, M.J., S.W. Krasner, C.J. Hwang, and G. Lzaguirre. 1981. "Closed-Loop Stripping Analysis as a Tool for Solving Taste and Odor Problems." *Journal AWWA*. 73(10):530-537.

McKie, M., L. Taylor-Edmonds, S.A. Andrews, and R.C. Andrews. 2015. "Engineered Biofiltration for the Removal of Disinfection By-product Precursors and Genotoxicity." *Water Research*. 81:196-207.

Nieminski, E. and S. Perry. 2015. "State and Provincial Perspectives on the Regulation of Biological Filtration." *Journal AWWA*. 107(12):31-38.

Nix, D.K. and J.S. Taylor. 2018 (2nd ed.) "Filter Evaluation Procedures for Granular Media." Denver: AWWA.

Newcombe, G. and B. Nicholson. 2004. "Water Treatment Options for Dissolved Cyanotoxins." *Journal of Water Supply: Research and Technology–AQUA*. 53(4):227-239.

Persson, F., G. Heinicke, T. Hedberg, M. Hermansson, and W. Uhl. 2007. "Removal of Geosmin and MIB by Biofiltration – An Investigation Discriminating Between Adsorption and Biodegradation." *Environmental Technology*. 28(1):95-104.

Schneider, O.D. and M.W. LeChevallier. 2017. "A Cost-Effective Treatment Process for Producing High-Quality Drinking Water." *Journal AWWA*. 109(3):39-47.

Upadhyaya, G., J. Brown, J. Nyfennegger, G. Pope, and V. Hart. 2021. "Optimizing Biofiltration and Integrating Biofiltration into Existing Treatment." Project #4731. Denver: WRF.

Williams, L., W. Persich, and K. Duncan. 2012. "Achieving Excellent Water Quality While Piloting at High Loading Rates." Toronto, Canada: Water Quality Technology Conf.

Zearly, T.L. and R.S. Summers. 2012. "Removal of Trace Organic Micropollutants by Drinking Water Biological Filters." *Environmental Science & Technology*. 46(17):9412-9419.

Chapter 3

Aerobic Biotreatment of Groundwater

Asher E. Keithley, Darren A. Lytle, and Amlan Ghosh

Aerobic biotreatment of groundwater relies on natural microorganisms to remove dissolved inorganic contaminants, such as ammonia, arsenic, manganese, and iron, in the presence of oxygen. This chapter discusses the target contaminants and regulatory concerns, biological removal mechanisms, treatment system configurations, and considerations for planning, process evaluation, design, and operation. This chapter also includes case studies on full-scale applications of the biotreatment of groundwater.

3.1 APPLICATION OVERVIEW

Aerobic biological treatment (biotreatment) of groundwater is defined as the utilization of naturally occurring microorganisms, generally bacteria, to oxidize dissolved contaminants, primarily inorganic in nature, in the presence of oxygen to a higher oxidation state that is relatively insoluble, more adsorbable, or in a less undesirable form. Often, biotreatment occurs in granular media filter beds designed to remove particles (e.g., iron), in which case it is often referred to as a biofilter. An aeration step is typically necessary to increase the amount of oxygen available to microorganisms when treating reduced contaminants in anoxic groundwaters.

The origins of aerobic biotreatment of groundwater are difficult to trace. It is reasonable to assume that such systems have unintentionally been in place for many decades in groundwater systems where aeration followed by filtration without chlorine or other disinfectant or strong oxidant addition in advance of the filter has been practiced. In the mid-2000's, the recognition of the detrimental effects of source water ammonia led to the development of engineered biotreatment approaches (de Vet et al. 2009, Lytle et al. 2007a, Štembal et al. 2005). Around the same time, the advantages of intentionally designed aerobic biotreatment to address manganese, arsenic, and iron gained recognition. Since that time, aerobic biotreatment of groundwater has gained recognition as a treatment option because of its effectiveness, reliability, and simplicity. Aerobic groundwater biotreatment differs from surface water biofiltration in important ways: target contaminants are typically reduced inorganic compounds, the raw water quality is relatively stable, upstream treatment is often limited to aeration, and backwash requirements are different. This section presents an overview of the benefits of aerobic biotreatment, discusses removal mechanisms, and summarizes regulatory considerations.

3.1.1 Benefits

Aerobic biotreatment of groundwater primarily focuses on the biological oxidation of reduced dissolved inorganic contaminants, such as free ammonia (the sum of ammonia [NH_3] and ammonium ion [NH_4^+], hereinafter referred to as NH_3), arsenic (As), manganese (Mn), and iron (Fe). In select cases, removal of total organic carbon (TOC) has also been observed. Advantages of aerobic biotreatment over conventional physical-chemical treatment processes (e.g., reverse osmosis, ion exchange softening, coprecipitation) include (1) efficient contaminant removal without the generation of concentrated waste streams, (2) low energy demand, and/or (3) reduced residual management disposal concerns. Generally, the final products of aerobic biotreatment of groundwater are acceptable at the produced concentrations or are physically removed.

There are further advantages to biotreatment relative to some conventional treatment techniques. Microorganisms associated with oxidizing inorganic groundwater contaminants in the presence of oxygen are generally autotrophic bacteria that use inorganic carbon (i.e., CO_2) as their carbon source for growth and have a low growth yield. This is important because groundwaters tend to be associated with low organic carbon levels as well as low levels of microbiological pathogens relative to surface waters. Biotreatment uses few chemicals, if any, with treatment additions being limited to aeration (i.e., oxygen augmentation), an acid or base for pH adjustment if necessary, and chlorine as a disinfectant. Unlike membrane systems, it also does not generate a waste brine stream. Additionally, iron and manganese oxides produced biotically are more crystalline and larger than those produced through physical-chemical processes, which may allow for higher loading rates and longer filter run times (Arturi et al. 2017, Bruins et al. 2015). The benefits and drawbacks of different treatment processes should be considered along with a cost–benefit analysis to decide the appropriate process for a particular site.

3.1.2 Contaminant Removal

The removal of targeted inorganic contaminants from groundwater by aerobic biotreatment is driven primarily by public health and aesthetic goals (see Section 3.1.4). Several studies have reported on the capability of aerobic biotreatment pilot- and full-systems to reduce such contaminants. In general, biotreatment was effective, achieving greater than 90 percent contaminant removal and lowering contaminants to below their respective maximum contaminant level (MCL) or secondary MCL (SMCL), and multiple contaminants can be treated simultaneously if the necessary conditions are provided (Table 3-1).

3.1.2.1 Ammonia. Many regions in the United States have NH3 in their drinking groundwater sources originating from naturally occurring processes and anthropogenic sources, such as agricultural and urban runoff, concentrated animal feeding operations, and municipal wastewater treatment plants. High free NH3 (e.g., greater than 1 mg N/L) is a concern because it negatively affects chlorination for contaminant removal (e.g., As and Mn) by forming chloramines, affects disinfection, and can lead to biological activity and resultant water quality challenges in the distribution system (Chen et al. 2018, Edwards and Dudi 2004, Odell et al. 1996). Biotreatment removes ammonia through the process of nitrification, in which microorganisms oxidize ammonia to nitrite (NO_2^-) and then nitrate (NO_3^-). Nitrite accumulation can be mitigated through appropriate design and operation, and posttreatment for nitrate would only be needed in instances in which the raw water ammonia concentration is very high (i.e., close to 10 mg N/L). Biotreatment is capable of consistently removing more than 90 percent of the raw water ammonia.

Table 3-1 Summary of selected pilot- and full-scale aerobic groundwater biotreatment systems

Reference	Scale	Treatment Process	Media	Loading Rate *gpm/ft²*	T °C	Influent Concentration NH₃ *mg/L*	As *µg/L*	Fe *mg/L*	Mn *mg/L*	Removal % NH₃	As	Fe	Mn
Bourgine et al. (1994)	Full	Open aeration, 3-stage filtration	Sand	10.6, 5.7, 5.7	11	0.6	—	4	0.6	>90	—	99	>95
Štembal et al. (2004)	Full	Pressure aeration, filtration	Sand (0.2–2 mm)	2.3	16.5	0.84	—	1.04	0.147	>40	—	>80	>65
	Full	Open aeration, closed filtration	Sand (0.2–2 mm)	2.3	12.9	0.018	—	0.4	0.425	100	—	>50	>88
Li et al. (2005)	Full	Cascade aeration, filtration	Sand (0.5–1.2 mm), Sand (0.9–1.0 mm)	1.6 and 3.3	9	—	—	0.01–0.5	0.57–3.0	—	—	100	100
Štembal et al. (2005)	Pilot	Pressure aerator to open reservoir, pressure filter	Sand (0.5–2 mm)	4.1–9.0	13.4	1.01	—	2.45	0.224	>90	—	>95	>78
	Pilot	Air injection, pressure filter	Sand (0.5–2 mm)	4.1–4.9	13.8	0.3	—	1.05	1.06	>67	—	>90	>95
	Pilot	Cascade aeration, pressure filter	Sand (0.5–2 mm)	4.1–9.8	11.2	0.02	—	<0.03	1.06	—	—	—	>95
	Pilot	Air injection, pressure filter	Sand (0.5–2 mm)	4.1	14.6	2.62	—	0.98	0.1	>96	—	>90	>50
Lytle et al. (2007b)	Full	Open aeration, retention basin, filtration	Sand	2.0	14	1.15	46	2.29	—	>90	88	>98	—
Burger et al. (2008)	Full	Pressurized aeration, pressure filtration	Sand	2.5–3.7	—	<0.06	<5	<0.2	0.93	—	—	—	>96
	Full	Pressurized aeration, pressure filtration	Sand	9.4	—	<0.06	<5	<0.2	0.86	—	—	—	>96
	Full	Pressurized aeration, pressure filtration	Sand	9.4	—	0.13	<5	<0.2	0.86	—	—	—	>96
	Full	Pressurized aeration, pressure filtration	Sand	9.4	—	<0.06	<5	<0.2	1.39	—	—	—	>96
Katsoyiannis et al. (2008)	Full	Aeration, up-flow filtration, coagulation, filtration	Anthracite (0.8–1.6 mm)/ sand (0.4–0.8 mm)	4.1	—	1.2	20	0.165	0.235	96	80	97	98
de Vet et al. (2009)	Full	Spray aeration, trickling filter	Anthracite (1.4–2.5 mm), sand (0.8–1.25 mm)	0.9	12	4.5	—	5.4	0.54	91	—	>99	>99
Li et al. (2013)	Full	Aeration, filtration	Sand (0.6–1.2 mm)	1.2–2.0	8	1.3	<2	15	1.2	>80	—	>98	>95

(continued)

Table 3-1 Summary of selected pilot- and full-scale aerobic groundwater biotreatment systems (*Continued*)

					Influent Concentration					Removal %			
Reference	Scale	Treatment Process	Media	Loading Rate *gpm/ft²*	T °C	NH_3 *mg/L*	As *μg/L*	Fe *mg/L*	Mn *mg/L*	NH_3	As	Fe	Mn
Cai et al. (2014)	Pilot	Cascade aeration, gravity filtration	Mn-coated sand (0.6–1.2 mm)	1.2–2.5	4	1.11	—	3.04	0.52	90	—	>96	80
Pacini et al. (2014)	Full	Aeration, 2-stage filtration	Gravel (6–12 mm)	1.8	—	<0.05	<20	1.2	0.3	—	—	>95	>90
Lytle et al. (2015)	Full	Continuous aeration contactor, gravity filtration	0.65 cm gravel, 0.1 cm anthracite/0.046 cm sand	2.3, 1.8	12	3.2	—	0.15	0.002	98	—	>95	0
McClellan (2015)	Full	2-stage pressurized air, pressure filtration		11.9, 5.9	—	— —	—	3	2.5	—	—	>96	>99
Richard et al. (2016)	Pilot	Pressurized aeration, filtration	Sand (3.15–5.6 mm)/ sand (1–2 mm)	0.9	16.5	3.5	—	0.022	0.006	99	—	>88	>90
Cheng et al. (2017)	Pilot	Aeration, gravity filtration	Mn-coated sand (0.8–1 mm)	2.5	8	1.27	—	11.27	1.20	92	—	>99	95
Dangeti et al. (2017)	Pilot	Filtration, aeration, filtration	Anthracite, sand	1.3	4–11	0.53	—	2.81	0.88	75	—	>99	97
Tatari et al. (2017)	Full	Aeration, 2-stage filtration	Anthracite (1.6–2.5 mm)/ sand (3–5 mm), calcined flint (1.5–2 mm)	—	—	0.51	—	2.9	—	>90	—	>96	—
	Full	Aeration, 2-stage filtration	Sand (3–5 mm), sand (0.8–1.4 mm)	—	—	1.08	—	1.9	—	>95	—	>95	—
	Full	Aeration, 2-stage filtration	Sand (2–3 mm), sand (0.8–1.4 mm)	—	—	1.09	—	3	—	>95	—	>96	—
	Full	Aeration, 2-stage filtration	Gravel (40–50 mm), sand (0.8–1.2 mm)	—	—	0.41	—	1.8	—	>98	—	>94	—
Gulay et al. (2018)	Full	Partial aeration, 2-stage filtration	Sand (1–5 mm)	—	8–10	—	—	12	0.3	—	—	>99	>98
	Full	Aeration, 2-stage filtration	Gravel (10–20 cm), sand (2–5 mm)	—	8–11	—	—	1.07	0.06	—	—	97	97
	Full	Aeration, 2-stage filtration	Sand (1–5 mm)	—	8–12	—	—	1.18	0.63	—	—	98	>99
Zeng et al. (2019)	Full	Aeration, filtration	Sand (0.5–1.2 mm)	2.0	12–14	—	—	0.2	1.2	—	—	—	>95

Source: Adapted from Tekerlekopoulou et al. (2013). *Blank cells indicate "field not applicable."*

3.1.2.2 Iron. Iron is a naturally occurring element commonly found in many groundwaters. Iron can originate from soils, rocks, and minerals in contact with groundwater, as well as from human activities. Iron can accumulate in the distribution system and has negative aesthetic effects, including staining, discoloration, and taste. Iron is most commonly found in the reduced and dissolved form in groundwaters and is oxidized chemically by aeration or biologically to a particulate, which is subsequently filtered from the water (Benefield et al. 1982, Faust and Aly 1981). Iron removal typically exceeds 90 percent.

3.1.2.3 Manganese. Manganese is a naturally occurring element that is widely geographically distributed in groundwaters. Elevated manganese concentrations (greater than 0.3 mg/L) have been associated with shallow, anoxic aquifers that also have high TOC concentrations (McMahon et al. 2019). Manganese can originate from soils, rocks, and minerals in contact with groundwater, as well as from human activities. Manganese can have negative aesthetic effects, including staining and discoloration, resulting in consumer complaints at concentrations of 0.01–0.02 mg/L (Dietrich and Burlingame 2015). It can accumulate in the distribution system where it can sorb and subsequently release other heavy metals, including lead and arsenic (Trueman et al. 2019). Although manganese is an essential nutrient, recent evidence suggests that it can cause neurotoxic effects, such as intellectual impairment and poorer neurobehavioral function, particularly in children (Iyare 2019). In most groundwaters, manganese is in the reduced and dissolved (or soluble) form. Dissolved manganese can be removed via biological oxidation and/or adsorption to a manganese oxide and subsequent oxidation, and the oxidized precipitate is then physically removed from the water. Removals exceeding 90 percent are often achieved with biotreatment.

3.1.2.4 Arsenic. Arsenic can have both natural and anthropogenic sources. In natural environments, arsenic may be leached from certain arsenic-containing minerals, such as arsenopyrite and other various arsenic sulfides and sulfosalts (USEPA 2003, Kim et al. 2002). In addition to arsenic-bearing host minerals, discharge from various industries, including mining, petroleum refining, glass, and ceramics manufacturing, can cause arsenic pollution. Pesticides, herbicides, and fertilizers are also known sources of arsenic release (Blakely 1984). The trivalent oxidation form of arsenic, arsenite, As(III) (As^{3+}), is normally found in groundwater. However, the pentavalent oxidation form, arsenate, As(V) (As^{5+}) has also been found in some groundwaters (USEPA 2003). As(III) (As^{3+}) is biologically oxidized to As(V) (As^{5+}) ions, which is a necessary step for effective arsenic removal. Additional treatment processes are required to then remove As(V) (As^{5+}) ions, such as iron coprecipitation and filtration, adsorption media processes, or anion exchange (USEPA 2003). Biotreatment can achieve arsenic removal rates of 25–50 percent to produce concentrations less than the MCL of 10 µg/L.

3.1.2.5 Organic carbon. Groundwaters tend to have a lower concentration of organic carbon than surface waters, but high TOC (e.g., TOC greater than 3 mg/L) sometimes occurs in groundwater, particularly in shallow aquifers (McMahon et al. 2019). Additionally, climate change and urbanization may cause groundwater TOC concentrations to increase in the future (McDonough et al. 2020). TOC is not regulated but is a precursor to regulated disinfection by-products.

Aerobic groundwater biotreatment systems are not typically designed around organic carbon removal. In biotreatment systems, TOC removal depends on how biodegradable the organic matter is. Surveys of groundwater facilities in the United States found that the biodegradable portion of organic carbon (i.e., measured as assimilable organic carbon) is typically less than 100 µg/L (Bradford et al. 1994, Kaplan et al. 1994). In such cases, upstream oxidation would be required; for example, sometimes ozone-biofiltration is used to treat groundwaters with high color (Rittmann et al. 2002). Ozone-biofiltration systems are discussed in more detail in Chapter 2. Because of the low organic carbon removal typically observed in aerobic biotreatment systems, the remainder of this chapter focuses on the treatment of inorganic contaminants.

3.1.3 Removal Mechanisms

Aerobic groundwater biotreatment typically relies on the growth and activity of autotrophic bacteria that use oxygen as an electron acceptor and carbon dioxide as a carbon source. Thus, because of the naturally low oxygen concentration in most groundwaters containing these target-reduced contaminants, oxygen addition is critical for biotreatment and for providing the appropriate water chemistry conditions for the produced oxidized compounds (McMahon et al. 2019, Schilling 2002). Ammonia by far has the greatest oxygen requirement on a mg/L basis as compared with other inorganic contaminants. Along with autotrophic bacteria, heterotrophic bacteria that grow on carbon sources other than carbon dioxide can coexist in bioreactors, using naturally present organic carbon or the organic by-products secreted by the autotrophic bacteria, particularly nitrifiers (Kindaichi et al. 2004). Although not responsible for the removal of contaminants of concern, heterotrophic bacteria can exert an oxygen demand and outgrow autotrophic bacteria, leading to increased biofilm on the media surface and potentially interfering with contaminant removal.

Table 3-2 gives an estimate for the reaction oxygen demand. It is important to note that (1) the desired dissolved oxygen (DO) concentration might be greater than the oxygen demand to provide the desired redox conditions, (2) other reaction mechanisms might contribute to contaminant transformation and removal, and (3) the stoichiometry does not consider the cell synthesis component of the biological oxidation reaction.

3.1.3.1 Ammonia. Nitrification is a two-step process in which ammonia is oxidized sequentially to nitrite and then to nitrate. This process may be performed by distinct ammonia-oxidizing bacteria microorganisms and nitrite-oxidizing bacteria (NOB), or by a single comammox (COMplete AMMonia Oxidation) bacteria (Gulay et al. 2019). The effectiveness of ammonia removal depends on the pH, alkalinity, and available oxygen. The optimal pH range for nitrification is 7–8. Nitrification has a high oxygen demand of 4.6 mg O_2/mg NH_3-N, and it consumes alkalinity at a rate of 7.1 mg $CaCO_3$/mg NH_3-N. Nutrients, including phosphorus and copper, are also required but should be added only if insufficient in the raw water (Wagner et al. 2018, Lytle et al. 2013).

3.1.3.2 Iron. Iron is removed through direct oxidation and precipitation or by sorption and oxidation (heterogeneous) processes. Iron may be oxidized both chemically by oxygen and biologically. The operation of biotreatment systems under lower DO (2–4 mg O_2/L) and pH (6–6.5) conditions has been proposed to favor biological over chemical iron oxidation (Gulay et al. 2018). The need for DO augmentation will depend on the raw water quality, such as DO, iron concentration, and presence of other contaminants. Multiple removal mechanisms likely occur concurrently in biotreatment systems. Precipitated iron is removed from the filter via regular backwashing, and the residuals in the backwash wastewater require appropriate disposal.

Table 3-2 Oxidation stoichiometry for reduced species common in drinking water

Oxidized Species	Chemical Oxidation Reaction	Oxygen Demand*
Ammonia	(1) $NH_4^+ + 1.5\ O_2 \rightarrow NO_2^- + H_2O + 2H^+$ (2) $NO_2^- + 0.5\ O_2 \rightarrow NO_3^-$	4.6 mg O_2/mg NH_4^+-N
Iron	$Fe^{2+} + 0.25\ O_2 + H^+ \rightarrow Fe^{3+} + 0.5\ H_2O$	0.14 mg O_2/mg Fe
Manganese	$Mn^{2+} + 0.5\ O_2 + 2\ H^+ \rightarrow Mn^{4+} + H_2O$	0.29 mg O_2/mg Mn
Arsenic†	$H_3AsO_3 + 0.5\ O_2 \rightarrow HAsO_4^{2-} + H^+$	0.21 mg O_2/mg As

*Oxygen requirements do not consider biological growth.
†As(III) and As(V) species are pH-dependent, but oxygen demand is the same.

3.1.3.3 Manganese. Manganese oxidation may proceed via three mechanisms in biofilters without the application of an oxidant such as free chlorine or permanganate (Tobiason et al. 2016). First, in homogenous oxidation, DO may oxidize Mn(II) to Mn(IV) in the water. This reaction is generally slow at pH less than 9 and not prominent in drinking water filters. Second, in heterogeneous oxidation, Mn^{2+} initially adsorbs to a metal (typically manganese) oxide on the filter media surface and subsequently is oxidized via a surface-catalyzed reaction. The third mechanism is biological oxidation in which bacteria directly oxidize manganese, potentially generating energy for growth (Yu and Leadbetter 2020). Bacteria also may contribute to manganese removal by producing biopolymers that adsorb manganese and/or catalyze its oxidation (Zhou and Fu 2020, Tebo et al. 2004). Biological manganese oxidation occurs faster under conditions in which the pH is greater than 7.4 and the DO is greater than 8 mg O_2/L, but may occur outside these conditions (Breda et al. 2019, Mouchet 1992). These conditions also are favorable for sorption and catalytic oxidation (Cerrato et al. 2011, Knocke et al. 1988).

All of these mechanisms produce manganese oxides, some of which become incorporated into the filter media coating while the other larger fraction gets removed during backwashing. Biogenic manganese oxides have a high adsorption capacity for Mn^{2+} and high autocatalytic activity, which enhance the overall removal performance of the filter (Bruins et al. 2015, Tebo et al. 2004). Manganese residuals in the backwash wastewater require appropriate disposal.

3.1.3.4 Arsenic. Biotreatment for arsenic removal requires first oxidizing As(III) (As^{3+}) to As(V) (As^{5+}). A range of bacteria are capable of enzymatically oxidizing As(III) to As(V), either as a detoxification mechanism or to generate energy for growth (Crognale et al. 2017). Then, As^{5+} must be removed, which is often accomplished via adsorption and coprecipitation with iron oxyhydroxides that are separated from the water via filtration (Lytle et al. 2007b, Lytle et al. 2005). This mechanism is common in biotreatment, particularly when the groundwater also contains iron. Removal of As(V) may be improved by maintaining the pH near 7 and supplementing iron if needed (Gude et al. 2018, Lytle et al. 2005). The arsenic-laden iron precipitates are removed from the filter via backwashing, and the generated residuals require appropriate disposal.

3.1.4 Regulatory Considerations

Although drinking water biotreatment has gained wide acceptance in recent years, misperceptions and concerns still exist. Informing and engaging the state regulatory agency whenever possible allows the opportunity for concerns to be raised and addressed early in the treatment design. Approval of biotreatment systems ultimately falls to the regulatory agency's discretion. Pilot testing will likely be required by the regulatory authority before approval is given to proceed with the design and construction of a biotreatment system, particularly in situations where limited past experiences exist. Other regulatory considerations include the following:

- particle removal and turbidity standards,
- backwash wastewater and residuals management,
- monitoring,
- data collection and reporting,
- treatment redundancy,
- specialized operator training and/or certification, and
- simultaneous compliance and unintended consequences of treatment, particularly in the drinking water distribution system (corrosion control, lead, copper, etc.).

Table 3-3 Existing US regulations for contaminants of interest

Contaminant	Existing Regulation	Notes
Ammonia	No US federal regulation	No direct health implications. Affects chlorination for treatment and disinfection. Can promote microbial growth in the distribution system, with residual loss, nitrite accumulation, and corrosion control concerns. By-products of oxidation, nitrite and nitrate, are regulated.
Nitrite	MCL = 1 mg N/L	Infants below the age of six months who drink water containing nitrate in excess of the MCL could become seriously ill and, if untreated, may die. Symptoms include shortness of breath and blue-baby syndrome.
Nitrate	MCL = 10 mg N/L	Infants below the age of six months who drink water containing nitrite in excess of the MCL could become seriously ill and, if untreated, may die. Symptoms include shortness of breath and blue-baby syndrome.
Iron	SMCL = 0.3 mg/L	Aesthetic issues.
Manganese	SMCL = 0.05 mg/L; lifetime HAL of 0.3 mg/L; acute HAL of 1.0 mg/L	Neurotoxic implications and aesthetic water quality issues.
Arsenic	MCL = 0.010 mg/L	Skin damage or problems with circulatory system and may have increased risk of getting cancer.

HAL—health advisory limit, MCL—maximum contaminant level, SMCL—secondary maximum contaminant level

USEPA (US Environmental Protection Agency). 2018. "Drinking Water Standards and Health Advisories Tables." EPA 882-F-18-001. Washington: Office of Water.

USEPA. 2004. "Drinking Water Health Advisory for Manganese." EPA-822-R-04-003. Washington: Office of Water.

Table 3-3 lists the existing US regulations for the contaminants of interests. Ammonia is not a regulated contaminant, but nitrification produces nitrite and nitrate, which are regulated with MCLs of 1 and 10 mg N/L, respectively. If nitrification in the distribution system is suspected, some states (e.g., Iowa) choose to enforce nitrite MCL in the distribution system in addition to the point of entry. When biotreatment is used to remove ammonia at the treatment plant, biofilter posttreatment may be required if the concentration of NO_2^- and/or NO_3^- generated exceeds their MCL (USEPA 2018). Nitrite accumulation can be resolved through treatment optimization, and chlorine applied for disinfection can oxidize residual NO_2^- to NO_3^-. Posttreatment for the generated NO_3^- would be needed if the raw water NH_3 concentration was greater than 10 mg N/L (or less if some NO_3^- is already present in the groundwater), but such instances are rare. Manganese guidelines have come under review as more information becomes available on the negative health effects caused by ingesting manganese in drinking water. Health Canada recently established a health-based maximum allowable concentration of 0.12 mg/L and an aesthetic objective of 0.02 mg/L, the World Health Organization established a provisional health-based guidance value of 0.08 mg/L, and the US Environmental Protection Agency (USEPA) has included manganese in the Draft Fifth Contaminant Candidate List for potential regulatory review (USEPA 2021, World Health Organization 2021, Health Canada 2019). Some states require public notice and restrict sequestration depending on the manganese concentration. A few states (e.g., New Hampshire) have established an arsenic MCL lower than the federal level. Water systems should engage with their state regulatory agencies to determine specific requirements.

3.2 PLANNING

Aerobic drinking water biotreatment is relatively well-documented. However, operating and design criteria can differ depending on the source water and treatment technology configuration. Therefore, the feasibility of a specific aerobic biotreatment application should be assessed with proper planning before considering full-scale implementation. Setting treatment goals and assessing the effectiveness of the technology are key planning steps. The following sections outline best practices for establishing treatment goals, conducting a desktop applicability evaluation, and developing a robust technology testing plan.

3.2.1 Treatment Goals

Establishing treatment goals is an important first step in evaluating biotreatment. Treated water production and finished water quality targets, source water quality, and applicable regulations will largely be the basis of treatment objectives and goals. In addition to the target contaminants, water quality goals for turbidity and disinfection by-products may be established. At the planning stage, expectations and treatment goals can be developed on the basis of general contaminant removal and system performance reported in the literature. Refer to Table 3-1 for a summary of contaminant removals reported in literature. Goals regarding system operations should also be considered, such as the biofilter head loss buildup rate and associated filter run times.

3.2.2 Applicability Evaluation

After treatment goals are established, a desktop analysis can evaluate the suitability of applicable treatment processes (e.g., ion exchange, chemical oxidation, and biotreatment) to address the targeted contaminant(s). The raw water quality from each well should be reviewed, and blending scenarios should be understood to establish the possible range of contaminant loadings. A range of water quality and performance data should be reviewed for multiple years to allow for identifying seasonal and temporal trends. When data are not available (e.g., new facilities), models or estimations of water quality may be required. Oxidation requirements, including oxidant concentration and reaction time, should be established. Table 3-4 summarizes potential processes that might be included in the treatment alternatives analysis.

3.2.2.1 Oxygen requirement. An applicability evaluation for aerobic groundwater biotreatment should begin with an assessment of oxygen requirements. For ammonia treatment, the oxygen requirement is estimated using the influent ammonia concentration and oxygen demand presented in Table 3-2. For iron, manganese, and arsenic treatment, the oxygen requirement is determined by the preferred redox conditions of the microorganisms and the stability of the oxidized precipitates, as discussed in Section 3.1.3. It is possible that the oxygen requirement may be greater than the saturation limit; for example, if the ammonia concentration is greater than 2 mg N/L, then the approximate oxygen demand would be 9.8 mg/L, which is greater than the saturation limit at a temperature of 16°C. Oxygen requirements for heterotrophic bacteria growth should also be considered and may constitute 0.5–2 mg/L O_2 depending on site-specific factors, such as the organic carbon in the groundwater, the microbial community within the biofilter, and kinetic constraints. Different process configurations are available to deliver the required oxygen, as discussed in Section 3.3.1.

3.2.2.2 Iron requirement for arsenic treatment. If arsenic removal is a treatment objective, assessing the iron speciation and concentration is necessary to determine whether enough iron naturally exists in the water or if iron might need to be supplemented. Arsenic removal can be optimized by having an iron-to-arsenic mass ratio (μg/μg) of at least 20:1, having iron and arsenic oxidation occur at the same time, and maintaining pH near 7 (Lytle et al. 2005).

3.2.2.3 Cyclical operation and seasonal shutdown. Groundwater supplies for small systems may be operated for only part of the day (i.e., cyclically), and some filters may be taken offline for extended periods during the winter when demand is lower. These operational practices are often raised as a concern when considering biotreatment, but aerobic biotreatment systems are generally robust and capable of achieving established water quality goals for ammonia, iron, and manganese under these different operating scenarios (Burger et al. 2008, Mouchet 1992). Different operating scenarios should be tested before full-scale implementation because performance may vary from system to system (Section 3.2.3).

3.2.2.4 Effect of temperature. Low groundwater temperatures can slow microbial metabolism and oxidation rates, and they might prolong the acclimation period. Still, biological systems have successfully treated groundwaters at temperatures as low as 4°C (Table 3-1), with the benefit of not experiencing large seasonal temperature variations. Another consideration is that the saturated DO concentration decreases as the temperature increases, so the system design should provide enough DO for contaminant transformation under anticipated warm temperatures. More consistent temperatures and other water quality parameters likely support a more consistent microbial community.

3.2.2.5 Effect on distribution system. The effect of any new treatment process or treatment change on finished and distributed water quality and the corrosion of distribution system materials, including lead, copper, and iron materials, must be considered. Biological aerobic treatment may change several water quality parameters, including pH, oxidation reduction potential (ORP), dissolved inorganic carbon (DIC), and alkalinity. Additionally, the disinfectant residual type (free chlorine and monochloramine) might change after implementing biotreatment if the source water contains ammonia because achieving nitrification at the plant might cause the system to change from running with chloramines to free chlorine. All of these water quality changes can affect metal corrosion and solubility. Each biotreatment applicability evaluation should consider anticipated water quality changes, and a corrosion control assessment study should be performed accordingly. As with any source water and treatment change, consultation with the respective state primacy agency is recommended, and appropriate testing to identify potential unintended consequences may be necessary.

Table 3-4 Treatment alternatives

Contaminant	Potential Treatment Processes
Ammonia	• Breakpoint chlorination • Cation exchange with zeolite resin • Reverse osmosis • Biotreatment
Iron	• Aeration and filtration • Cation exchange (maintain reducing conditions) • Biotreatment
Manganese	• Oxidation with free chlorine or permanganate and greensand filtration • Cation exchange (maintain reducing conditions) • Biotreatment
Arsenic	• Adsorptive media • Reverse osmosis • Oxidation with a strong oxidant (free chlorine, permanganate, or ozone) followed by iron coagulation and removal • Oxidation with a strong oxidant followed by anion exchange • Biotreatment followed by iron coagulation and removal

All treatment processes have benefits and drawbacks, so the treatment effectiveness, cost–benefit analysis, and site-specific factors should be considered to determine the appropriate process.

3.2.3 Testing

After the applicability evaluation, testing needs to be conducted to validate process performance, identify any limitations or unintended consequences, and develop design criteria for the full-scale system. Treatment performance is site-specific and influenced by criteria such as local water quality or requirements to retrofit existing pre- or posttreatment. Appropriate time should be allotted to perform testing, and resource and staffing needs must be considered. Experience has found that communicating with and addressing questions of the state primacy agency during the planning and operation of studies can assist in facilitating a timely full-scale plant approval process. Recommended testing objectives, scales, design, and outcomes are described in the following subsections. Specific design configurations are presented in Section 3.3.1.

3.2.3.1 Testing objectives. While testing objectives can be site- and contaminant-specific, there are generally three broad testing objectives.

- **Demonstrate treatment process performance:** Testing can be used to demonstrate the effectiveness of the selected treatment process train to remove individual or multiple contaminants of concern. Testing can also be used to compare design components and assess enhancement strategies, such as different engineered aeration or oxygenation configurations, filter media (depth, type, and size), and effects of nutrient supplementation.
- **Develop design and operational criteria for full-scale system:** Pilot-testing results are typically used to develop the design and operational criteria for the full-scale system. Criteria may include oxygen augmentation rates, media type and size, filter loading rates, empty bed contact times (EBCTs), and filter run times (backwash frequency). Pilot testing may also be used to characterize residuals quantity and quality to inform the design of residuals management processes. Preliminary estimates of capital and operational costs for the full-scale system may also be developed using the test results.
- **Obtain regulatory approval:** Aerobic groundwater biotreatment is still considered by some as an emerging technology, and full-scale installations have not yet been permitted by many regulatory agencies. As such, demonstration of technology through testing may be required by the regulatory agency before permitting. As part of the required testing before permitting, it may be necessary to demonstrate the duration of the startup and acclimatization phase (i.e., period between system startup and acclimatization of the biological community when treatment targets are achieved), demonstrate the operation during the least-favorable conditions (e.g., cold temperatures or higher loading rates), or demonstrate the effects of intermittent operations (i.e., periodic shutdown periods between continuous runs).

3.2.3.2 Testing scales. Testing can be performed at bench, pilot, and/or full scale. The appropriate scale of testing is determined by considering the initial desktop applicability evaluation, testing objectives, available budget, and/or consultation with state primacy agencies. As needed, limited process optimization testing may also be conducted at full scale following implementation. The three different testing scales are described in detail in Section 1.5.

Bench-scale testing uses small quantities of water with flow rates in the mL/min range. Bench-scale testing can be used to support pilot- or full-scale demonstration testing in several ways, including the following:

- detailed characterization of water quality and oxygen demand of water;
- investigation of factors affecting simultaneous removal of multiple contaminants (e.g., capacity of iron to sorb arsenic);

- "proof-of-concept" testing of novel oxygenation technologies;
- preliminary evaluation of filter media and bio-growth support media (e.g., gravel);
- characterization of microbial communities growing on biological filters; and
- pre- and posttreatment design (disinfection needs, corrosion control needs, etc.).

Pilot-scale testing is most frequently used for the evaluation of aerobic groundwater biotreatment. Pilot-scale testing generally has a flow rate in units of gpm or L/min. Pilot-scale testing is recommended from 6 to 16 months depending on the operating scenario and target contaminants. Pilot testing can be used to address multiple objectives, including the following:

- demonstrate treatment process performance;
- develop design criteria for unit treatment processes;
- develop operational conditions for optimal contaminant removal;
- assess startup and acclimatization durations;
- test the effect of intermittent operations;
- test challenge conditions (increase in contaminant concentrations, temporary loss of oxygen supply, seasonal changes in water temperature, production rates, etc.); and
- support obtaining regulatory approval for a given process.

Full-scale testing is typically performed to demonstrate that the installed equipment performs to meet specifications, achieves water treatment goals, and is sufficiently robust for changes in real-world conditions. Full-scale demonstration testing has flow rates in the hundreds to thousands of gallons per minute. Successful full-scale demonstration testing may lead to certification by a regulatory agency, implementation by the water utility, and be used as a model for nearby communities needing to address similar source water concerns.

3.2.3.3 Testing design. Regardless of the scale of testing, the testing design should be tailored to achieve the test and evaluation objectives. Routine monitoring of key performance indicators and the collection of accurate and extensive data are critical for the success of testing at any scale. USEPA (2015) provides valuable guidance in designing, constructing, and operating pilot systems. Although it is geared toward surface water conventional treatment, many of the operating principles and design equations are applicable to testing aerobic biotreatment systems for groundwater. The following parameters and considerations need to be taken into account when developing a testing plan. Detailed discussion on other important aspects (resource planning, data management and interpretation, quality control, safety, etc.) to be considered when designing a testing plan are provided in Section 2.2.3.3. Process design criteria are discussed in detail in Section 3.3.

3.2.3.3.1 Process configuration and media type. Typically, a process configuration is selected before testing, although it is possible to test and compare multiple biotreatment configurations. Multiple filter columns may be set up and operated in parallel to test aspects of the design, such as media type, bed depth, number of filtration stages, and aeration/oxygenation system. Common media types for an aerated contactor bioreactor include gravel because of its larger size and pore volumes and a low-density ceramic media that allows for bed expansion during backwashing, which is particularly helpful when treating iron. Media with a manganese oxide coating may be preferred when treating manganese because it can help shorten the startup period. The media size should be selected to achieve the contaminant removal and turbidity goals when particle removal is a treatment objective.

3.2.3.3.2 Biological acclimation. It is critical to assess biological acclimation (see Section 1.7), which refers to the time required to establish a consistent, mature, and active microbial community on the support media before achieving steady-state targeted contaminant removal. Multiple parameters, including source water quality, the contaminant of interest, temperature, and media type affect the duration required for biological acclimation. The testing duration and budget should allow for full acclimation in which the system reaches a pseudo-steady state before adjusting operational conditions, which may take from one week to three to four months. General ranges in acclimation time observed under continuous, routine operating conditions in pilot- and full-scale systems include the following:

- **ammonia:** one to three months (Ramsay et al. 2018, Richard et al. 2016, Lytle et al. 2007a);
- **arsenic:** one to three months (Lytle et al. 2007b, Lytle et al. 2020);
- **iron:** 1–14 days (Dangeti et al. 2017, McClellan 2015, Cai et al. 2014, Tekerlekopoulou et al. 2013); and
- **manganese:** two weeks to four months (Breda et al. 2019, 2017; Dangeti et al. 2017; McClellan 2015; Pacini et al. 2014; Tekerlekopoulou et al. 2013; Štembal et al. 2005).

3.2.3.3.3 Monitoring and process optimization. Routine monitoring of key parameters is essential for evaluating process performance and effectiveness and understanding operation conditions and constraints. Important parameters include flow rate, influent, and effluent contaminant concentrations (and potentially speciation), pH, DO, temperature, turbidity, effluent biomass concentrations, and head loss. Process optimization may begin after acclimation and a period of baseline steady-state performance. Operational parameters to optimize include the DO concentration, chemical feed (e.g., nutrient supplementation and pH adjustment) location and dose, and contact time. If arsenic is a target contaminant, improving its removal may include adjusting the pH to improve sorption to Fe, supplementing Fe, or adding an As-removal process downstream of the biotreatment system.

3.2.3.3.4 Challenge testing. Challenge tests should be performed to assess system performance under conditions that are outside of normal operations but are still likely to be experienced. Challenging conditions to test may include intermittent spikes in contaminant concentrations, increased hydraulic load, such as when a redundant unit is offline, and periods of extended shutdown.

3.2.3.3.5 Posttreatment requirements. Posttreatment requirements for the biotreatment system effluent before entering the distribution system should be evaluated during testing. Posttreatment will include disinfection and may include corrosion control practices such as adjusting the pH, increasing the DIC concentration, or adding phosphate for corrosion control. Aeration practices and nitrification affect the pH and alkalinity, which can affect corrosion in a distribution system. Evaluating these parameters during testing will help mitigate unintended consequences.

3.2.3.3.6 Residuals characterization. When As, Fe, and/or Mn are target contaminants in the source water, biotreatment will produce residuals that require appropriate disposal. Therefore, the residuals in the backwash water should be characterized during testing to determine their quantity, composition, and settling characteristics. These parameters will inform the design of the full-scale residuals handling processes.

3.2.3.4 Testing outcomes. The success of testing is dependent on its ability to achieve the testing objectives. As such, the outcomes of a well-designed and operated testing approach would include data and information to support the subsequent design process and outline operational conditions/criteria. The collected data should identify any differences between the anticipated theoretical and actual findings, and how those influence the design or operations of the system. The need for any additional testing and/or other next steps to fill data gaps should also be identified through the testing. For example, if the targeted contaminant removal is not achieved, additional monitoring during testing, such as metal speciation (i.e., oxidation state, soluble versus particulate, particle size), may be warranted to better understand process performance and inform treatment modifications.

3.3 DESIGN

Design of full-scale aerobic biotreatment systems for groundwater should be based on criteria identified through testing at the appropriate scale, which typically is piloting. This section is organized with the following subsections: process configurations and design components. Design components include aeration/oxygenation design systems, filtration systems, backwashing, nutrient addition, pH adjustment, posttreatment, and residuals handling. Key design considerations for each of these processes are described in the following subsections.

3.3.1 Process Configurations

Process configurations differ according to the aeration/oxygenation delivery systems, the amount of provided DO, pressurization, and number of filtration stages. A list of advantages and disadvantages of each of the three configurations discussed is presented in Table 3-5.

Table 3-5 Comparison of process configurations

Configuration	Advantages	Disadvantages
Open aeration followed by filtration	• Simple • Common • Robust • Efficient in producing DO saturated influent water	• Not able to meet oxygen demand greater than saturation equilibrium state of water • May require cover/venting if aeration is indoors • System needs to be repressurized • Freezing concerns of aeration system in cold climates
Air/oxygen injection and pressurized filtration system	• Compact system • Can meet higher oxygen demand by producing supersaturated influent water • DO can be controlled across different filtration stages • May not need to repressurize system	• Air bubble entrapment within pressurized system • May need to introduce air break to minimize bubble formation/cloudy water and repressurize system
Aeration contactor through a coarse bed followed by filtration	• Can meet higher oxygen demand by producing saturated water throughout the bed • Simple and robust • Suitable to address multiple contaminants	• May require cover/venting indoors • System needs to be repressurized • Relatively new approach

DO—dissolved oxygen
Finally, guidance on the number of filtration stages is provided.

3.3.1.1 Open aeration followed by filtration. Open aeration followed by filtration configuration is illustrated in Figure 3-1. Water first passes through an aeration system that is open, such as a spray or packed-tower aerator, and then to the atmosphere to increase the DO. The maximum DO leaving the aerator is the saturation limit, which is approximately 8–11 mg O_2/L depending on the water temperature. The water then flows through the biologically active filter. The water may be pumped through a pressurized filter or flow by gravity through a filter that is open to the atmosphere. A second filtration stage in series may be provided, if necessary, driven by water quality and the design of the first filtration stage. Nutrients, such as phosphate, may be added ahead of the filter to support microbial growth. Chemical feeds for disinfection and corrosion control may be added to the filter effluent.

3.3.1.2 Air/oxygen injection followed by pressure filtration. Air/oxygen injection followed by the pressure filtration configuration is illustrated in Figure 3-2. Compressed air or liquid oxygen is injected into the water under pressure, for example through a Venturi-type device, and the oxygenated water then flows through the filter. Often the filter is closed and pressurized so that supersaturated DO levels can be achieved in the filter influent. A second or subsequent filtration stages in series may be provided, depending on the concentrations of the target contaminants, and the DO may be increased ahead of each stage. Possible pretreatment includes nutrient addition and pH adjustment, and posttreatment includes disinfection and corrosion control. If the system is pressurized from the wells through the filters, a clearwell open to the atmosphere should be considered so that dissolved gases may be released before pumping finished water to the distribution system to avoid cloudy water at the customer tap.

3.3.1.3 Continuous bubble aeration through a coarse bed (aeration contactor) followed by filtration. Figure 3-3 presents a schematic of this configuration, which consists of two packed bed reactors in series that both support biological growth and activity. The first reactor, termed the aeration contactor, is packed with coarse media, and air is continuously injected up through the water column through diffusers at the bottom of the bed. Thus, saturated DO levels are maintained throughout the contactor. Water may flow up (co-current) or down (countercurrent) through the contactor. Water then flows down through a granular media filter, which may be pressurized or open to the atmosphere. The first stage is not intended to remove resulting solids, whereas the second stage serves to remove particles (e.g., precipitated iron and manganese) and could provide further biological oxidation if necessary. Nutrients and pH adjustment may be provided ahead of the contactor, and posttreatment may include disinfection and corrosion control.

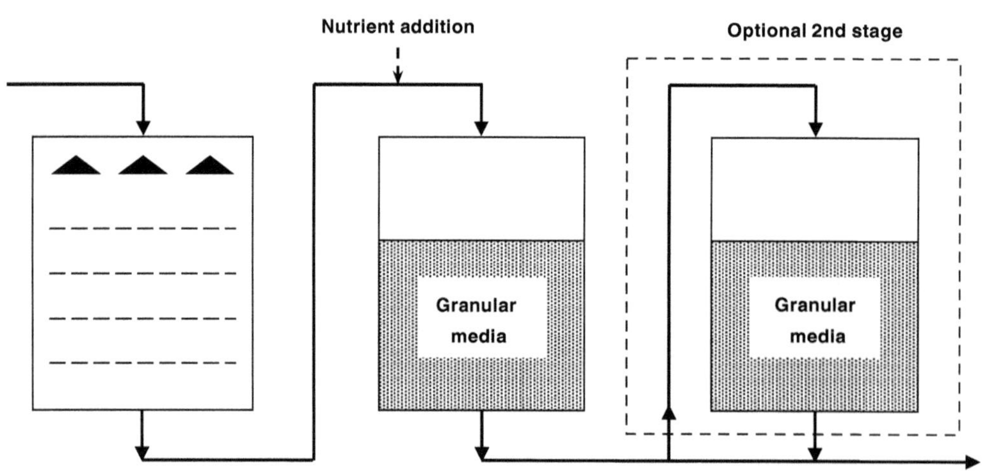

Figure 3-1 Schematic of open aeration followed by filtration system

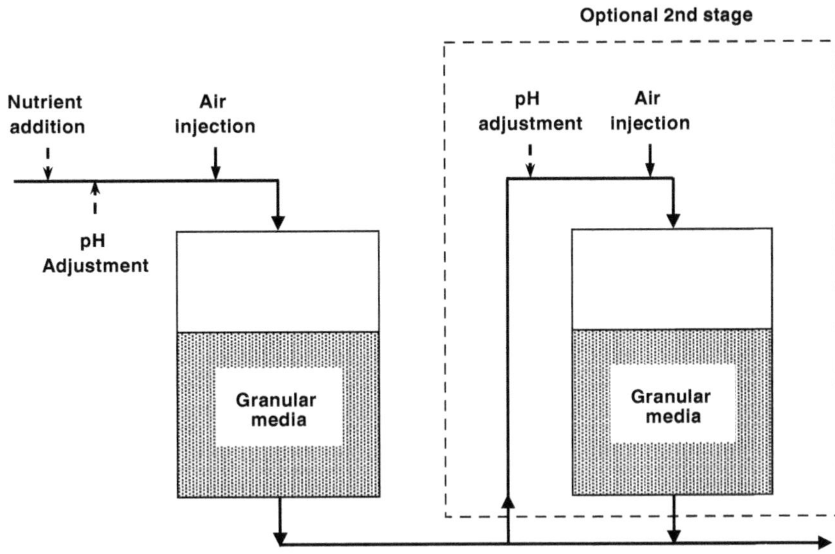

Figure 3-2 Schematic of air injection and pressurized filtration system

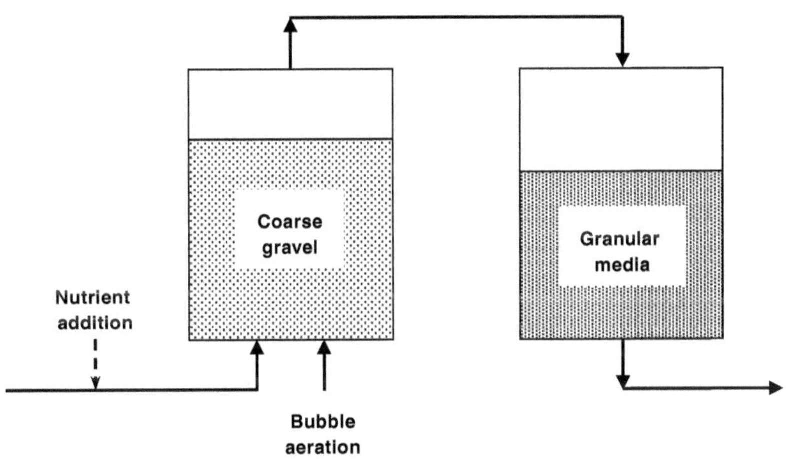

Figure 3-3 Schematic of aeration contactor followed by filtration

3.3.1.4 Number of reactor stages. Multiple reactor stages (i.e., filters in series) might be needed when treating multiple contaminants, although several examples of treating co-occurring contaminants within one column exist (e.g., Cheng et al. 2017, Štembal et al. 2005; Table 3-1). The need for a second stage depends on (1) raw water contaminant concentrations, (2) the contact time provided in the first stage, (3) whether the first stage is designed to act as both a biological reactor and a filter, and (4) whether a separate arsenic removal step (if arsenic is present at higher than its regulatory level) is required.

Contaminant oxidation and removal occur within different preferred pH and ORP ranges. It is possible to establish different zones within one column, but it might be beneficial to have two stages and adjust the DO and pH between them. This separation may be particularly important for iron removal. Biological iron oxidation occurs at lower (potentially ambient) DO levels (2.5–4 mg/L) and lower pH (6–7) as compared with chemical iron oxidation by air or biological oxidation of manganese or ammonia. Iron oxides formed biologically have a different morphology than those formed chemically and tend to lead to slower head loss accumulation (Arturi et al. 2017), so promoting biological iron removal in the first stage could lead to longer filter run times and less backwash wastewater. Another advantage of separating iron removal from that of manganese or ammonia is it allows for longer filter runs in the second stage, causing less disruption to the microbial community and better treatment performance.

Arsenic removal depends on the biological oxidation of As(III) to As(V) and the subsequent coprecipitation and/or adsorption onto iron oxyhydroxides. Arsenic removal with iron occurs best when the iron solids are freshly formed because they have a greater active surface area (Lytle et al. 2005). Thus, if arsenic is a target contaminant, it is important that arsenic oxidation occurs along with iron oxidation and that effective particle removal is accomplished. If arsenic is oxidized but effluent concentrations are above the MCL, then treatment enhancements such as adding an iron coagulant or installing a downstream arsenic removal process should be considered.

3.3.2 Design Components

The primary design components for an aerobic biotreatment for groundwater applications are the aeration/oxygenation delivery system, microbiological support system, filtration system (can also support microbiological growth and activity), pre- and posttreatment requirements, and residuals handling. Because of limited full-scale installations, relationships among important engineering design and water quality factors (e.g., pH, nutrients) and treatment performance need further research and refinement. Piloting (see Section 3.2.3), therefore, is necessary in defining site-specific conditions and engineering criteria necessary to design a new biological treatment system. The design principles and considerations for each unit treatment process that affects performance and outcomes are summarized in the following subsections.

3.3.2.1 Aeration/oxygenation delivery systems. A variety of aeration technologies are available that can add oxygen to water (Table 3-6). Open aeration and air/oxygen injection processes are widely used in the drinking water industry for several different applications. Detailed descriptions of these processes as well as design considerations can be found in AWWA Manual of Practice, M64 (Gas Transfer Applications in Water: Addition and Removal) (AWWA 2021). Design considerations of these two treatment processes are not discussed here.

If continuous bubble aeration through a coarse bed (aeration contactor) is selected as the aeration configuration to meet a necessary oxygen demand with the intent of achieving biotreatment, the design details of the aeration system and the aeration contactor need to be carefully considered. As shown in Lytle et al. (2020), most of the target contaminant can be oxidized within the aeration contactor, even before the filtration step. Aeration contactors are typically constructed as gravel-filled (or alternate less dense attachment material) atmospheric vessels that can be concrete, carbon steel, or aluminum. Aeration contactors are operated in an up-flow mode, with co-current (or countercurrent) introduction of air and water. The following parameters need to be evaluated as part of the testing and design process:

- **contactor size and loading rate:** driven by DO augmentation and minimum contact time requirements, typically less than 2.5 gpm/ft^2;
- **airflow rate:** determined by water quality (oxidant demand) and DO augmentation needs;

Table 3-6 Comparison of aeration systems (Lytle et al. 1998)

Aeration Configuration	Aeration Type	Advantages	Disadvantages
Open	Packed tower	• Efficient delivery of oxygen.	• Large packing depths required. • Freezing in colder climates if not enclosed.
Open	Cascade tray tower	• 50 percent higher air-to-water ratio than traditional packed tower. • Lower packing depths than traditional packed-tower aerators. • Relatively easy modification of existing stripping towers. • Accommodates high water loading rates.	• Supplemental air source required; increased cost and maintenance. • Freezing in colder climates if not enclosed.
Open	Spray	• High air-to-water ratio.	• Large areas may be required. • Duration of exposure of water to air is short. • Nozzle prone to clogging by iron, manganese or hardness. • Freezing in colder climates if not enclosed.
Injection	Diffuser-type fine-bubble	• Package units available. • Smaller space requirements than spray or cascade. • Improved gas exchange compared with packed-tower aerators. • Not subject to freezing. • Negligible water head loss.	• Higher cost than most other aerators. • Longer contact time. • Blowers required; increased operational cost and maintenance. • Harder to adapt to existing aerators.
Injection	Inline LOX injection	• Easy to implement on an existing process train. • Capital improvements limited to storage of LOX. • Does not rely on oxygen transfer from gas to liquid phase.	• Operational costs are high due to the use of LOX. • Inline mixer needed.
Aeration contactor	Diffuser-type fine-bubble	• Can operate at atmospheric pressure. • Attachment material typically gravel (inexpensive media). • Significant contaminant removal may be achieved through the contactor before the downstream filtration process.	• Has not been tested or implemented as widely as open aeration or injection systems. • Only a few commercial suppliers available currently for this technology.

LOX—liquid oxygen

- **contactor media:** typically, round gravel between 0.25- and 0.5-in. diameter to support microbiological growth;
- **air diffuser:** needs to be designed considering potential of periodic biofouling and cleaning and maintenance needs; and
- **backwash needs:** typically, backwashes are infrequent (ranging from once per several weeks to months) and performed at a high rate to remove accumulated solids.

3.3.2.2 Filtration systems. Aerobic biotreatment of groundwater can be achieved using a fixed bed that could also function as a filter by achieving particle removal. Fixed-bed approaches can use natural granular media, engineered media, or membranes to support biofilm growth and activity. With the exception of a retrofit to an existing membrane facility, membrane filtration is typically not used for aerobic biotreatment of groundwater when the objectives of treatment are the contaminants discussed in Section 3.1.2. As such, this section discusses only media filtration.

Media filtration can be configured as either pressure filters (closed to atmosphere) or gravity filters (open to atmosphere), both of which are widely used in drinking water applications. In pressure filters, the media is contained in a pressure vessel (such as a steel tank), and filters are operated by a pump or from a high-pressure water source such as a groundwater well pump. Pressure filters can be configured either vertically or horizontally. Vertical pressure filters are typically used for smaller facilities, allowing for modular operation and future incremental plant expansion. Horizontal pressure filters are used in larger plants and allow maximization of the filter surface area.

Gravity filters are more commonly used in surface water plants but can also be used for groundwater treatment, particularly in the case of iron removal plants where aeration is practiced or lime softening plants where pressure breaks are already created. Gravity filters are typically operated at lower loading rates compared with pressure filters but can be designed to operate at higher loading rates, if necessary. Capital costs for the implementation of gravity filters are likely higher than pressure filters for smaller water treatment plants, but for larger plants, gravity filters may provide cost savings over pressure filters.

As discussed in Section 3.3.1, both pressurized and gravity filters can be used for the aerobic biotreatment of groundwater. There are some important design factors that should be considered when determining which approach is most suitable to meet the biotreatment objectives that are listed in this section.

- **Number of filters and filter stages:** As discussed in Section 3.3.1, biologically active filters can be configured in a single stage or multiple stages depending on the raw water quality (e.g., high iron) and treatment needs. The number of filters should also consider local process redundancy requirements.
- **Filter size and hydraulic loading rate (HLR):** Sizing should be determined on the basis of influent and target effluent water qualities. HLRs of biological filters for aerobic groundwater treatment can vary significantly and depend on the target contaminant, filter depth, and EBCT, and potentially on the media size because gravel has a lower specific surface area than anthracite or sand. Table 3-1 lists HLRs used in pilot- and full-scale systems. Reported HLR ranges are as follows: 1–4 gpm/ft^2 for ammonia, with less than 2 gpm/ft^2 common; 2–12 gpm/ft^2 for iron and manganese; and 2–4 gpm/ft^2 for arsenic. Pressure filters may have greater HLRs as compared with gravity filters, however, final design will most likely be dictated by the kinetics of biological processes. Ultimately pilot testing will be necessary to identify site and process specific operating conditions necessary to achieve treatment objectives.

- **Filter media:** A variety of filter media can be used, including anthracite, sand, manganese-coated silica sand, granular activated carbon, and greensand for filters, and larger ceramic and gravel for aerated contactors. Ceramic media allows for bed expansion during backwashing, which helps remove accumulated iron, whereas gravel does not. When manganese-based media are used, appropriate pH and ORP conditions and an optimal pH range must always be maintained within the filters so that manganese and other contaminants are not released from the media. Selection of filter media is dependent on target contaminants and concentrations, co-contaminants and overall water quality, and site-specific constraints. Local design and related experiences should also be considered, and pilot testing should be used to establish full-scale materials of construction.
- **Filter underdrains:** Underdrains need to be designed carefully to support the filter media and maintain uniform flow distribution through the filter. Higher head loss due to biological clogging of filter underdrains needs to be considered in the design of biologically active filters. With the increasing use of biotreatment applications, tailored underdrain designs are now available from several manufacturers that minimize biological fouling, prevent filter media loss, and can also be efficiently cleaned through air scouring and filter backwash processes.

3.3.2.3 Filter backwash. The rates, frequency, and design of filter backwash processes are designed on the basis of raw water quality and treated water quality objectives. The design of the filter backwashing process for aerobic biotreatment is similar to the design of backwash in the non-biotreatment of contaminants, such as iron and manganese, in groundwater. Similar backwash rates and media expansion targets can be adopted according to guidance set by the local regulatory and permitting agencies. Backwash frequency may range from three days to two weeks, depending on metals concentrations, loading rates, and media characteristics.

The backwash water source is an important design decision. Table 3-7 summarizes four possible sources along with their associated advantages and disadvantages. Backwashing with unchlorinated, filtered water is typically preferred, although backwash using chlorinated water is also possible while still achieving treatment goals (Lytle et al. 2007b).

Table 3-7 Backwash water source comparisons

Source	Advantages	Disadvantages
Raw water	• No storage. • No chlorine residual.	• Depends on capacity of raw water pumps. • May be less effective if iron concentrations are high.
Filter effluent from other on-line filters (i.e., without storage)	• Treated water. • No storage. • No chlorine residual.	• May require additional filters be installed to ensure adequate flow.
Backwash water storage tank filled with filter effluent without a chlorine residual	• Treated water. • No chlorine residual.	• Requires additional water storage tank that should be cleaned routinely to prevent microbial growth.
Treated water clearwell with a chlorine residual	• Treated water. • No storage. • Has been used in past aerobic biotreatment applications without issue.	• Chlorine residual may harm the microorganisms and negatively influence treatment effectiveness.

3.3.2.4 Nutrient addition. The availability of nutrients, such as phosphorus, can influence microbial growth and process performance. Lytle et al. (2013) reported slow acclimation and incomplete nitrification in the absence of phosphorus addition and observed improved treatment performance immediately following initiation of a phosphate feed. Nevertheless, several biotreatment applications achieve target treated water quality without supplementing phosphorus (e.g., Case Studies #1 and #3). When phosphorus is added, a target dose of 0.3 mg/L phosphate has been reported (Lytle et al. 2020, 2013), and the actual dose can be optimized during testing as it may be affected by the contaminant concentrations, sorption to iron, and other site-specific factors.

3.3.2.5 pH adjustment. pH adjustment may be beneficial depending on the contaminants present. Arsenic sorption to iron occurs best at pH near 7 (Lytle et al. 2005). Microorganisms involved in nitrification prefer pH 7–8 (Zhang et al. 2009). Additionally, nitrification consumes approximately 14 mg/L (as $CaCO_3$) of alkalinity per 1 mg N/L of ammonia oxidized and may lower the pH, depending on the initial ammonia concentration and alkalinity of the raw water (Zhang et al. 2009). On the basis of metal speciation and precipitation, a pH of 6.5–7 may be beneficial for iron removal to encourage biological rather than chemical iron oxidation, while a pH of 7.5–8 may be preferred for manganese oxidation (Mouchet 1992). Nevertheless, several pilot- and full-scale systems successfully operate without pH adjustment, possibly because many groundwaters naturally fall within this range and aeration can produce water with pH 7.5–8.2, depending on its design and operation. The need for pH adjustment for optimal performance should be tested through bench- or pilot-scale treatment. Typically, in drinking water applications, pH is adjusted using acids (carbon dioxide, sulfuric acid, etc.) or bases (caustic soda, lime, etc.).

3.3.2.6 Posttreatment. Biological processes foster the growth of microorganisms, and as such, a higher concentration of microorganisms could be observed in the filter-effluent water compared with raw water. Heterotrophic plate count can have a wide range of concentrations in aerobic biotreatment of groundwater ranging from less than 100 to greater than 10,000 CFU/mL in a steady-state biofilter effluent (Lytle et al. 2020, 2015). Effective disinfection is critical to inactivate microbes in a biofilter effluent. Disinfection requirements for biotreatment vary by location, and the local regulatory agency should be contacted for guidance and compliance requirements. In some cases, disinfection requirements for aerobic groundwater treatment may be as much as required for surface water treatment.

Additionally, aerobic biotreatment may change multiple water quality parameters, such as pH, alkalinity, DO, and ORP, which affect corrosion in distribution systems. For the biological oxidation of ammonia, the total consumption of alkalinity by nitrification is 7.1 mg, as calcium carbonate per mg of ammonium ion oxidized (USEPA 1993). Additional treatment may be required given the site's treated water quality targets and corrosion control strategy.

3.3.2.7 Residuals handling. Aerobic biological filters will generate wastes from both the aeration contactor and the biological filter as part of the backwash process. Characteristics of backwash wastes from these processes are similar to backwash wastes from similar drinking water treatment processes with higher concentrations of contaminants (iron, manganese, arsenic, etc.) that are removed from water, and higher microbial loads. Because biological filtration does not require the addition of coagulation or softening chemicals, the volume of waste generated through the biological processes is generally anticipated to be lower. Typically, residuals handling systems for aerobic biotreatment should be designed similarly to residuals handling systems for oxidation/filtration systems for inorganic contaminants (e.g., iron and manganese) in groundwater. If any water used in the backwash process is planned to be returned to the head of the plant following treatment, detailed evaluations are needed to ensure that the returned backwash water does not introduce any challenges to the biotreatment process. Especially if contaminants such as arsenic, manganese, or low levels of radioactive elements are present in the backwash water, treatment systems should be designed to remove these contaminants before any portion of the backwash water stream is returned to the head of the plant. Residuals disposal will depend on the chemical composition and local regulations and may include beneficial applications such as iron for phosphorus removal in wastewater treatment.

3.4 OPERATIONS

The operation of aerobic groundwater biotreatment systems needs to be carefully managed through the development of a detailed monitoring and operations and maintenance (O&M) plan. The monitoring plan should clearly identify monitoring parameters, methods, sampling frequency and location, and sample collection and handling protocols. The plan should also detail data management, analysis, and reporting procedures. Additionally, it should also be developed so that it outlines expectations and protocols for the various phases of operation: startup and biological acclimation, optimization, steady-state operation, shutdown and restart, and troubleshooting.

3.4.1 Monitoring

Operational, water quality, and microbiological parameters are key in assessing performance of the biotreatment system, determining effectiveness in achieving desired treated water quality, and adjusting process control variables to optimize performance. A monitoring plan should be developed with the key monitoring parameters and monitoring frequency. For an accurate assessment of system performance, a combination of inline and grab sample analyses should be included in the plan. Table 3-8 summarizes recommended monitoring parameters and their relevance.

A clear and organized data management plan should also be developed, including a schedule and well-defined protocol for inputting and analyzing data. Many of the parameters listed in Table 3-8 can be effectively monitored and recorded using supervisory control and data acquisition (SCADA) for real-time and long-term performance metrics. Grab sample results, generated either through in-house testing or testing at an external lab, should be compiled using software, such as Microsoft Excel or Microsoft Access. Performance tracking tools, such as Proformance (Nyfennegger et al. 2016), can be used to compile and manage data received from multiple sources (i.e., inline analyzers, SCADA system, field data, and lab reports). These tools can autogenerate graphs, tables, trend plots, and statistical summary tables to assess treatment in a streamlined fashion.

3.4.2 Initial Startup/Acclimation

Once the system is started, microorganisms indigenous to the source water gradually establish on the growth support media (see Section 1.3.3). As a general rule for plants that operate for only part of the day, the acclimation time will be delayed by approximately the reciprocal of the fraction of operating time per day (Lytle et al. 2020). For example, if a plant operates for 12 hours per day (i.e., half of the day), its acclimation time will be approximately twice as long as if it operated continuously. When treating ammonia, ammonia-oxidizing microorganisms typically develop before NOB, which may lead to a period of high nitrite in the effluent before complete nitrification is achieved. Staging and blending the bioreactors along with dosing chlorine downstream can be used to achieve acceptable finished water quality during acclimation, or it may be necessary to operate the plant to waste (i.e., not connected to the distribution system) until the bioreactors are acclimated. Using manganese oxide–coated media may shorten the time before the target

Table 3-8 Recommended monitoring parameters

Category	Parameter	Sampling Location	Sampling Method	Sampling Frequency	Performance Indication
Operating parameters	Flow/hydraulic loading rate	Inf	Inline	Continuous	Dictates contact time, which is directly related to biological contaminant degradation.
	Head loss/ pressure drop	Across the system	Inline	Continuous	Allows determining clogging/ biofouling. Clean-bed head loss signifies the efficacy of backwashing protocol, and head loss accrual trend may help determine preferential channel formation.
	Air injection pump pressure	At air injection inlet	Inline	Continuous	Indicates obstruction of air feed line.
	Backwash pressure	Downstream of the underdrain	Inline	Continuous	Indicates underdrain clogging if backwash pressure is increased.
Water quality	DO	Inf, Eff, downstream of aeration contactor	Inline	Continuous	DO is essential for aerobic biotreatment. Effluent DO monitoring allows rapid detection of performance anomaly. Monitoring DO downstream of the aeration contactor allows determination of the consumption of DO through the aeration contactor.
	Temperature	Inf	Inline; grab	Continuous; daily	Affects biological growth and activity.
	pH	Inf, Eff	Inline; grab	Continuous; daily	Affects biological growth and activity. Several processes (e.g., aeration and contaminant transformation) can cause the pH to change. Constant finished water pH is important for water quality in the distribution system.
	Contaminant and intermediate product concentration	Inf, Eff	Inline, grab	Continuous; regular (e.g., two times per week)	Monitoring contaminant and intermediate product (e.g., nitrite) concentrations is critical for determining treatment process effectiveness. Total and dissolved fractions of metals and the oxidation state of arsenic should be analyzed during startup and troubleshooting.
	Turbidity	Inf, Eff	Inline	Continuous	Indicates biomass/solids present in treated water. Influent turbidity is an indicator of particle loading. Effluent turbidity monitoring helps assess the disinfection requirements.

(continued)

Table 3-8 Recommended monitoring parameters (*Continued*)

Category	Parameter	Sampling Location	Sampling Method	Sampling Frequency	Performance Indication
	Phosphorus	Inf	Grab	Occasional (e.g., every month)	Affects biological growth and activity. Can affect arsenic removal with iron. Monitoring influent phosphorus is critical for tailoring nutrient dose.
	Alkalinity	Inf	Grab	Regular (one time per month)	Indicates buffering capacity of the water. pH decrease or increase can be significant in water with low alkalinity.
	AOC/BDOC	Eff	Grab	Occasional (e.g., every two to three months)	Indicates regrowth potential in distribution system.
Microbial	HPC	Eff after disinfection	Grab	Regular (e.g., two times per week)	Indicates disinfection efficacy.
	Total coliform/ *Escherichia coli*	Eff after disinfection	Grab	Regular (e.g., two times per week)	Indicates disinfection efficacy.
	EPS	Filter media	Grab	When hydraulic issues are observed	Excess EPS may be attributed to increased head loss.
	Microbial community structure	Filter media	Grab	When sustained performance issues are experienced	Provides information on which bacteria and functional genes are present in the system. Requires the use of molecular microbial methods; typically performed at specialty labs and universities.
	ATP	Filter media	Grab	During biological acclimation or system recovery	Indicates biological activity. Because biological growth and activity are intentionally enhanced in biotreatment, ATP monitoring is recommended only during the initial biological acclimation or during reacclimation.

AOC—assimilable organic carbon, ATP—adenosine triphosphate, BDOC—biodegradable organic carbon, DO—dissolved oxygen, Eff—effluent, EPS—extracellular polymeric substance, HPC—heterotrophic plate count, Inf—influent

effluent manganese concentration is achieved because the loaded manganese initially will adsorb onto the media, but the effluent concentration over time will depend on the media adsorption capacity, manganese loading rate, and how quickly the manganese-oxidizing bacteria acclimate (Breda et al. 2019, 2017; Cai et al. 2014). Additionally, "seeding," in which microorganisms from a functioning biofilter are added to a new biofilter during startup, has been considered to shorten acclimation time. Results of seeding strategies have been mixed, sometimes yielding a benefit (Tekerlekopoulou et al. 2013) and other times not (Lytle et al. 2007a). Local regulatory agencies should be consulted to determine if "seeding" can be performed and to determine the type and volume of "seeding" performed. Piloting studies may be useful to assess startup expectations and enhancement options (e.g., seeding, continuous operation, and nutrient addition).

3.4.3 Steady-State Operation

In general, after complete biological acclimation, steady-state system performance can be assumed if variability in contaminant removal remains within ±10% across three to five consecutive data points over a week and consistent filter run times are observed. Routine operation and maintenance activities include monitoring water quality parameters, monitoring process operation parameters, backwashing the fixed-bed bioreactors, and performing maintenance per manufacturer recommendations.

Operational (e.g., filter head loss, chemical feed rate, others), water quality, and microbiological parameters should be monitored routinely during steady-state operation to establish a baseline of the average and range of typical values (Table 3-8). Tracking these parameters over time will allow for identification of any subsequent performance deviations and may assist with troubleshooting efforts. Additionally, standard operating procedures should be developed and implemented for maintaining and calibrating chemical feed systems and analyzers.

3.4.4 Shutdown and Restart

The robustness of biotreatment with respect to operations and effectiveness is sometimes questioned over concerns regarding the sensitivity of established biofilm to downtime and the ability to rapidly recover, particularly after prolonged shutdown. Furthermore, treatment systems in many small communities operate intermittently for a fraction of the day (as little as 4–6 hours per day). Such considerations should be raised early in the treatment assessment phase when shutdown and restart scenarios can be evaluated during piloting after the system has stabilized. For example, collecting a time series of water samples for measuring contaminants of concern after the pilot has been shut down for an extended time period is recommended. Any time delay before returning to optimal treatment operations should be noted and planned for.

Previous work reported that once an aerobic biological ammonia, manganese, and arsenic treatment system was acclimated, the system quickly returned to normal operation. No negative effects on ammonia oxidation rates were observed after a nine-day shutdown of the aeration contactor and biological filter (Lytle et al. 2020). The redox conditions in a bioreactor typically shift toward being more reducing when offline, which may lead to the desorption of accumulated metals. Therefore, a specialized backwashing and filter-to-waste program may be beneficial after an extended shutdown period.

3.4.5 Optimization

Optimization refers to relatively fine adjustments made to the full-scale operation of a biotreatment system once steady-state operation has been achieved to optimize reduction of targeted contaminants. Such adjustments might include adjustment to airflow rate, nutrient feed rate, backwash frequency and/or duration change, or filter loading rates. Although optimization adjustments should be relatively minor, system monitoring of relevant water quality and operational parameters should increase in frequency shortly after the optimization process to evaluate system response, including quantifying improvements and identifying unintended consequences.

3.4.6 Common Challenges

Two common challenges with the biotreatment process are discussed here: (1) contaminant effluent concentrations greater than the targets and (2) shortened filter run times. Treatment performance issues may be caused by insufficient DO, pH outside the optimal range, microbial kinetics limitations, precipitate removal, and nutrient limitation. These

Table 3-9 Strategies to address suboptimal treatment performance

Parameter to Investigate	Why It Is Important	Possible Improvement Strategies
DO	• Needed for microbial metabolism • Important for metals speciation/ precipitate stability	• Increase aeration rate to achieve target influent and effluent DO concentrations • Investigate ways to increase microbial activity (e.g., decrease flow rate, add nutrients) if DO consumption is less than expected
pH	• Microbial preference • Precipitate stability	• Feed acid or base to adjust pH
Metals speciation	• Oxidation state and particle size help identify what aspect of the treatment process should be focused on for improvement	• If reduced metals are present in the effluent, evaluate DO, pH, and microbial activity • If oxidized iron and manganese are present in the effluent, evaluate particle formation, stability, and removal • If oxidized arsenic is present in the effluent, supplement iron, adjust pH to near 7, and/or assess competition with phosphate
Temperature	• Microbial metabolism kinetics	• Decrease flow rate, if possible
Loading rate/EBCT	• Microbial metabolism kinetics	• Decrease flow rate, if possible
Nutrient concentrations	• Phosphorus, in particular, is an essential nutrient that might be present at low concentration and limiting microbial growth	• Dose low concentrations of nutrients (e.g., orthophosphate)
Filter run time	• Decreased filter run times could limit plant production	• Investigate biomass and/or EPS accumulation on media • Evaluate backwash system • Perform solids retention analysis on filter media

DO—dissolved oxygen, EBCT—empty bed contact time, EPS—extracellular polymeric substance

parameters, as well as metals speciation, should be investigated and possibly adjusted to improve effluent water quality (Table 3-9). Shortened filter run times may be caused by excessive biofouling and high EPS concentrations or by particle accumulation (e.g., formation of cake on the surface of the filter). The backwash system, including pumps, air scour equipment, and underdrains, should be checked to assess whether they are operating correctly and delivering the target flow rates. Additionally, investigating the biofilm and metals on the media surface may help identify the cause of the problem. Possible improvement strategies include optimizing the backwash program, supplementing nutrients, and applying an oxidant.

3.5 FEATURED CASE STUDIES

The case studies presented in this chapter represent a variety of applications regarding process design, specific contaminants targeted for removal, scale, and geographic location. Descriptions of these facilities are presented in this chapter, such as process flow diagrams, raw water quality, treatment goals, and performance of each facility. These facilities are listed in Table 3-10, which outlines location, treatment type, and targeted contaminants.

Table 3-10 Aerobic biotreatment of groundwater case studies

Facility, Location	Treatment	Targeted Contaminants
Greene County Southern Plant, Xenia, Ohio	Conventionally biologically active filters	Ammonia, arsenic, iron
City of Gilbert, Gilbert, Iowa	Aeration contactor with polishing biofilter	Ammonia, arsenic, iron, manganese
Putnam Water Pollution Control Authority, Putnam, Conn.	Biological filtration with aeration and pH adjustment	Iron, manganese

Table 3-11 Raw water quality

Parameter	Units	Value
pH		7.48±0.1
Temperature	°C	14±2
DO	mg/L	1.22±0.3
ORP	mV	–114±19
Alkalinity	mg/L as $CaCO_3$	373±15
Iron	mg/L	2.289±0.114
Manganese	µg/L	11.6±0.8
Arsenic	µg/L	46±8
Ammonia	mg N/L	1.15±0.04
TOC	mg/L	1.2

DO—dissolved oxygen, ORP—oxidation reduction potential, TOC—total organic carbon. *Blank space indicated field not applicable.*

3.5.1 Conventional Biologically Active Filters, Greene County Southern Plant, Xenia, Ohio

Adapted from Lytle, D. A., et al. (2007). "Biological nitrification in a full-scale and pilot-scale iron removal drinking water treatment plant." Journal of Water Supply: Research and Technology–AQUA. 56(2):125-136.

The Greene County Sanitary Engineering Department operates and maintains the drinking water facilities for much of Greene and serves more than 18,000 water customers. The system is supplied by two groundwater treatment plants and water purchased from neighboring communities.

3.5.1.1 Facility description. Greene County's "Southern" Plant (Xenia, Ohio) treats an average of 0.6 million gallons per day (mgd) and a maximum of 1.2 mgd. The plant was originally designed for Fe removal from a hard, high alkalinity groundwater (Table 3-11). The groundwater had a low DO and low ORP, indicating reducing conditions in which soluble Fe, soluble As(III), and NH_3 were the dominant iron, arsenic, and nitrogen species, respectively.

The treatment plant draws groundwater from three wells. The treatment process includes aeration, retention, biologically active sand filtration (three filter cells), and postchlorination (Figure 3-4, Figure 3-5). The major treatment train components are listed in Table 3-12.

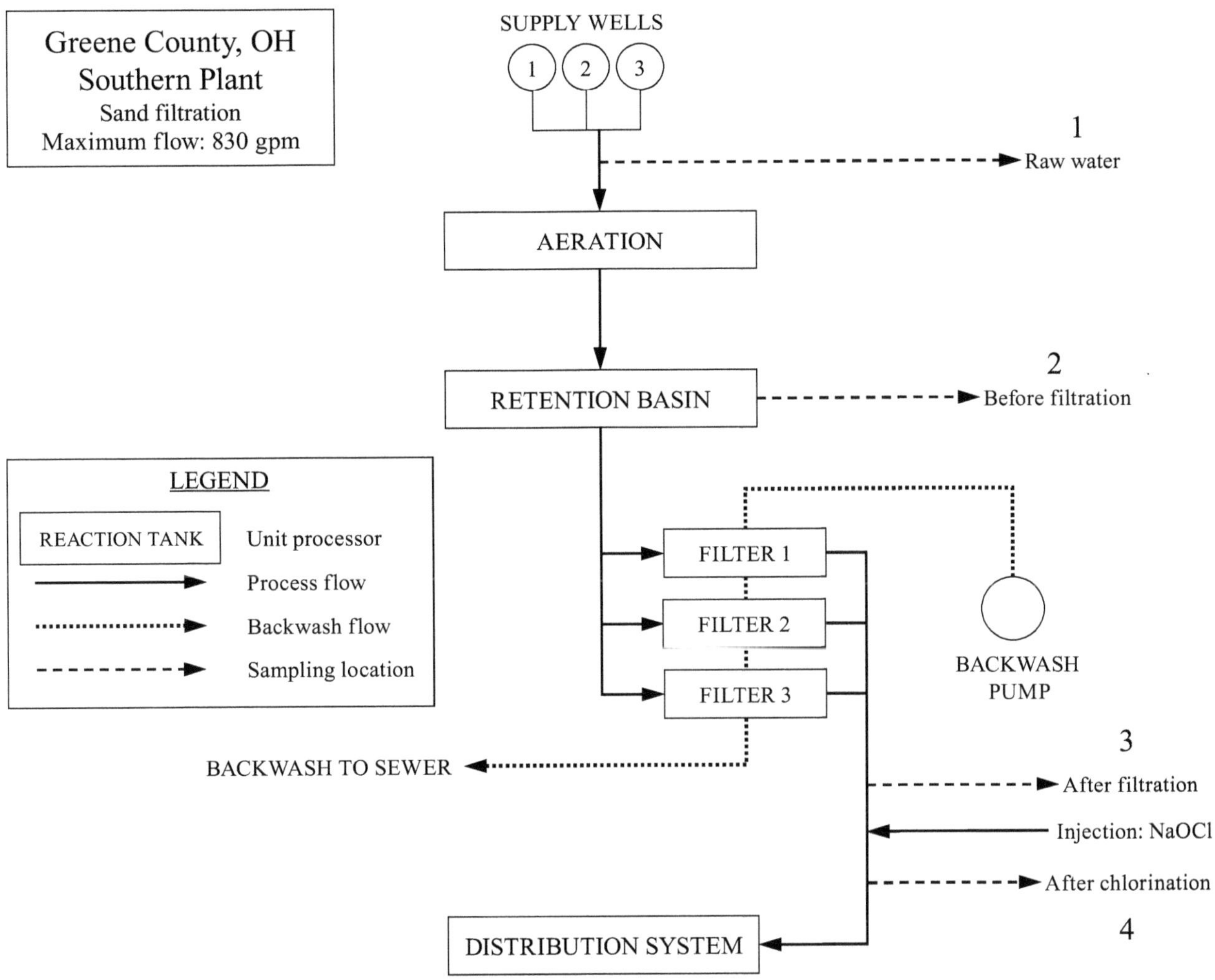

Figure 3-4 Process flow diagram for Greene County Southern Plant

3.5.1.2 Facility treatment goals. The primary treatment objective was iron removal to achieve a finished water iron concentration less than the SMCL of 0.3 mg/L. When the EPA lowered the arsenic MCL from 50 to 10 μg/L, several treatment plants had to apply a treatment process. The raw water at the Greene County Southern Plant contained an average arsenic concentration of 46±8 μg/L, but it consistently achieved finished water arsenic concentrations less than 10 μg/L despite not applying a strong oxidant. The objectives of this study were to document the removal of arsenic and conversion of ammonia in the full-scale water treatment plant and document the role of microorganisms in arsenic removal.

3.5.1.3 Operation and performance. The performance of the full-scale plant was investigated over a six-month period by collecting monthly samples of the raw water before the filters, after the filters, and of the finished water after chlorination.

Figure 3-5 Aeration blowers (A) and gravity sand biologically active filters at Greene County Southern Plant (B) (Lytle et al. 2007a)

Table 3-12 Design criteria

Parameter	Value
Aeration type	Forced draft
Retention basin volume	5,000 gallons
Average retention basin detention time	12 min
Filtration type	Gravity
Number of biofilters	3
Filter surface area	148 ft^2
Filter media type	Sand/gravel
Filter HLR	2 gpm/ft^2
Filter run time	3 days
Filter backwash rate	14 gpm/ft^2
Target finished water Cl_2 residual	1.0 mg/L

HLR—hydraulic loading rate

Figure 3-6 shows the DO and pH levels through the plant. The source water had low DO, and aeration increased the DO from 1.2 mg/L to 6.74 mg/L on average. Aeration also increased the pH slightly, from 7.48 to 7.76, presumably due to stripping of CO_2. The DO decreased through the filter by 1.34 mg/L on average, which is less than the stoichiometric amount of 5.1 mg/L that theoretically is required to oxidize ammonia to nitrate. It is suspected that air was introduced to the sample when measuring the DO, thereby causing an erroneously high reading. The pH did not change substantially through the filter or chlorination, likely because of the high alkalinity of the water.

Figure 3-7 shows the average concentration and speciation of the inorganic contaminants iron, arsenic, and ammonia. The average iron concentration in raw water was 2.28 mg/L and was present in the soluble form. Most of the iron was chemically oxidized and precipitated during aeration before the filters, and the filters removed the iron particles. The total iron concentration in the finished water was less than 0.025 mg/L.

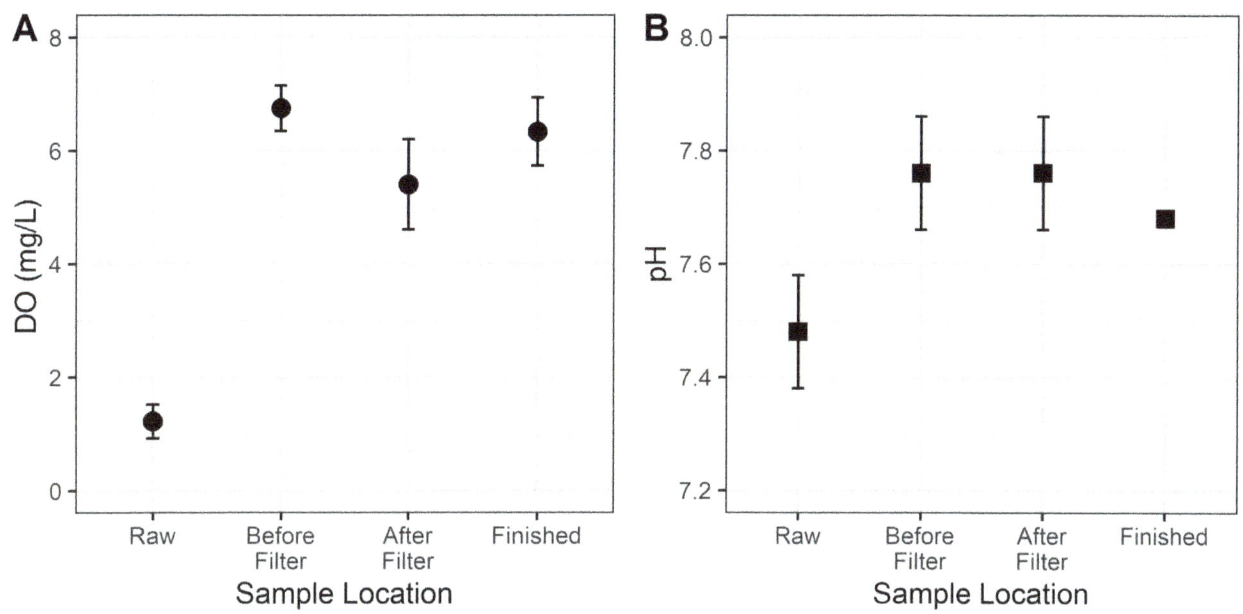

DO—dissolved oxygen

Figure 3-6 DO (A) and pH (B) profiles through the treatment plant. Points show the average value, and error bars represent one standard deviation.

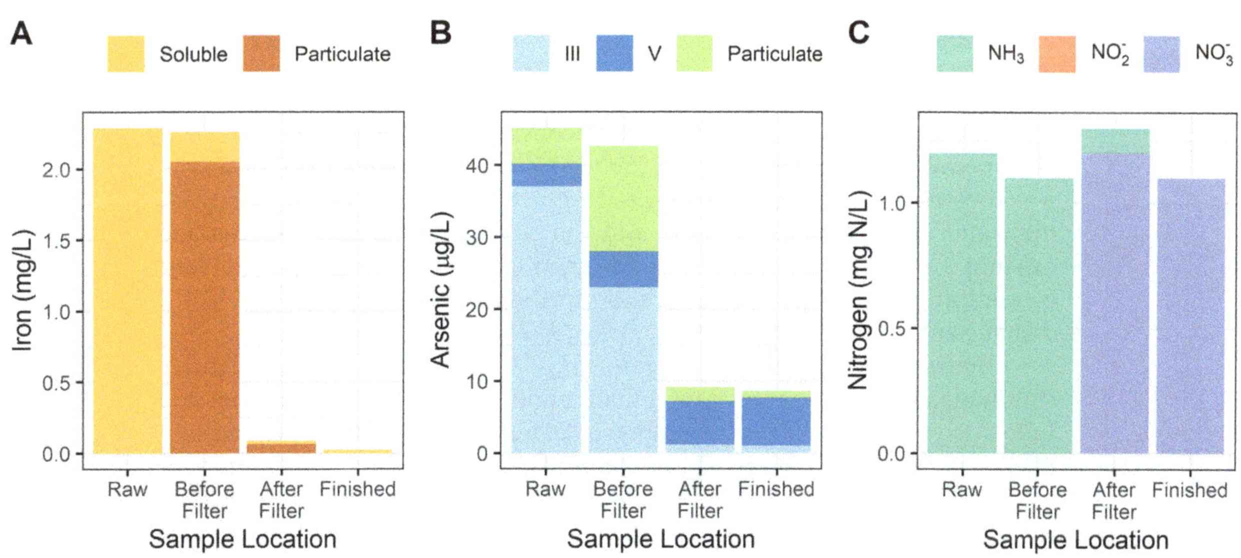

Figure 3-7 Average iron (A), arsenic (B), and inorganic nitrogen species (C) through the treatment plant

Total arsenic concentrations in raw water ranged from 34.9 to 54.8 μg/L, most of which existed as As(III). Although the water was not preoxidized with a strong oxidant like NaOCl or $KMnO_4$, there was almost complete conversion of As(III) to As(V) through the filters, which was unexpected. The oxidation likely was biologically mediated, and subsequent investigations of the filter media identified arsenic-oxidizing bacteria on the media (White et al. 2012). The filter removed most arsenic-laden solids, and most of the remaining arsenic in the filter effluent existed as As(V). Arsenic concentrations in the finished water consistently were less than the MCL of 10 μg/L, ranging between 7.3 and 9.7 μg/L.

Figure 3-7C shows the inorganic nitrogen speciation through the plant. The raw water contained ammonia at an average concentration of 1.15 mg/L. Most of the ammonia was oxidized to nitrate in the filters, with no detectable nitrite accumulation. Like the arsenic, it is likely that the ammonia oxidation through the filters was biologically mediated. Removing ammonia allowed the plant to produce a stable free chlorine residual in the finished water that ranged from 0.3 to 1.2 mg/L (as Cl_2).

3.5.1.4 Key success factors and lessons learned. This case study demonstrates that a conventional iron removal treatment process can operate with biologically active filters if a strong oxidant like chlorine or permanganate is not added ahead of the filters. Microorganisms in the filters helped facilitate arsenic removal by oxidizing As(III) to As(V), which was then removed along with the iron, thus allowing the utility to meet the MCL of 10 μg/L without adding a strong chemical oxidant. Additionally, the biofilters further improved finished water quality by converting ammonia to nitrate, which produces a more stable disinfectant residual and lowers the nitrification potential in the distribution system (AWWA 2013).

Key success factors include the following:

- providing aeration to increase the DO and ORP for microbial arsenic oxidation and complete nitrification of the ammonia present in the raw water; and
- placing chlorine addition downstream of the filters.

3.5.2 Pilot Study on Arsenic, Ammonia, Manganese, and Iron Removal, Gilbert, Iowa

Adapted from Lytle, D. A., et al. (2020). "The removal of ammonia, arsenic, iron and manganese by biological treatment from a small Iowa drinking water system." Environmental Science: Water Research & Technology ***6****(11): 3142-3156.*

The City of Gilbert, Iowa is a rural community with a population of approximately 1,150 people. Water for the City is supplied by two wells. The existing water treatment plant was constructed in the 1970s and was designed to remove iron and manganese and soften the water. The groundwater also contains elevated levels of ammonia and arsenic. The existing treatment process was not designed to remove ammonia, and elevated ammonia concentrations can lead to nitrification in the distribution system, which can lower the disinfectant residual and generate nitrite (AWWA 2013).

Biotreatment was selected for the new water treatment plant because of its ability to simultaneously address the four contaminants (ammonia, arsenic, iron, and manganese). In 2016, a year-long pilot demonstration of an innovative biotreatment system was initiated.

Table 3-13 Raw water quality

Parameter	Units	Raw Water Quality
Temperature	°C	13.3
pH	SU	7.63
Alkalinity	mg/L as $CaCO_3$	410
Hardness	mg/L as $CaCO_3$	280
As	µg/L	23
Fe	mg/L	2.91
Mn	mg/L	0.08
NH_3	mg N/L	2.91
NO_2^-	mg N/L	0.01
NO_3^-	mg N/L	0.02
TOC	mg/L	2.68
DO	mg/L	1.1

DO—dissolved oxygen, TOC—total organic carbon

3.5.2.1 Facility description. Table 3-13 summarizes the raw water quality from the two wells. The oxidation state of iron and manganese in the source water was not determined, but the reduced forms Fe(II) and Mn(II) were likely predominantly based on water chemistry, low DO, and local geology. Arsenic was predominantly present as As(III) in the raw water.

The aeration contactor followed by a polishing biofilter was selected to treat the raw water because it can meet the high oxygen demand (13 mg/L) required for nitrification of the 2.91 mg N/L of ammonia in the raw water. The gravel in the contactor allows bacteria attachment and growth yet reduces the potential for "clogging" of the media, reduces backwashing frequency, and allows air bubbles to move up through the contactor. The purpose of the filter, which also supports microbial growth, is to remove arsenic-containing iron particles, precipitated manganese, and biomass flocs. It also provides additional capacity for biological oxidation of any remaining ammonia and/or nitrite in the contactor effluent.

The pilot system was based on the USEPA patented design (US 8,029,674 B2 awarded on Oct. 2, 2011) and consisted of one pair of 3-in.-diameter columns in series built from clear polyvinyl chloride and other common plumbing materials. The first column, the aerated contactor, was packed with medium gravel with a nominal ½-in. diameter. The second column, the filter, was packed with anthracite on top of ADGS$^+$ silica sand. This sand is a proprietary media that is coated with Mn-oxide, but it is unlikely that the coating improved Mn removal in this biofilter application. The contactor operated co-current and was aerated from the bottom using a diffuser (US Patent Application Serial 14/459,277) connected to a gas pump. Figure 3-8 shows a process flow diagram and photograph of the pilot system. Table 3-14 summarizes the design criteria for the pilot system.

The pilot was operated approximately seven hours a day, seven days a week, matching the water treatment plant operating schedule. The system was backwashed with nonchlorinated effluent that was stored in a 50-gallon effluent storage tank that overflowed to the sewer.

Table 3-14 Pilot design criteria

Parameter	Units	Contactor	Filter
Filter loading rate	gpm/ft^2	2.2	1.8
Airflow rate	cfm/ft^2	2.86	--
Empty bed contact time	min	15.6	13.9
Backwash duration	min	5	15
Backwash bed expansion	%	0	50
Backwash loading rate	gpm/ft^2	51	17
Filter media type	--	Gravel	Anthracite/ADGS$^+$ silica sand
Bed depth	in.	55	10/30
Effective size	in.	0.5 (0.25–1.25)	0.04/0.012–0.014
Effective size	mm	12.7 (6.35–31.8)	0.97/0.30–0.35

Dashes indicate field not applicable.

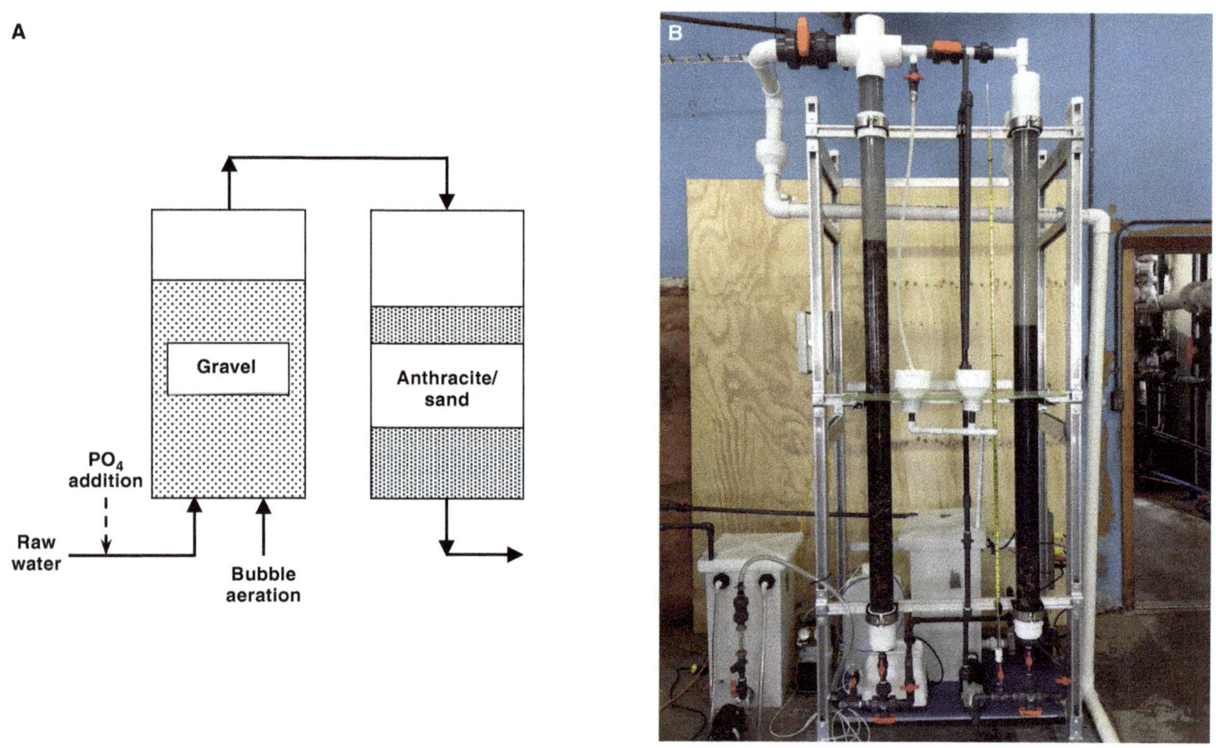

Figure 3-8 Process flow diagram (A) and photograph of pilot system (B)

3.5.2.2 Treatment goals. Although ammonia in water is not regulated, Iowa Department of Natural Resources (IDNR) can require water systems in the state to monitor nitrite and nitrate at their points of entry to the distribution system and in their distribution systems if they suspect that nitrification of the source water ammonia is occurring. Nitrite and nitrate have MCLs of 1 (in Iowa 1.0) and 10 mg N/L, respectively, measured at the point of entry to the distribution system. The IDNR extends the MCL sampling location definition to drinking water in the distribution when nitrification is a concern. Particularly worrisome are waters with ammonia levels greater than 1 mg N/L where incomplete nitrification can lead to exceedances of the lower nitrite MCL.

The finished water quality goals to be achieved by the biotreatment process were as follows:

- As less than 10 μg/L;
- NH_3 less than 0.1 mg N/L;
- Fe less than 0.1 mg/L; and
- Mn less than 0.02 mg/L.

Additional goals for the City were ease of operation, reduced operating costs, and reliability.

3.5.2.3 Operation and performance. Acclimation, which was defined in terms of nitrification performance, took longer than expected. Nitrification acclimation time was expected to take 90 days on the basis of previous experience at a similar site that took 30 days to acclimate when operated continuously and previous work suggesting that acclimation time is inversely proportional to the daily run time (Lytle et al. 2007a, 2007b). Upon investigation, an issue with the aeration system was discovered, which limited the average DO in the contactor to approximately 7 mg/L. Once the issue was resolved on day 114, DO in the biofilter influent increased to greater than 9 mg/L. NH_3 removal in the contactor increased to 60 percent within 14 days of increasing the airflow. It continued to increase up to 80 percent over the next 105 days, and acclimation was considered complete on day 220.

Figure 3-9 shows NH_3, NO_2^-, and NO_3^- in the raw water, contactor effluent, and biofilter effluent during the pilot study. Most (60–80 percent) of the nitrification occurred in the contactor, and additional nitrification occurred in the filter such that the treated water NH_3 concentration was less than 0.1 mg N/L at steady state. NO_2^- accumulation in the filter effluent occurred during the acclimation period in the filter with a peak NO_2^- concentration of 2.3 mg N/L. Such a spike is typical because there is a lag in the growth of NOB until significant nitrite levels are present to trigger their activity. While it is possible for nitrite to peak above the MCL of 1 mg N/L, for this study, nitrite was only above the MCL for 20 days. Additionally, if needed, nitrite can be oxidized with chlorine with a theoretical chlorine dose of 5 mg Cl_2 per mg NO_2-N.

Figure 3-10 shows the total iron, manganese, and arsenic concentrations in the raw water, contactor effluent, and biofilter effluent during the pilot study. For all three contaminants, removal occurred in both the contactor and biofilter. The contactor was designed to be the main point where iron, manganese, and arsenic oxidation occurred and was not intended to remove particles; however, considerable, although variable, iron (59 percent), manganese (36 percent), and arsenic (60 percent) removal occurred in the contactor. No flow restriction was observed in contactor.

Iron removal occurred quickly, within one week of startup (Figure 3-10A). Chemical oxidation of Fe(II) by air is rapid under the pilot conditions, but it is also possible that some biological iron oxidation occurred. Iron removal was consistent and complete with an average filter-effluent concentration of 0.02 mg/L over the study period.

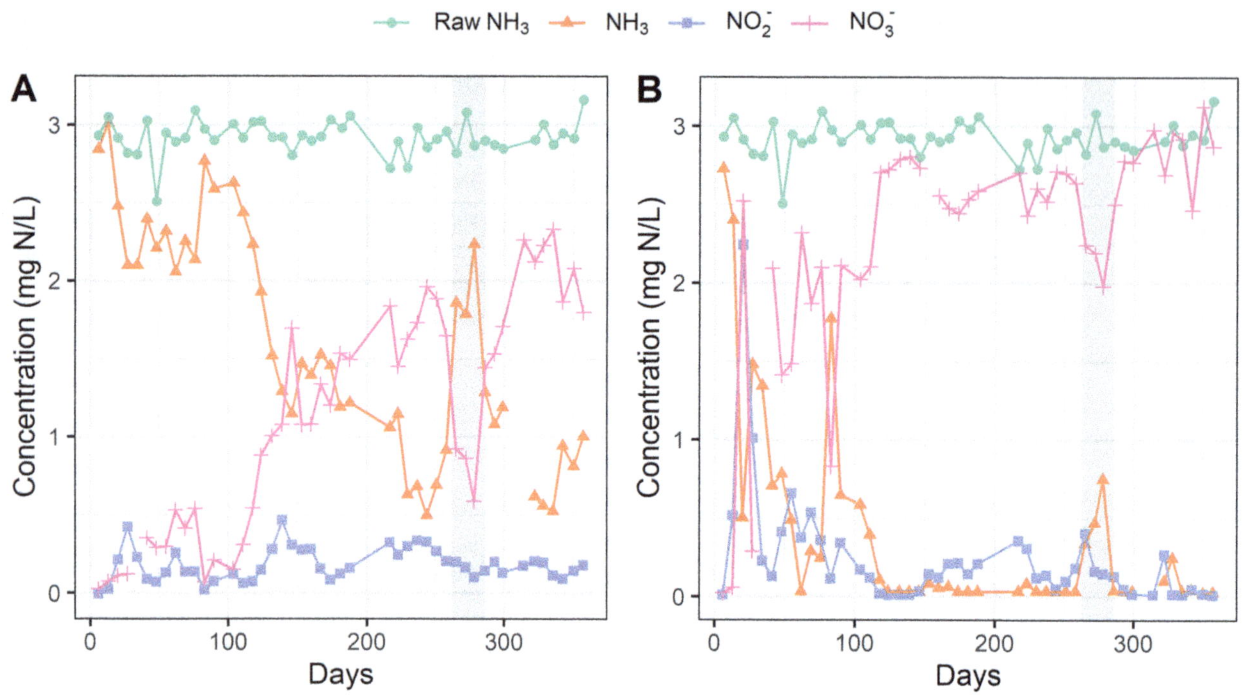

Figure 3-9 Nitrogen species through the contactor (A) and the biofilter (B) over the course of the pilot study

Excellent Mn removal occurred through the pilot system, with the filter-effluent concentration being less than 0.005 mg/L at steady state (Figure 3-10B). Mn removal in the contactor continued to improve over the one-year pilot, accounting for approximately 75 percent of the total removal by the end of the study. It is likely that Mn oxidation occurred biologically. Although Mn(II) adsorbs to MnO_x surface coatings, such as those on the sand in the filter, a chemical oxidant such as Cl_2 is required to oxidize the sorbed Mn(II) (Tobiason et al. 2016). If it is not oxidized, then the adsorption capacity becomes exhausted, and no Mn would be removed. The consistent, long-term removal of Mn in the pilot indicates that biological oxidation of Mn occurred. Although oxygen alone will not oxidize soluble Mn(II) under the pilot conditions, DO did appear to affect Mn removal because consistent removal was observed in the filter after the DO was increased on day 114, whereas it previously had been inconsistent. DO affects Mn-oxide solubility as well as the activity of Mn-oxidizing microorganisms, and thus is a critical operating parameter.

Biological activity in the pilot completely oxidized As(III) to As(V), and the total arsenic concentration in the filter effluent averaged 8.0 μg/L and was at or below the MCL of 10 μg/L for most of the study except for a few samples (Figure 3-10C). Adding acid to decrease the pH in the contactor where the average effluent pH was 8.07 might improve arsenic removal because the optimal range for As(V) sorption to iron occurs best at pH 7.0–7.2 (Lytle et al. 2005). Arsenic oxidation was evident shortly after startup, and after 50 days, as much as 93 percent of the arsenic in the contactor effluent was present as oxidized As(V) (Figure 3-10D). Higher arsenic values might have been due to operational issues, such as floc carryover when the airflow rate in the contactor was increased or extending filter run time past the designated backwash point.

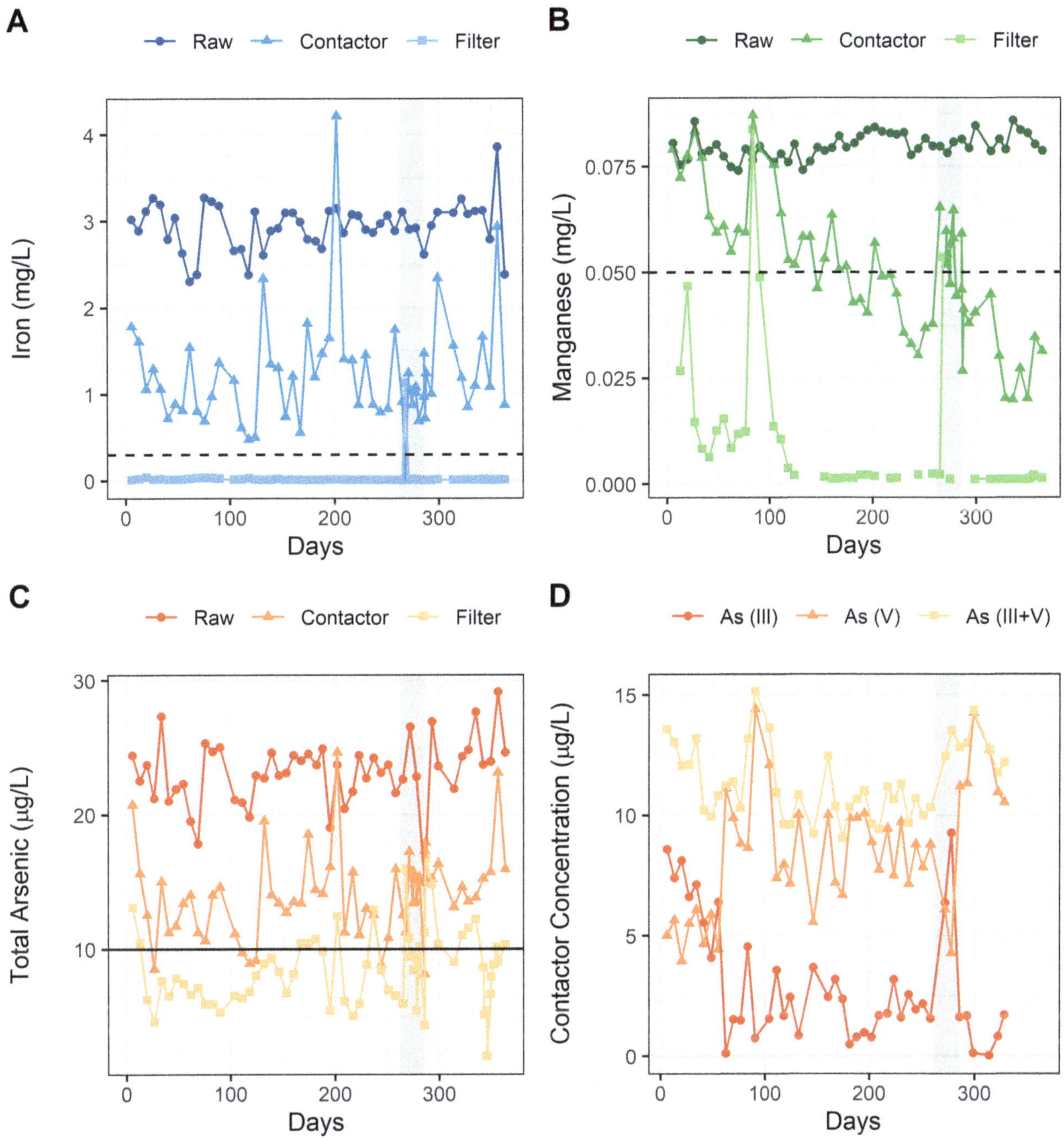

Figure 3-10 Iron (A), manganese (B), and total arsenic concentrations (C) in the raw water, contactor effluent, and filter effluent, and soluble arsenic speciation in the contactor effluent (D) over the course of the pilot study

The loading rate to the contactor was doubled to 4.4 gpm/ft^2 for 25 days to day 260 to simulate having one of the two redundant contactors in the full-scale design out of service. This corresponded to a decrease in the EBCT from 16 to 8 min. Ammonia removal decreased to approximately 75 percent; iron and manganese spiked to 1.13 mg/L and 0.053 mg/L, respectively, and then dropped back down to similar removal, as observed at the lower loading rate; and the arsenic concentration in the filter was greater than the MCL of 10 µg/L for two of the three sampling events. The filter helped mitigate the lower treatment performance through the contactor. Upon returning to the original loading rate, water quality parameters rapidly returned to previous levels.

3.5.2.4 Key success factors. The piloted biotreatment system effectively addressed the four target contaminants (ammonia, arsenic, iron, and manganese) and achieved the City's water quality goals. Therefore, the City proceeded with a full-scale design and construction of the piloted system. The plant began startup in December 2020.

- Both the aerated contactor and filter contribute to contaminant transformation and removal.
- Supplying enough air to maintain saturated DO throughout the contactor is critical for nitrification, Mn oxidation, and As oxidation in both the contactor and filter. Maintaining saturated DO also improved Mn removal.
- Good arsenic removal (65 percent) was observed throughout the study, but the total arsenic concentration in the biofilter effluent was above the MCL of 10 µg/L on a few occasions. Arsenic removal via sorption to iron particles is affected by the pH and iron concentration. The capability to adjust pH and supplement iron ahead of the contactor was included in the full-scale plant to improve arsenic removal if necessary.

3.5.3 Biological Iron and Manganese Removal, Putnam, Conn.

Derek R. Belanger, Tighe & Bond, Worcester, Mass.
Asher E. Keithley, USEPA, Cincinnati, Ohio

The Putnam Water Pollution Control Authority (WPCA) provides drinking water to a large portion of the Town of Putnam, Conn., and to small areas in the neighboring towns of Woodstock and Thompson. The average and maximum daily demand from the WPCA's system are approximately 0.90 and 1.8 mgd, respectively. The WPCA obtains its water supply from three sources, one of which is the Park Street Wellfield, which has a total yield capacity of 1.8 mgd.

In the past, polyphosphate for sequestration of Mn was used at the wellfield. Accumulation of Mn in the distribution system and customers' water quality concerns prompted the WPCA to investigate opportunities for removal of Fe and Mn from the groundwater. At the start of this project, the wellfield contained a single building that housed chemical treatment systems for pH adjustment, chlorine disinfection, and polyphosphate addition. The purpose of this project was to select and construct an Fe and Mn removal system at the Park Street Wellfield water treatment plant. First, pilot tests were performed to evaluate physical/chemical and biotreatment approaches. Then, a full-scale biotreatment plant was designed and constructed.

In 2014, the WPCA commissioned a pilot study to assess the feasibility of Mn removal using three different technologies:

- chemical/physical using GreensandPlus and anthracite with sodium hypochlorite and potassium permanganate feeds;
- chemical/physical using pyrolusite with sodium hypochlorite and potassium permanganate feeds; and
- biological filtration with aeration and pH adjustment.

All three treatment processes achieved the finished water quality goal of Mn less than 0.03 mg/L, and biotreatment was selected as the treatment process for the full-scale plant. Pilot testing demonstrated that biotreatment had several advantages as compared with physical/chemical treatment:

- low effluent Mn concentration;
- higher unit filter run volumes (approximately 25,000 gal/ft^2 versus 5,000 gal/ft^2);
- fewer residuals because of lower backwash frequency and no added permanganate; and
- reduced chemical feed requirements.

3.5.3.1 Facility description. The wellfield contains six gravel pack wells and four bedrock wells. Table 3-15 summarizes the range of raw water quality among the 10 wells. The WPCA operates the wellfield under a variety of different pumping scenarios, so the Fe and Mn concentrations in the treatment plant raw water can vary over a wide range.

The treatment scheme consists of two pressure filtration stages in series with compressed air injection for aeration. The first stage is intended to remove Fe, and the second stage is intended to remove Mn. Iron removal was included in the full-scale design in case the raw water iron concentration increased over time. Potassium hydroxide is added ahead of the second filtration stage to increase pH so that the pH in the filter effluent is approximately 7.5. Filter effluent is discharged to a filtered water storage tank that serves as an atmospheric break and the backwash water source. Sodium hypochlorite is added for disinfection, and zinc orthophosphate is added for corrosion control in line just ahead of the finished water pumps that serve the distribution system. The CT for 4-log virus inactivation credit is achieved in line between the plant and the first customer tap.

The filters are designed to allow for backwash on the basis of differential pressure, filter run time, or filter run volume. Backwashes are typically performed at a preset filter run time of 500 hours because the water treatment plant's average flow is less than the design capacity of the filter system. A regular backwash consists of water only wash at a rate between 7 and 10 gpm/ft^2. Backwash with air scour is provided every three to four backwashes to further clean the media. Backwash with air scour is a multistep process that consists of air scour (3.5 scfm/ft^2), air + water wash (3.5 scfm/ft^2 + 6 gpm/ft^2), and high-rate rinse (8 gpm/ft^2). Spent backwash wastewater and the released Fe and Mn residuals are discharged to a holding tank that discharges to the sanitary sewer. Although the design allows for recycle of settled backwash wastewater, the recycle system is not used.

Figure 3-11 presents the process flow diagram. Table 3-16 summarizes the design criteria.

Table 3-15 Raw water quality

Parameter	Units	Range of Raw Water Quality
Iron	mg/L	0.072–0.9
Manganese	mg/L	0.002–1.070

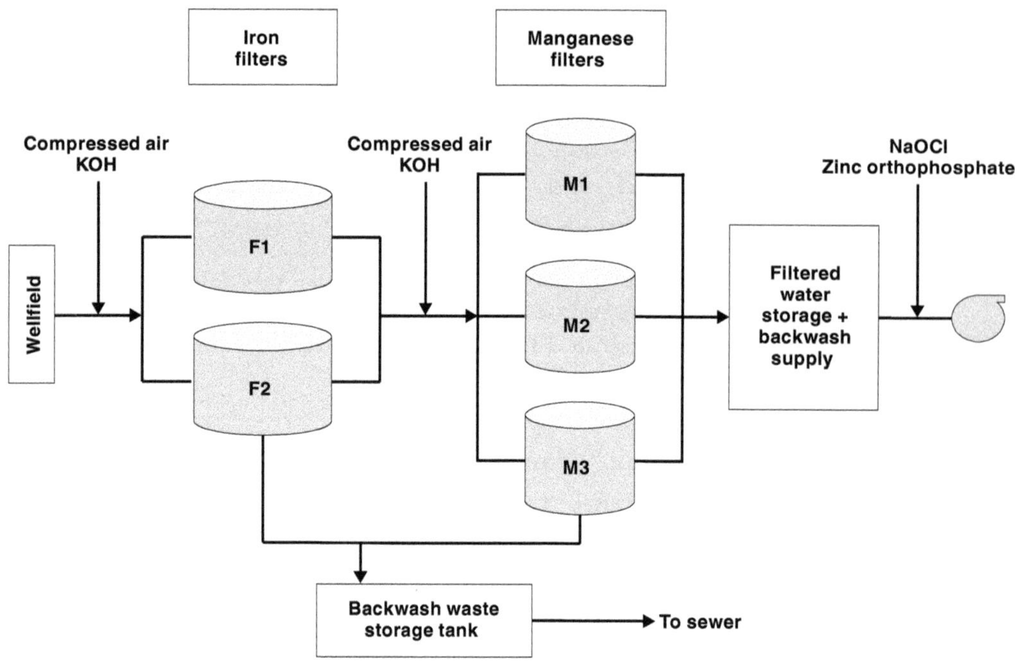

Figure 3-11 Process flow diagram

Table 3-16 Biotreatment design criteria

Parameter	Unit	Design Criteria
Minimum design flow	mgd	0.3
Maximum design flow	mgd	1.77
Maximum raw water iron	mg/L	2.0
Maximum raw water manganese	mg/L	1.5
Iron removal design filter loading rate	gpm/ft^2	8.1
Iron removal average filter loading rate	gpm/ft^2	2–4
Number of iron removal filters	--	2
Diameter of iron removal filters	ft	10
Iron removal filter media type	--	Sand
Iron removal filter media size	mm	0.95
Manganese removal design filter loading rate	gpm/ft^2	8.1
Manganese removal average filter loading rate	gpm/ft^2	2–4
Number of manganese removal filters	--	3
Diameter of manganese removal filters	ft	8
Manganese removal filter media type	--	Sand
Manganese removal filter media size	mm	0.95
Backwash water pump capacity	gpm	630
Backwash residuals disposal location	--	Sanitary sewer

mgd—million gallons per day. Blank space indicates field not applicable

3.5.3.2 Treatment goals. The Putnam WPCA has the following goals for the Park Street Wellfield water treatment plant:

- consistently achieve a finished water iron concentration less than 0.15 mg/L;
- consistently achieve a finished water manganese concentration less than 0.030 mg/L;
- operate effectively over a wide range of flow rates and influent chemistry;
- maximize energy efficiency, equipment resiliency, and ease of operation; and
- minimize O&M costs.

3.5.3.3 Operation and performance. Startup of the filtration system began in August 2018. Figure 3-12 shows the acclimation period in the manganese filters. The biological filter system was initially seeded with backwash solids from a pilot filter to assist in the acclimation of biological activity on the media surface. However, one month after startup, the filters showed little progress in lowering effluent manganese. At this point, plant operators identified a partially opened underground valve that allowed chlorinated water with a chlorine residual of 0.1–0.3 mg/L to pass through the filters, which hindered biofilm development. After fixing the cross connection, media from mature biologically active filters for Fe and Mn removal at a water treatment plant in Cavendish, Vt., was added to the top of the filters. The filter-effluent Mn concentration was less than the SMCL of 0.05 mg/L approximately one month after reseeding the filters, and it achieved the Mn treatment goal of 0.03 mg/L within two months. In total, startup lasted approximately four months. Mn acclimation can take two to five months, so it is unclear the extent to which the chlorine residual hindered and the seeding practices aided acclimation at this facility (Tobiason et al. 2016, Tekerlekopoulou et al. 2013).

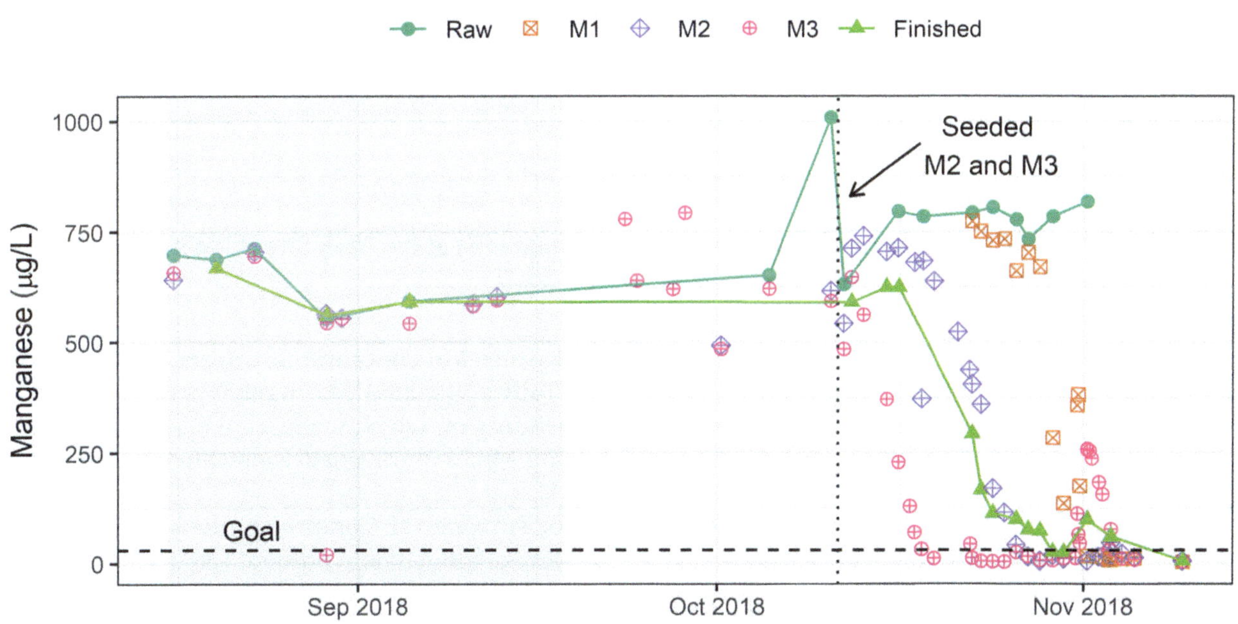

Figure 3-12 Manganese acclimation at the Park Street Wellfield water treatment plant

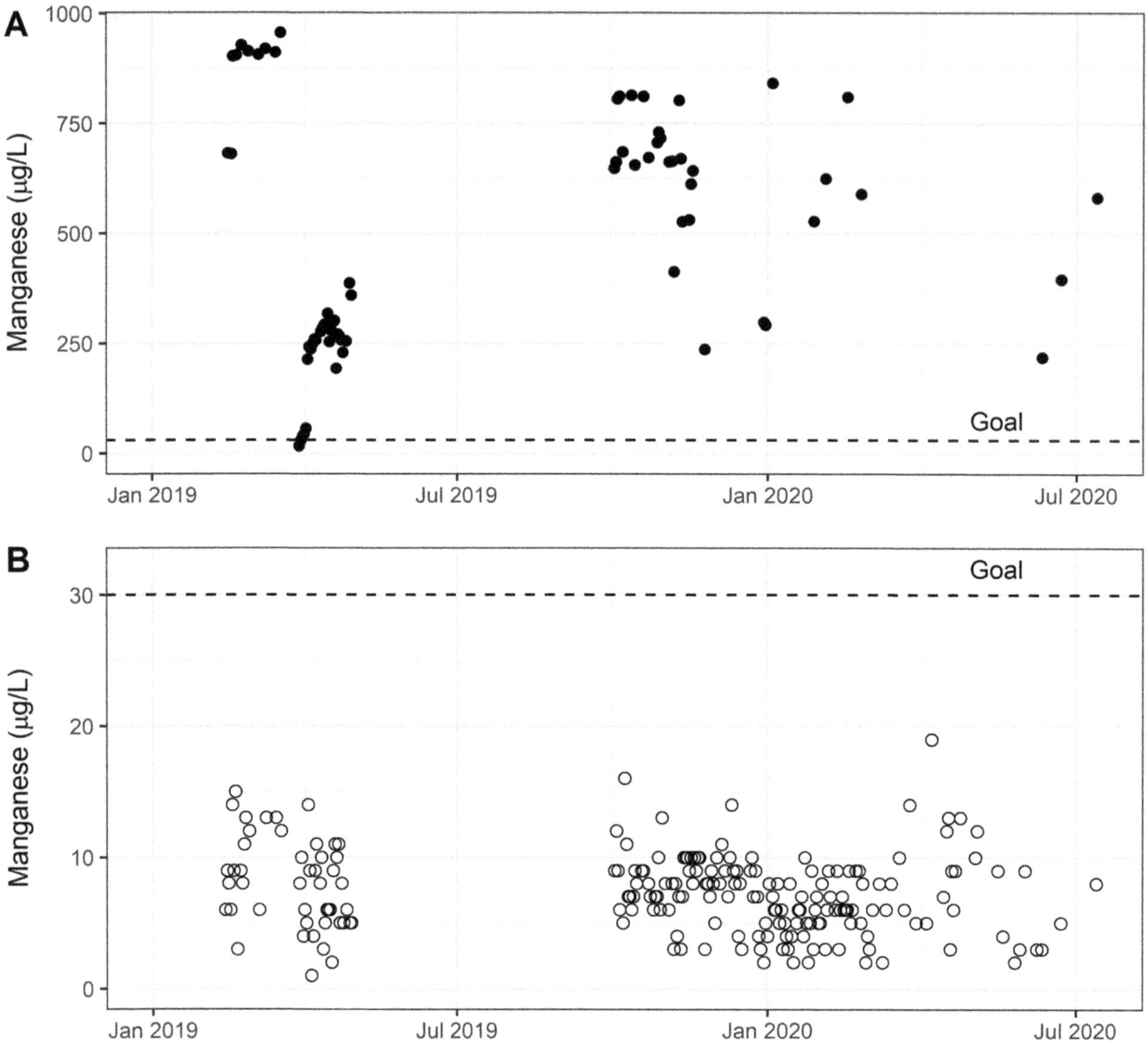

Figure 3-13 Raw (A) and finished manganese concentrations (B) at the Park Street Wellfield water treatment plant from February 2019 to July 2020

Figure 3-13 shows raw and finished water Mn concentrations from February 2019 through July 2020, after the biofilters were fully acclimated. During that period, raw water Mn ranged from 0.016 to 0.956 mg/L, depending on which wells were on-line. The treatment system was very effective at removing Mn, with finished water Mn concentrations averaging 0.007±0.003 mg/L (97% removal).

DO is a critical operating parameter for Mn removal. Compressed air is injected ahead of the Mn filters to increase the DO. The average influent and effluent DO in the Mn filters are 7.8 and 4.7 mg/L, respectively, under typical operating conditions. The decrease in DO through the Mn filters was higher than expected, and it is possible that biological oxidation of other reduced contaminants in the water contributed to the oxygen consumption.

Raw water iron concentrations are generally low, and the plant does not monitor iron frequently, but limited data suggest that the plant effectively removes iron. One sample showed that the Fe filters lowered the Fe concentration from 0.18 mg/L in the raw water to 0.03 mg/L in the Fe filter effluent. Although it is possible to aerate the raw water before the Fe filters, this option is not exercised to promote biological instead of chemical oxidation of iron.

Table 3-17 Filter operating parameters, June 2019 through June 2020

Filtration Stage	Filter Run Time d	Volume Treated mil gal	Pressure Increase psi
1st Iron	20.2 (17.2–23.0)	5.31 (3.46–7.40)	1.10 (0.06–4.38)
2nd Manganese	20.9 (18.2–21.6)	3.82 (3.17–4.40)	0.40 (0–5.96)

Note: Data presented as median (10th–90th percentile).

Biotreatment was selected because of its superior operational performance observed during the pilot study, and full-scale operating data support that decision (Table 3-17). All filter vessels were typically backwashed on the basis of reaching the maximum filter run time of 500 hours (20.8 days), which is much longer than what is usually achieved in physical/chemical Fe and Mn removal processes. The volume treated in the Mn filters was lower than in the Fe filters because there is one more filter vessel in the second Mn removal stage than in the first Fe removal stage. The pressure increased over a filter run was generally very low, with median increases of 1.10 psi in the Fe filters and 0.40 psi in the Mn filters. These results demonstrate the operational advantages of biotreatment for Fe and Mn removal.

3.5.3.4 Key success factors. Pilot testing demonstrated that both biological and physical/chemical treatment systems effectively removed iron and manganese and achieved the finished water quality goals, and biotreatment was better operationally because it treated greater unit filter run volumes and produced less residual waste. Therefore, a full-scale biotreatment system was constructed and has been in operation for two years. The plant treats raw water manganese concentrations from 0.016 to 0.956 mg/L and consistently achieves a finished water manganese concentration of 0.007±0.003 mg/L, which is below WPCA's goal of 0.03 mg/L.

Lessons learned include the following:

- Biotreatment is stable and consistent at treating high Mn concentrations;
- Seeding new filters using media from other biologically active filters may shorten the acclimation time; and
- If treatment goals are not achieved within the expected time period, it may be beneficial to investigate potential sources of stress on the microbial community, such as the unintended chlorine residual in this study.

3.6 REFERENCES

Arturi, K.R., C.B. Koch, and E.G. Søgaard. 2017. "Characterization and Comparison of Iron Oxyhydroxide Precipitates from Biotic and Abiotic Groundwater Treatments." *Journal of Water Supply: Research and Technology–AQUA*. 66(2):96-104.

AWWA (American Water Works Association). 2021. M64, *Gas Transfer Applications in Water: Addition and Removal*. Denver: AWWA.

Benefield, L.D., J.F. Judkins, and B.L. Weand. 1982. *Iron and Manganese Removal from Process Chemistry for Water and Wastewater Treatment*. Englewood Cliffs, N.J.: Prentice-Hall.

Blakely, N.C. 1984. "Behavior of Wastes Co-disposed with Domestic Solid Wastes." *Journal of the Water Pollution Control Federation*. 56(1):69-75.

Bourgine, F., M. Gennery, J. Chapman, H. Kerai, J. Green, R. Rap, S. Ellis, and C. Gaumard. 1994. "Biological Processes at Saints Hill Water-Treatment Plant, Kent." *Water and Environment Journal*. 8(4):379-391.

Bradford, S.M., C.J. Palmer, and B.H. Olson. 1994. "Assimilable Organic Carbon Concentrations in Southern California Surface and Groundwater." *Water Research*. 28(2):427-435.

Breda, I.L., L. Ramsay, and P. Roslev. 2017. "Manganese Oxidation and Bacterial Diversity on Different Filter Media Coatings During the Start-up of Drinking Water Biofilters." *Journal of Water Supply: Research and Technology–AQUA*. 66(8):641-650.

Breda, I.L., L. Ramsay, D.A. Søborg, R. Dimitrova, and P. Roslev. 2019. "Manganese Removal Processes at 10 Groundwater Fed Full-Scale Drinking Water Treatment Plants." *Water Quality Research Journal*. 54(4):326-337.

Bruins, J.H., B. Petrusevski, Y.M. Slokar, K. Huysman, K. Joris, J.C. Kruithof, and M.D. Kennedy. 2015. "Biological and Physico-chemical Formation of Birnessite During the Ripening of Manganese Removal Filters." *Water Research*. 69:154-161.

Burger, M.S., C.A. Krentz, S.S. Mercer, and G.A. Gagnon. 2008. "Manganese Removal and Occurrence of Manganese Oxidizing Bacteria in Full-Scale Biofilters." *Journal of Water Supply: Research and Technology–AQUA*. 57(5):351-359.

Cai, Y., D. Li, Y. Liang, H. Zeng, and J. Zhang. 2014. "Operational Parameters Required for the Start-up Process of a Biofilter to Remove Fe, Mn, and NH3-N from Low-Temperature Groundwater." *Desalination and Water Treatment*. 57(8):3588-3596.

Cerrato, J.M., W.R. Knocke, M.F. Hochella Jr., A.M. Dietrich, A. Jones, and T.F. Cromer. 2011. "Application of XPS and Solution Chemistry Analyses to Investigate Soluble Manganese Removal by MnO(x)(s)-Coated Media." *Environmental Science & Technology*. 45(23):10068-10074.

Chen, A.S., L. Wang, D.A. Lytle, and T.J. Sorg. 2018. "Arsenic/Iron Removal from Groundwater with Elevated Ammonia and Natural Organic Matter." *Journal AWWA*. 110(3):E2-E17.

Cheng, Q., L. Nengzi, L. Bao, Y. Wang, J. Yang, and J. Zhang. 2017. "Interactions Between Ammonia, Iron and Manganese Removal Using Pilot-Scale Biofilters." *Journal of Water Supply: Research and Technology–AQUA*. 66(3):157-165.

Crognale, S., S. Amalfitano, B. Casentini, S. Fazi, M. Petruccioli, and S. Rossetti. 2017. "Arsenic-Related Microorganisms in Groundwater: A Review on Distribution, Metabolic Activities and Potential Use in Arsenic Removal Processes." *Reviews in Environmental Science and Bio/Technology*. 16(4):647-665.

Dangeti, S., B. Roshani, B. Rindall, J.M. McBeth, and W. Chang. 2017. "Biofiltration Field Study for Cold Fe(II)- and Mn(II)-Rich Groundwater: Accelerated Mn(II) Removal Kinetics and Cold-Adapted Mn(II)-Oxidizing Microbial Populations." *Water Quality Research Journal*. 52(4):229-242.

de Vet, W.W.J.M., L.C. Rietveld, and M.C.M. van Loosdrecht. 2009. "Influence of Iron on Nitrification in Full-Scale Drinking Water Trickling Filters." *Journal of Water Supply: Research and Technology–AQUA*. 58(4):247-256.

Dietrich, A.M. and G.A. Burlingame. 2015. "Critical Review and Rethinking of USEPA Secondary Standards for Maintaining Organoleptic Quality of Drinking Water." *Environmental Science & Technology*. 49(2):708-720.

Edwards, M.A. and A. Dudi. 2004. "Role of Chlorine and Chloramine in Corrosion of Lead-Bearing Plumbing Materials." *Journal AWWA*. 96(10):69-81.

Faust, S.B. and O.M. Aly. 1981. *Chemistry of Natural Waters*. Ann Arbor, Mich.: Ann Arbor Science Publishers.

Gude, J.C.J., L.C. Rietveld, and D. van Halem. 2018. "As(III) Removal in Rapid Filters: Effect of pH, Fe(II)/Fe(III), Filtration Velocity and Media Size." *Water Research*. 147:342-349.

Gulay, A., Y. Cekic, S. Musovic, H.J. Albrechtsen, and B.F. Smets. 2018. "Diversity of Iron Oxidizers in Groundwater-Fed Rapid Sand Filters: Evidence of Fe(II)-Dependent Growth by Curvibacter and Undibacterium spp." *Frontiers in Microbiology*. 9:2808.

Gulay, A., S.J. Fowler, K. Tatari, B. Thamdrup, H.J. Albrechtsen, W.A. Al-Soud, S.J. Sorensen, and B.F. Smets. 2019. "DNA- and RNA-SIP Reveal Nitrospira spp. as Key Drivers of Nitrification in Groundwater-Fed Biofilters." *mBio*. 10(6):e01870-19.

Health Canada. 2019. "Guidelines for Canadian Drinking Water Quality: Guideline Technical Document – Manganese." Ottawa: Health Canada.

Hopenhayn-Rich, C., M.L. Biggs, and A.H. Smith. 1998. "Lung and Kidney Cancer Mortality Associated with Arsenic in Drinking Water in Córdoba, Argentina." *International Journal of Epidemiology*. 27(4):561-569.

Iyare, P.U. 2019. "The Effects of Manganese Exposure from Drinking Water on School-age Children: A Systematic Review." *Neurotoxicology*. 73:1-7.

Kaplan, L.A., D.J. Reasoner, and E.W. Rice. 1994. "A Survey of BOM in US Drinking Waters." *Journal AWWA*. 86(2):121-132.

Katsoyiannis, I.A., A. Zikoudi, and S.J. Hug. 2008. "Arsenic Removal from Groundwaters Containing Iron, Ammonium, Manganese and Phosphate: A Case Study from a Treatment Unit in Northern Greece." *Desalination*. 224(1-3):330-339.

Kim, M.-J., J. Nriagu, and S. Haack. 2002. "Arsenic Species and Chemistry in Groundwater of Southeast Michigan." *Environmental Pollution*. 120(3):379-390.

Kindaichi, T., T. Ito, and S. Okabe. 2004. "Ecophysiological Interaction Between Nitrifying Bacteria and Heterotrophic Bacteria in Autotrophic Nitrifying Biofilms as Determined by Microautoradiography-Fluorescence in situ Hybridization." *Applied and Environmental Microbiology*. 70(3):1641-1650.

Knocke, W.R., J.R. Hamon, and C.P. Thompson. 1988. "Soluble Manganese Removal on Oxide-Coated Filter Media." *Journal AWWA*. 80(12):65-70.

Li, D., J. Zhang, H. Wang, H. Yang, and B. Wang. 2005. "Operational Performance of Biological Treatment Plant for Iron and Manganese Removal." *Journal of Water Supply: Research and Technology–AQUA*. 54(1):15-24.

Li, X.K., Z.R. Chu, Y.J. Liu, M.T. Zhu, L. Yang, and J. Zhang. 2013. "Molecular Characterization of Microbial Populations in Full-Scale Biofilters Treating Iron, Manganese and Ammonia Containing Groundwater in Harbin, China." *Bioresource Technology*. 147:234-239.

Lytle, D.A., A.S. Chen, T.J. Sorg, S. Phillips, and K. French. 2007b. "Microbial As(III) Oxidation in Water Treatment Plant Filters." *Journal AWWA*. 99(12):72-86.

Lytle, D.A., M.R. Schock, J.A. Clement, and C.M. Spencer. 1998. "Using Aeration for Corrosion Control." *Journal AWWA*. 90(3):74-88.

Lytle, D.A., T.J. Sorg, and V.L. Snoeyink. 2005. "Optimizing Arsenic Removal During Iron Removal: Theoretical and Practical Considerations." *Journal of Water Supply: Research and Technology–AQUA*. 54(8):545-560.

Lytle, D.A., T.J. Sorg, L. Wang, C. Muhlen, M. Rahrig, and K. French. 2007b. "Biological Nitrification in a Full-Scale and Pilot-Scale Iron Removal Drinking Water Treatment Plant." *Journal of Water Supply: Research and Technology–AQUA*. 56(2):125-136.

Lytle, D.A., C. White, D. Williams, L. Koch, and E. Nauman. 2013. "Innovative Biological Water Treatment for the Removal of Elevated Ammonia." *Journal AWWA*. 105(9):E524-E539.

Lytle, D.A., D. Williams, C. Muhlen, M. Pham, K. Kelty, M. Wildman, G. Lang, M. Wilcox, and M. Kohne. 2015. "The Full-Scale Implementation of an Innovative Biological Ammonia Treatment Process." *Journal AWWA*. 107(12):E648-E665.

Lytle, D.A., D. Williams, C. Muhlen, E. Riddick, and M. Pham. 2020. "The Removal of Ammonia, Arsenic, Iron and Manganese by Biological Treatment from a Small Iowa Drinking Water System." *Environmental Science: Water Research & Technology*. 6(11):3142-3156.

McClellan, J. 2015. "Biological Iron and Manganese Treatment: 5 Years of Operating Experience in Cavendish VT." *Journal of the New England Water Works Association*. 129(4):245-248.

McDonough, L.K., I.R. Santos, M.S. Andersen, D.M. O'Carroll, H. Rutlidge, K. Meredith, P. Oudone, J. Bridgeman, D.C. Gooddy, J.P.R. Sorensen, D.J. Lapworth, A.M. MacDonald, J. Ward, and A. Baker. 2020. "Changes in Global Groundwater Organic Carbon Driven by Climate Change and Urbanization." *Nature Communications*. 11(1):1279.

McMahon, P.B., K. Belitz, J.E. Reddy, and T.D. Johnson. 2019. "Elevated Manganese Concentrations in United States Groundwater, Role of Land Surface-Soil-Aquifer Connections." *Environmental Science & Technology*. 53(1):29-38.

Mouchet, P. 1992. "From Conventional to Biological Removal of Iron and Manganese in France." *Journal AWWA*. 84(4):158-167.

Odell, L.H., G.J. Kammeyer, A. Wilczak, J.G. Jacangelo, J.P. Marcinko, and R.L. Wolfe. 1996. "Occurrence of Nitrification in Chloraminated Distribution Systems." *Journal AWWA*. 88(7):86-98.

Pacini, V., A.M. Ingallinella, R. Vidoni, G. Sanguinetti, and R. Fernández. 2014. "Transformation of an Existing Physicochemical Plant for Iron and Manganese Removal by the Application of Biological Processes." *Journal of Water Supply: Research and Technology–AQUA*. 63(6):507-517.

Ramsay, L., I.L. Breda, and D.A. Søborg. 2018. "Comprehensive Analysis of the Start-up Period of a Full-Scale Drinking Water Biofilter Provides Guidance for Optimization." *Drinking Water Engineering and Science*. 11(2):87-100.

Richard, L., E. Mayr, M. Zunabovic, R. Allabashi, and R. Perfler. 2016. "Experiences with the Practical Implementation and Operation of Biological Nitrification for a Small-Scale Drinking Water Treatment Plant." *Water Supply*. 16(4):922-930.

Rittmann, B.E., D. Stilwell, J.C. Garside, G.L. Amy, C. Spangenberg, A. Kalinsky, and E. Akiyoshi. 2002. "Treatment of a Colored Groundwater by Ozone-Biofiltration: Pilot Studies and Modeling Interpretation." *Water Research*. 36(13):3387-3397.

Schilling, K.E. 2002. "Occurrence and Distribution of Ammonium in Iowa Groundwater." *Water Environment Research*. 74(2):177-186.

Štembal, T., M. Markic, F. Briški, and L. Sipos. 2004. "Rapid Start-up of Biofilters for Removal of Ammonium, Iron and Manganese from Ground Water." *Journal of Water Supply: Research and Technology–AQUA*. 53(7):509-518.

Štembal, T., M. Markić, N. Ribičić, F. Briški, and L. Sipos. 2005. "Removal of Ammonia, Iron and Manganese from Groundwaters of Northern Croatia—Pilot Plant Studies." *Process Biochemistry*. 40(1):327-335.

Tatari, K., S. Musovic, A. Gulay, A. Dechesne, H.J. Albrechtsen, and B.F. Smets. 2017. "Density and Distribution of Nitrifying Guilds in Rapid Sand Filters for Drinking Water Production: Dominance of Nitrospira spp." *Water Research*. 127:239-248.

Tebo, B.M., J.R. Bargar, B.G. Clement, G.J. Dick, K.J. Murray, D. Parker, R. Verity, and S.M. Webb. 2004. "Biogenic Manganese Oxides: Properties and Mechanisms of Formation." *Annual Review of Earth and Planetary Sciences*. 32(1):287-328.

Tekerlekopoulou, A.G., S. Pavlou, and D.V. Vayenas. 2013. "Removal of Ammonium, Iron and Manganese from Potable Water in Biofiltration Units: A Review." *Journal of Chemical Technology & Biotechnology*. 88(5):751-773.

Tobiason, J.E., A. Bazilio, J. Goodwill, X. Mai, and C. Nguyen. 2016. "Manganese Removal from Drinking Water Sources." *Current Pollution Reports*. 2(3):168-177.

Trueman, B.F., B.S. Gregory, N.E. McCormick, Y. Gao, S. Gora, T. Anaviapik-Soucie, V. L'Herault, and G.A. Gagnon. 2019. "Manganese Increases Lead Release to Drinking Water." *Environmental Science & Technology*. 53(9):4803-4812.

USEPA (US Environmental Protection Agency). 2021. Drinking Water Contaminant Candidate List 5 - Draft, Federal Register.

USEPA. 2018 "Drinking Water Standards and Health Advisories Tables." Washington: Office of Water.

USEPA. 2015. "Water Treatment Pilot Plant Design Manual: Low Flow Conventional/ Direct Filtration Water Treatment Plant for Drinking Water Treatment Studies." EPA/600/R-15/045. Cincinnati: USEPA.

USEPA. 2003. "Removal of Arsenic from Drinking Water by Adsorptive Media." Cincinnati: National Risk Management Research Laboratory.

USEPA. 2003. "Arsenic Treatment Technology Evaluation Handbook for Small Systems." EPA 816-R-03-014. Washington: Office of Water.

USEPA. 1993. "Process Design Manual: Nitrogen Control." EPA/625/R-93/010. Washington: USEPA.

Wagner, F.B., P.B. Nielsen, R. Boe-Hansen, and H.J. Albrechtsen. 2018. "Remediation of Incomplete Nitrification and Capacity Increase of Biofilters at Different Drinking Water Treatment Plants Through Copper Dosing." *Water Research*. 132:42-51.

White, C.P., R.W. Debry, and D.A. Lytle. 2012. "Microbial Survey of a Full-Scale, Biologically Active Filter for Treatment of Drinking Water." *Applied and Environmental Microbiology*. 78(17):6390-6394.

WHO (World Health Organization). 2021. "Manganese in Drinking-Water. Background document for Development of WHO Guidelines for Drinking-Water Quality." Geneva: WHO.

Yu, H. and J.R. Leadbetter. 2020. "Bacterial Chemolithoautotrophy via Manganese Oxidation." *Nature*. 583(7816):453-458.

Zeng, H., C. Yin, J. Zhang, and D. Li. 2019. "Start-Up of a Biofilter in a Full-Scale Groundwater Treatment Plant for Iron and Manganese Removal." *International Journal of Environmental Research and Public Health*. 16(5):698.

Zhang, Y., N. Love, and M. Edwards. 2009. "Nitrification in Drinking Water Systems." *Critical Reviews in Environmental Science and Technology*. 39(3):153-208.

Zhou, H. and C. Fu. 2020. "Manganese-Oxidizing Microbes and Biogenic Manganese Oxides: Characterization, Mn(II) Oxidation Mechanism and Environmental Relevance." *Reviews in Environmental Science and Bio/Technology*. 19:489-507.

Chapter 4

Anoxic Biological Treatment

Giridhar Upadhyaya, Jess Brown, Brian Noma, and Bruce Rittmann

Anoxic biological treatment is used primarily in groundwater applications and relies on natural microorganisms to remove inorganic (e.g., nitrate and perchlorate) and organic contaminants (e.g., trichloroethylene [TCE] and dichloroethylene [DCE]). Removal occurs in the absence of oxygen and typically requires adding a substrate (i.e., carbon source or inorganic compound). This chapter discusses the target contaminants, regulatory concerns, biological removal mechanisms, applicable system configurations, and considerations for the planning, process evaluation, design, and operation of anoxic biological treatment processes. This chapter also includes case studies on four full-scale anoxic biological treatment applications.

4.1 APPLICATION OVERVIEW

Biological processes have been widely used for nitrate removal from drinking water sources in Europe for several decades (Rittmann and Huck 1989). Extensive information is available on denitrification processes because they have been used for tertiary wastewater treatment and in water reuse applications for many years (Tchobanoglous et al. 2002). The first full-scale drinking water biological denitrification plant came on-line in the United States (in Coyle, Okla.) in 1998 and treated around 40,000 gallons of groundwater per day (Silverstein and Carlson 1999). Since then, multiple full-scale biological drinking water plants have been commissioned for treating nitrate (Ali 2017, Brown et al. 2016, Gudal 2016). Though gaining traction, anoxic biotreatment applications are still relatively limited compared with aerobic biological surface water treatment.

4.1.1 Benefits

In an anoxic environment, oxidized contaminants can be removed through biologically-mediated reduction reactions. Anoxic biological treatment shares the inherent benefits of biological treatment processes, including effective contaminant removal without the generation of concentrated waste streams, removal of multiple contaminants in a single unit process, and low energy demand. The end products of anoxic biological treatment are either innocuous or less toxic, and easily separable solids and no concentrated waste streams are generated. In general, anoxic biological treatment may provide a cost-effective and sustainable treatment approach, especially when removing multiple oxidized contaminants is a primary objective. However, treatment costs must be assessed for a specific application and site and should consider substrate chemical costs as well as posttreatment costs (reoxygenation, electron donor removal, turbidity removal, and/or disinfection).

4.1.2 Contaminants

With sufficient contact time, anoxic biological treatment can remove several chemical compounds of public health concern to below the maximum contaminant level (MCL) or method detection limit. Table 4-1 summarizes health risks associated with the contaminants of interest for anoxic biological treatment and the existing regulations.

4.1.3 Removal Mechanisms

Microorganisms mediate electron transfer from an electron donor (e.g., acetic acid) to the electron acceptor (e.g., nitrate) and use the energy generated during the process for growth. During the electron transfer process, the electron donor is oxidized (e.g., acetate to carbon dioxide [CO_2] and water [H_2O], and hydrogen gas [H_2] to H_2O), whereas the oxidized contaminant is converted to its reduced form (e.g., nitrate [NO_3^-] to nitrogen gas [N_2]). Figure 4-1 illustrates the typical conversions in an anoxic biological process.

Because most groundwater sources are typically oligotrophic (i.e., contain low organic carbon and nutrients), supplying a bioavailable electron donor is the most critical component of an anoxic biological treatment system. Anoxic biological treatment can occur in either autotrophic or heterotrophic conditions. In an autotropic system, an inorganic chemical (e.g., H_2) is used as the electron donor, and the microbial populations use inorganic carbon as their carbon source. In contrast, in a heterotrophic system, organic carbon (e.g., acetate) is fed as the electron donor and carbon source.

An initial estimate of an electron donor to be added can be obtained using stoichiometric electron transfer reactions (Table 4-2). These equations do not consider the electron donor/carbon source requirement for cell synthesis and estimate the minimal electron donor amount to be added to achieve the target oxidant reduction. The optimal dose, including the requirement for cell synthesis, should be determined during the testing phase. At minimum, the concentration of an electron donor should be close to the stoichiometric demand of the oxidized contaminants to be reduced. Adding insufficient electron donor will result in incomplete reduction of the contaminants. Adding excess electron donor could lead to negative unintended consequences, such as an effluent residual of readily biodegradable organic carbon and/or undesired reduction products like sulfide or methane.

Table 4-1 Health risks and existing regulations for oxidized contaminants of interest

Contaminant	Health Risks	Existing Regulation
Nitrate	• Causes methemoglobinemia in infants	MCL 10 mg/L N
Nitrite	• Causes methemoglobinemia in infants	MCL 1 mg/L N
Perchlorate	• Interferes with iodide uptake by thyroid gland • Can adversely affect hormone production	• No federal regulation • State standards ranging from 1 to 18 μg/L
Bromate	• Genotoxic • Carcinogen	• MCL 10 μg/L
Cr(VI)	• Carcinogen • Can damage kidney and liver	• No federal regulation for Cr(VI) • California MCL 10 μg/L • MCL for chromium 100 μg/L
Se(VI)	• Causes circulatory problems • Causes hair and fingernail loss	• MCL for selenium 50 μg/L
U(VI)	• Can damage kidney • Can raise cancer risk	• MCL for uranium 30 μg/L
Radium 226 and 228	• Can cause cataract, teeth decay, and anemia • Can cause cancer	Combined MCL 5 pCi/L
Gross alpha*	• Can cause cancer	15 pCi/L for all alphas except radon and uranium
VOCs		
PCE	• Can raise cancer risk	• MCL 5 μg/L
TCE	• Can raise cancer risk	• MCL 5 μg/L
1,1-DCE	• Causes liver problems	• MCL 7 μg/L
1,2-DCE	• Can raise cancer risk	• MCL 5 μg/L
cis-1,2-DCE	• Causes liver problems	• MCL 70 μg/L
trans-1,2-DCE	• Causes liver problems	• MCL 100 μg/L
DBCP	• Can cause cancer	• MCL 0.2 μg/L
1,2,3-TCP	• Potential carcinogen	• California MCL 5 ng/L

Cr(VI)—hexavalent chromium, DBCP—dibromochloropropane, DCE—dichloroethylene, MCL—maximum contaminant level, PCE—tetrachloroethylene, Se(VI)—hexavalent selenium, TCE—trichloroethylene, TCP—trichloropropane, U(VI)—hexavalent uranium, VOC—volatile organic compound

*Gross alpha = overall radioactivity in water

Competition for the electron donor is important for determining whether an oxidized contaminant is reduced in an anoxic process. Although reducing the target contaminant is the primary goal of an anoxic treatment process, sulfate reduction is highly undesirable. Thermodynamics and kinetics work together to dictate which acceptors are used first in the presence of an electron donor. In general, though not always, microorganisms preferably use electron acceptors in which the energy reaction produces more free energy, and a sequential use of $O_2 > NO_3^- > ClO_4^- > SeO_4^{2-} > CrO_4^{2-} > SO_4^{2-}$ is observed. However, it is not uncommon to observe overlapping contaminant-reduction zones (e.g., O_2 and NO_3^- reductions occurring together). Because nearly complete O_2 removal needs to occur before the reduction of other oxidized pollutants, the electron donor demand for oxygen reduction must always be included in calculating the concentration of electron donor to be added. Furthermore, because the electron donor is used as the carbon source for cell

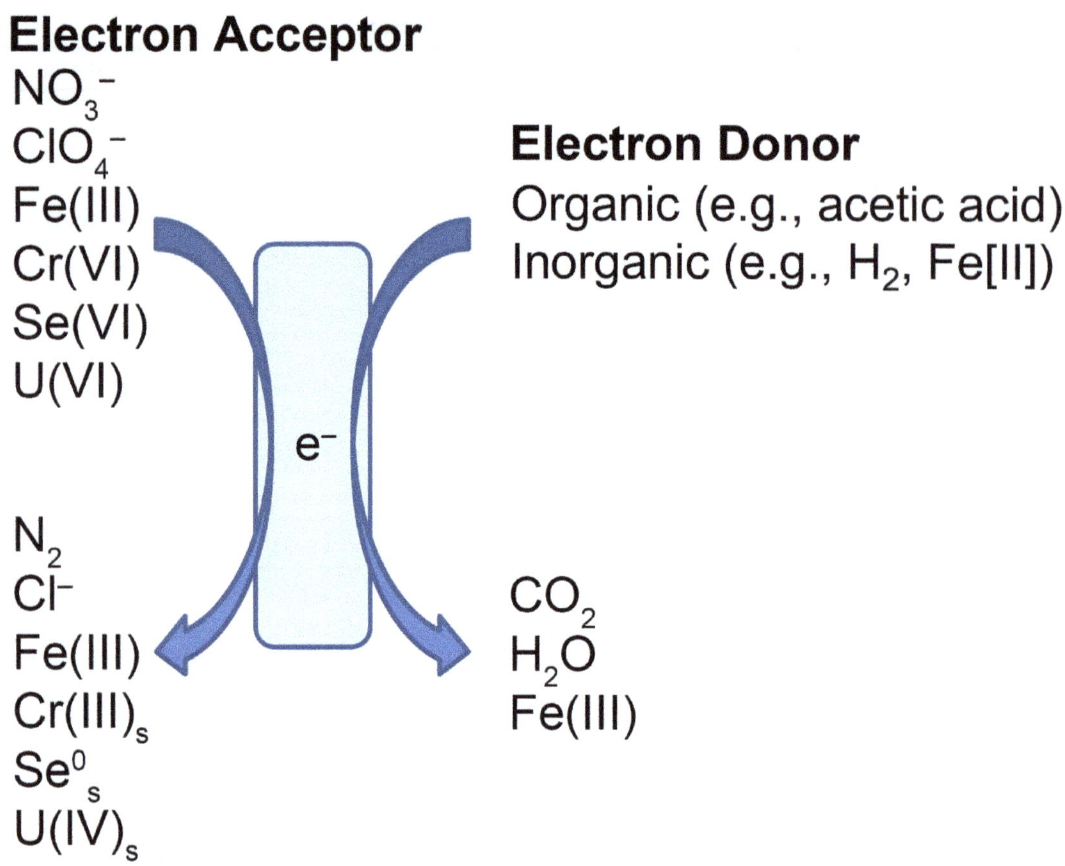

Figure 4-1 Simplified schematic for the fate of oxidized contaminants and electron donors during anoxic biological treatment

Table 4-2 Reduction stoichiometry for oxidized species common in drinking water

Oxidized Species	Reduction Reaction	Donor Demand
Oxygen	$O_2 + 4H^+ + 4e^- \rightarrow 2H_2O$	1 g OD/g O_2
Nitrate	$NO_3^- + 5H^+ + 5e^- \rightarrow 0.5N_2 + OH^- + 2H_2O$	2.86 g OD/g N
Nitrite	$NO_2^- + 3H^+ + 3e^- \rightarrow 0.5N_2 + OH^- + H_2O$	1.71 g OD/g N
Fe(III)	$Fe^{3+} + e^- \rightarrow Fe^{2+}$	0.14 g OD/g Fe(III)
Perchlorate	$ClO_4^- + 8H^+ + 8e^- \rightarrow Cl^- + 4H_2O$	0.64 g OD/g ClO_4^-
Selenate	$SeO_4^{2-} + 6H^+ + 6e^- \rightarrow Se° + 2OH^- + 2H_2O$	0.34 g OD/g SeO_4^{2-}
Chromate	$CrO_4^{2-} + 3H^+ + 3e^- + H_2O \rightarrow Cr(OH)_3 + 2OH^-$	0.41 g OD/g CrO_4^{2-}
Uranate	$UO_2^{2+} + 2e^- \rightarrow UO_2$	0.07 g OD/g U
Bromate	$BrO_3^- + 6H^+ + 6e^- \rightarrow Br^- + 3H_2O$	0.25 g OD/g BrO_3^-
Oxygen	$O_2 + 4H^+ + 4e^- \rightarrow 2H_2O$	1 g OD/g O_2
Bicarbonate	$HCO_3^- + 8H^+ + 8e^- \rightarrow CH_4 + 2H_2O + OH^-$	1.05 g OD/g HCO_3^-
Sulfate	$SO_4^{2-} + 8H^+ + 8e^- \rightarrow HS^- + 3H_2O + OH^-$	0.67 g OD/g SO_4^{2-}

OD—oxygen demand

Note: OH^- represents the generation of base.

synthesis in heterotrophic systems, the electron donor calculations should also include the organic carbon requirement for the cell synthesis. However, the carbon requirement for cell synthesis is typically minimal compared with that required for the contaminant removal, and approximate estimation of the donor requirement can be calculated in terms of oxygen demand (OD) as shown below. The OD represents mass of electron donor in terms of oxygen equivalents.

$$O_2 + 4H^+ + 4e^- = 2H_2O$$

$$NO_3^- + 5H^+ + 5e^- = 0.5N_2 + OH^- + 2H_2O$$

Therefore,

Oxygen Demand per Electron = 32 g of DO/4 electrons = 8 g OD/electron

OD per g N = (8*5)/14 = 2.86 g

Calculations for converting the OD to actual electron donor requirements are presented in Section 4.2.

Table 4-3 and Table 4-4 summarize previous studies on the anoxic biological treatment of individual or combinations of inorganic and organic contaminants, respectively. Thus, these tables highlight, in part, simultaneous removal of multiple contaminants. The contact times evaluated in previous studies for volatile organic compound (VOC) removal differed by orders of magnitude (Table 4-4) depending on treatment goals and type and concentrations of contaminants evaluated and cannot be used for designing a treatment system. There are three sets of data for three different types of reactors. For each application and type of VOC, it is recommended to characterize the process performance and determine the minimum contact time through pilot testing.

Table 4-3 Inorganic contaminants removal observed with anoxic biological treatment

Contaminant	Raw Water Concentration	Unit	Removal %	Configuration	Scale	Operating Conditions: Electron Donor	Contact Time *min*	Temp °C	pH *SU*	References
Nitrate	10–12	mg/L NO3	>99	FXB (two-stage)*	Bench	Acetic acid	10 (1st stage), 20 (2nd stage)	22	7.2±0.5	Upadhyaya et al. (2012, 2012a, 2010)
Arsenic	200–250	µg/L	88–92							
Nitrate	10–12	mg/L NO3	100	FXB (two-stage)	Bench	Acetic acid	7–10 (each stage)	NA	7.1±0.5	Upadhyaya et al. (2012)
Uranium	300	µg/L	90–95							
Nitrate	1–2	mg/L N	100	FXB (two-stage)	Bench	Acetic acid	25–50	21–24	7.5–8.2	Upadhyaya and Brown (2018)
Selenium	35–45	µg/L	91–93							
Nitrate	150–160	mg/L NO3	>99	FXB (two-stage)	Bench	Acetic acid	25–50	21–24	8.0–8.3	Upadhyaya and Brown (2018)
Hex. chrom.	16–18	µg/L	>70							
Uranium	51–53	µg/L	>90							
Selenium	220–260	µg/L	>95							
Nitrate	1	mg/L N	>90	MBfR	Bench	H_2	200‡	25	~7.5	Ontiveros-Valencia et al. (2016)
Selenium	12	µg/L	>90							
Nitrate	4–5	mg/L N	>99	FXB	Pilot	Acetic acid	30	NA	NA	Logan et al. (2004)
Perchlorate	70–80	µg/L	>99							
Nitrate	5.6	mg/L N	>99	FXB	Pilot	Acetic acid	18	18–22	7.6–7.8	Li et al. (2010)
Perchlorate	50	µg/L	>99							
Nitrate	0.9±0.09	mg/L N	>99	FXB	Pilot	Acetic acid	5.8	19.9±0.8	6.74±0.1	Min et al. (2004)
Perchlorate	77±13	µg/L	>95							
Nitrate	12	mg/L N	>99	MBfR	Pilot	H_2	33	27	7.6	Tang et al. (2010)
Nitrate	9	mg/L N	>99	MBfR, two stages	Pilot	H_2	15 per stage	18.5	~7.6	Zhao et al. (2014)
Nitrate	20	mg/L N	100	FBR	Pilot	Ethanol	126	19–21	NA	Hatzinger et al. (2000)
Perchlorate	400,000	µg/L	>99.9							

(continued)

Table 4-3 Inorganic contaminants removal observed with anoxic biological treatment (*Continued*)

Contaminant	Raw Water Concentration	Unit	Removal %	Configuration	Scale	Operating Conditions: Electron Donor	Contact Time *min*	Temp °C	pH *SU*	References
Nitrate	0.7±0.3	mg/L N	100	FXB (two-stage)	Pilot	Acetic acid	10	18.7±3.7	8.0±0.3	Upadhyaya et al. (2019)
Hex. chrom.	75–100	µg/L	91–93							
Nitrate	35–40	mg/L N	>99	FXB (two-stage)	Pilot	Acetic acid	10–15	10–25	7.5–7.8	Upadhyaya et al. (2022)
Arsenic	9–14	µg/L	>70†							
Nitrate	0.9±0.27	mg/L N	>97	FXB	Pilot	Acetic acid	71.5	10–30	7.6±0.26	Dugan et al. (2009)
Perchlorate	50±0.39	µg/L	>96							
Nitrate	7.1–9.0	mg/L N	>87	FXB	Pilot	Acetic acid	10	17.5–25.4	7.2–7.7	Najm et al. (2014)
Hex. chrom.	20–29	µg/L	>80							
Nitrate	1.9	mg/L N	NA	FBR	Full	Acetic acid or ethanol	61.5	21–24	NA	Polk et al. (2001)
Perchlorate	14,700	µg/L	>99							
Nitrate	1.5	mg/L N	100	FBR	Full	Ethanol	NA	NA	NA	Hatzinger et al. (2000)
	6,000–8,000	mg/L N	>99.9							
Nitrate	6.1–6.3	mg/L N	>99.5	FBR	Full	Acetic acid	12.2	18.5	7.7	Webster and Togna (2009)
Perchlorate	50–53	µg/L	>99.5							
Nitrate	19.7	mg/L N	65–79	FXB	Full	Ethanol	22–38.3	10	7.2	Richard (1989)
Nitrate	0.34	mg/L N	>99.4	FBR	Full	Ethanol	10	19	7.2	McCarty and Meyer (2005)
Perchlorate	2,600	µg/L	>99.9							

Source: Adapted from Brown et al. (2017) *Blank cells in table indicate field not applicable.*

FBR—fluidized-bed bioreactor, FXB—fixed-bed, hex. chrom.—hexavalent chromium, MBfR—membrane biofilm reactor, MBR—membrane bioreactor, SU—standard unit

*Arsenic removal achieved through interaction with sulfides. First stage dedicated for nitrate removal; second stage dedicated for sulfate reduction and arsenic removal.

†Arsenic removal achieved through adsorption on or coprecipitation with iron hydroxides.

‡The specific surface area of membrane is low in a bench-scale system. The specific surface increases about six-fold for full-scale units, and this reduces the contact time by a factor of about six. So, a contact time of 200 min in a bench-scale unit translates to about 33 min in a full-scale unit.

Table 4-4 Organic contaminant removal in anoxic biological treatment

Contaminant	Influent Conc. *μg/L*	Percent Removal	Configuration	Scale	Environmental/Operating Conditions: Electron Donor	Contact Time *min*	Temp *C°*	pH *SU*	References
PCE	1,490	95–98	FXB	Bench	Lactate	1,440	20	NA	De Bruin et al. (1992)
1,2-DCP	3,000	>90	FBR	Bench	Methanol Acetic acid	1,440	25–27	NA	Hauck and Hegemann (2000)
PCE	500–1,000	80	FXB, two-stage	Bench	Acetic acid	7,360	22	NA	Fathepure and Vogel (1991)
CF	500–1,000	32							
PCE	18±3	86±7	FXB	Bench	Acetic acid	3,600	NA	NA	Bouwer and Wright (1988)
CF	17±4	95±3							
1,1,1-TCA	10±2	>99							
CT	7±2	>99							
DBCP	17±2	>99							
BDCM	13±1	89±8							
BF	18±3	>99							
1,2-DBA	20±4	>99							
HCA	4±1	>99							
PCE	8±4	13±45	FXB	Bench	Acetic acid	3,600	NA	NA	Bouwer and Wright (1988)
CF	17±3	−39±37							
1,2-DBA	19±3	63±9							
1,1,1-TCA	9±2	72±8							
BDCM	16±2	96±5							
DBCP	12±4	98±7							
CT	7±1	>99							
BF	19±3	>99							
HCA	7±2	>99							
PCE	7±2	0±40	FXB	Bench	Acetic acid	3,600	NA	NA	Bouwer and Wright (1988)
CF	28±2	−13±16							
DBCP	37±3	14±11							
1,2-DBA	30±2	23±7							
TCA	27±3	30±7							
HCA	4±2	80±17							
BDCM	24±2	90±8							
BF	36±2	>99							
CT	17±2	>99							

(continued)

Table 4-4 Organic contaminant removal in anoxic biological treatment *(Continued)*

Contaminant	Influent Conc. *μg/L*	Percent Removal	Configuration	Scale	Environmental/Operating Conditions: Electron Donor	Contact Time *min*	Temp C°	pH *SU*	References
PCE	8,000–12,000	>98	AAFEB	Bench	Sucrose	108–240	15	~7	Carter and Jewell (1993)
TCE	300	100	FXB	Bench	Lactate	1,080	14	7	Schollhorn et al. (1997)
		74			Formate				
		70–90			Acetic acid				
		>80			Methanol				
		>75			H_2/CO_2				
PCE	3,000	38±7	UASB	Bench	Lactate	1,440	NA	7.1–7.4	Hwu and Lu (2008)
		76±4			Sucrose	5,760		7.1–7.4	
PCE	7.9±3.7	99±1	FXB, two-stage	Bench	Acetic acid	15	22	6.9–7.3	Upadhyaya and Brown (2016)
TCE	11.8±2.9	99±4							
1,1-DCE	3.9±0.9	100							
cis-1,2-DCE	0.48±0.11	98±14							
DBCP	0.22±0.07	98±4							
CF	0.32±0.06	100							
CT	0.17±0.03	100							
TCE	2.4±3.4	70±44	FXB, two-stage	Bench	Acetic acid	25	20–24	6.9–7.5	Upadhyaya et al. (2013)
PCE	87.6±180.8	94±16							
TCP	82.86±171.9	76±63							
TCE	189±25	93±4	FXB, two-stage	Pilot	Acetic acid	7–12	20–24	6.9–7.5	Brown et al. (2015)
1,1-DCE	4.4±0.5	67±16							
TCP	6.4±0.7	93±7							
1,2-DCA	2,000–15,000	>99	FXB	Full	1,2-DCA	NA	8–12	7.1–7.6	Stucki et al. (1995)

Source: Adapted from Brown et al. (2017). *Blank cells in table indicate field not applicable.*

AAFEB—anaerobic attached-film expanded-bed (similar to FBR), BDCM—bromodichloromethane, BF—bromoform, CF—chloroform, CT—carbon tetrachloride, DBA—dibromoethane, DBCP—dibromochloropropane, DCA—dichloroethane, DCE—dichloroethylene, DCP—dichloropropane, FBR—fluidized-bed reactor, FXB—fixed-bed bioreactor, HCA—hexachloroethane, PCE—tetrachloroethylene, SU-standard unit, TCA—tetrachloroethane, TCE—trichloroethylene, TCP—trichloropropane, UASB—upflow anaerobic sludge blanket, VOC—volatile organic compound

4.1.4 Regulatory Considerations

Before designing a full-scale anoxic biological treatment system, it is strongly recommended that the project owner or owner's representative contact the state and/or local primacy agency regarding the approval process for these systems. Furthermore, all chemical additives must be certified by the American National Standard Institute/National Sanitation Foundation (NSF) Standard 60. Similarly, all materials, including equipment, that come in contact with drinking water must be NSF standard 61-certified. Waste residual volume and toxicity must be considered in accordance with local, state, and/or federal discharge and/or disposal requirements. Toxicity associated with the residuals is of greater concern when target contaminants include metals (e.g., chromium and selenium), metalloids (e.g., arsenic), and VOCs. Pilot testing might be required by the administrative authority before approval is given to proceed with the design and construction of a full-scale biological treatment system. Other regulatory considerations include the following:

- data collection and reporting for specific system operating parameters;
- treatment redundancy (i.e., dual train parallel treatment);
- water quality requirements not directly related to the target contaminants (e.g., turbidity);
- downstream treatment requirements; and
- specialized operator training and/or certification.

4.1.4.1 Contaminant-specific regulations. The proposed treatment goal should be to consistently remove the target contaminant to or below a predetermined level (e.g., 50 percent of the MCL), health risk limit, or health-based value without creating any harmful by-products or other contaminant compliance issues. Applicable federal, state, and local regulatory requirements should be consulted for regulated limits. State and local health departments should be consulted for health risk limits or health-based values. In cases with no federal or state regulatory requirements, the World Health Organization's guidelines should be consulted. Regulatory provisions for the contaminants of interest for anoxic biological treatment are presented in Table 4-1.

4.1.4.2 Posttreatment. Treatment downstream of anoxic biological treatment often includes reoxygenation, filtration for turbidity removal, electron donor removal, and disinfection. As discussed in Section 4.1.3, DO is preferred by bacteria because of the amount of energy generated while mediating aerobic redox reactions. Therefore, anoxic systems require near-complete DO removal before the target oxidized contaminants are biologically reduced, which leads to lower redox conditions in the treated water. Lower redox conditions can promote corrosion in the distribution system, leading to metal release. Therefore, the treated water should be reoxygenated. Considerable biological growth occurs in anoxic processes, and it is not uncommon to observe biomass release into the treated water, requiring downstream filtration. If an electron donor is fed in excess of the required amount, the electron donor remaining in the bioreactor effluent should be removed in a downstream treatment process (e.g., an aerobic bioreactor) to minimize potential biological growth in the distribution system. Disinfection and turbidity requirements for anoxic biological treatment systems are similar to those for surface water treatment processes. In general, a 4-log virus inactivation carbon tetrachloride (CT) ensures adequate disinfection of anoxic biotreatment system effluent.

4.2 PLANNING

Applications of anoxic biological drinking water treatment are expanding. However, operating and design criteria can vary depending on the source water quality, seasonal variability (including temperature), and treatment technology configuration. Therefore, the feasibility of a specific anoxic biological treatment application should be assessed with proper planning before considering full-scale implementation. Setting treatment goals and assessing the effectiveness of the technology are key planning steps. The following sections outline best practices for establishing treatment goals, conducting desktop applicability evaluations, and developing a robust technology testing plan.

4.2.1 Treatment Goals

Assessing possible implementation of anoxic biological treatment for a given contaminant(s) starts with establishing treatment goals. Raw water quality, target contaminants, and applicable regulations will determine overall contaminant-specific treatment goals. The contaminants listed in Section 4.1.2 can be removed to below detection via anoxic biological treatment; however, the contact time required to achieve low effluent levels must be evaluated during the testing phase (Section 4.2.3) because it drives reactor volume requirements and thus capital costs. At the planning stage, expectations and treatment goals can be developed on the basis of general contaminant removal and system performance reported in the literature. Table 4-3 and Table 4-4 provide examples of contaminant removals expected with a given contact time. In anoxic biological treatment, hydraulic performance might affect contaminant removal (e.g., too frequent and aggressive biomass cleaning could result in occasional reacclimation requirements), and therefore should be considered when developing treatment goals.

4.2.2 Applicability Evaluation: Example

A simple desktop evaluation can make it clear whether anoxic biological treatment is a viable option for meeting the treatment goals established for a given water source. A desktop evaluation generally involves two steps. The first step is to assess whether the water source contains oxidized contaminants at concentrations of concern (Section 4.1) by reviewing historical raw water quality data, if available. If data are not available, raw water quality monitoring should be conducted. As much as possible, the water quality assessment should focus on delineating seasonal variability. If the water source contains one or more oxidized contaminants above target concentrations, the next step is to perform a simple quantitative analysis of electron donor requirements and sizing, which can be used for a preliminary economic analysis. A preliminary estimate of the supply rate of an electron donor can be estimated from the stoichiometry in Table 4-3. This is illustrated by the following example:

> A flow rate of 1,000 m^3/d (0.264 million gallons per day [mgd]) contains 20 mg/L NO_3^--N, 100 μg/L ClO_4^-, and 8 mg/L O_2. NO_3^- and ClO_4^- are the oxidized contaminants to be removed. Though O_2 is not a contaminant, it needs to be nearly completely removed to achieve nitrate and perchlorate removals. For preliminary estimates of applicability, assume that DO, nitrate, and perchlorate are completely removed (i.e., 100 percent removal). Anoxic biotreatment is typically applied to groundwater, which tends to have less than 6 mg/L DO. When an equalization tank is used upstream of an anoxic biotreatment system, DO concentrations can be 8 mg/L or higher. We used 8 mg/L as an example for the calculations.

On the basis of the stoichiometry of Table 4-1, the mg/L OD demands are as follows:

NO_3^-: 20.0 mg/L N × 2.86 g OD/g N = 57.20 g OD/L
ClO_4^-: 0.100 mg/L ClO_4^- × 0.640 g OD/g ClO_4^- = 0.064 mg OD/L
O_2: 8.00 mg/L O_2 × 1.00 g OD/g O_2 = 8.00 mg OD/L
Total: 57.20 + 0.064 + 8.00 = 65.3 mg/L OD

Using the total OD, the mass of electron donor required per day can be calculated using the following relation:

Mass of Electron Donor (kg OD/d) = Total OD Demand (mg OD/L) × Flow Rate (m^3/d) × 1,000 (L/m^3) × 1/10^6 (kg/mg)

Therefore, for the given example,

Electron Donor Requirement per day = 65.3 (mg OD/L) × 1,000 (m^3/d) × 1,000 (L/m^3) × 1/10^6 (kg/mg) = 65.3 kg OD/d

Two key electron donors are acetate ($C_2H_3O_2^-$) and H_2 gas. The OD rate can be converted to demands of acetate and H_2 by stoichiometry:

Acetate: 65.3 kg OD/d × 64 g OD/59 g Acetate = 70.8 kg Acetate/d
H_2 gas: 65.3 kg OD/d × 16 g OD/2 g H_2 = 522.4 kg H_2/d

The next step is to estimate the volume of the process. Anoxic biological treatment processes are biofilm processes, for which the fundamental design criterion is the substrate flux, J, which represents the rate of electron acceptor (i.e., contaminant of interest) use per unit biofilm–surface area. A simple mass balance on a steady-state biofilm reactor is as follows:

$$0 = Q(S_i - S_e) - JaV$$

or, $$V = Q(S_i - S_e)/aJ$$

where,

Q = flow rate through the reactor
S_i = influent electron acceptor concentration
S_e = effluent electron acceptor concentration
J = electron acceptor flux
a = biofilm specific surface area
V = reactor volume

J values are needed for all the oxidized contaminants, and the contaminant giving the largest V controls the size of the reactor. Previous publications can be referenced for guidance on J and a values (Rittmann and McCarty 2020, Zhou et al. 2014). For the given example, reasonable J values for NO_3^- and ClO_4^- are 0.6 g N/m^2–d and 0.01 g ClO_4^-/m^2–d, respectively. With a value of 200 m^2/m^3 (i.e., 200 m^{-1}), the V values for nitrate and perchlorate are as follows:

NO_3^-: V = 1,000 m^3/d (20–0 g N/m^3)/[(200 m^{-1})(0.6 g N/m^2–d)] = 167 m^3
ClO_4^-: V = 1,000 m^3/d (0.1–0 g ClO_4^-/m^3)/[(200 m^{-1})(0.01 g ClO_4^-/m^2–d)] = 50 m^3

Therefore, for the given example, NO_3^- removal requires 167 m^3 (approximately 0.44 *million gallons*) and controls the reactor volume.

The reactor volume and substrate dose requirements largely drive capital and operations and maintenance (O&M) costs, respectively. Therefore, the values obtained through these fundamental calculations must be confirmed and refined through pilot testing for assessing economic feasibility.

4.2.3 Testing

Though desktop modeling may estimate the applicability of anoxic biological treatment for a given application, site-specific testing is typically performed to rigorously assess process efficacy. Robust testing programs improve design criteria, improve capital and operating cost estimate accuracy, and improve full-scale performance. Best practices are summarized in the following sections for establishing testing objectives, determining testing scale(s), designing the test, and evaluating testing outcomes.

4.2.3.1 Testing objectives. Although testing objectives can be site- and contaminant-specific, there are generally five broad testing objectives:

- **Demonstrate proof of concept:** Testing for proof of concept is typically conducted with a bench-scale system. In general, results obtained during a proof of concept evaluation might not be scalable for full-scale implementation.
- **Develop design and operating criteria for a full-scale system:** Pilot-scale testing is generally conducted to verify the results obtained during the proof of concept testing and to develop design and operating criteria for a full-scale application.
- **Optimize system design and operation:** Pilot-scale testing is generally used for optimizing process performance (i.e., hydraulic and water treatment performance). Full-scale testing of an existing treatment system could also be performed, with limited flexibility, for process improvement.
- **Identify any process limitations and potential unintended consequences:** Pilot-scale testing can identify process limitations and potential unintended consequences, which can then be used to develop mitigation strategies.
- **Familiarize operators and local regulators with the treatment system:** Though gaining traction, anoxic biotreatment applications are still relatively limited; therefore, pilot testing can provide invaluable opportunities to address operators' and regulators' questions and concerns.

4.2.3.2 Testing scales. The selection of a testing scale is driven by the testing objectives. Depending on the objective, testing might need to progress through multiple scales. For a new contaminant of interest, evaluation should start at the bench-scale, in which multiple parameters/factors can be economically assessed under controlled conditions. The results from this testing should be further verified at the pilot-scale system with more representative operational and monitoring conditions that support accurate scaling of the results for demonstration-scale implementation. As needed, limited process optimization testing can also be conducted at the full-scale system. The three different testing scales are described in detail in Section 1.5.

Bench-scale testing uses small quantities of water with flow rates in the mL/min range. Bench-scale testing is often considered when the treatment setting differs considerably from what has been well-studied for anoxic biological treatment. Unique attributes that favor bench-scale testing are water that has a mixture of oxidized contaminants, the possibility of inhibitory constituents, or that which is atypical in terms of temperature, pH, and/or the dissolved solids concentration. Bench-scale testing is used to determine whether oxidized contaminants can be reduced in the unique setting and to gain initial information on workable substrate fluxes, microbial ecology, and special challenges, such as maintaining a good pH or preventing mineral scaling. Bench-scale tests are commonly conducted for three to six months, which may need to be extended depending on the contaminant of interest because biological acclimation itself may require up to six months. When granular activated carbon (GAC) is used as the growth support media for removing contaminants that can be removed through adsorption (e.g., VOCs), exhaustion of the media can delay biological acclimation.

Extensive pilot testing might not be required if the treatment scenario (e.g., treatment technology and raw water quality) has already been studied rigorously. The goals of pilot testing are to document workable loading rates/contact times, to uncover and remedy operational challenges (e.g., too little or too much biofilm accumulation), and to provide technical data for process design and economic analysis. Pilot testing should also answer questions associated with seasonal temperature and water quality variations, and anticipated system operating conditions, including intermittent operation and intentional/accidental shutdowns. Pilot-scale testing is recommended from 3 to 12 months depending on the target contaminant. The pilot testing should be long enough to allow time for complete media exhaustion (if the contaminant of interest can be removed through adsorption on the media), biological acclimation, process optimization, evaluation of a range of operating conditions, uncovering and solving challenges, and meeting primacy agency requirements.

Full-scale demonstration testing employs commercial-scale equipment and has flow rates in the hundreds to thousands of gallons per minute. The ultimate goal of a demonstration test is to verify that process performance meets specifications, determined by pilot testing and treatment goals, for the long term and under actual conditions. Success with a full-scale demonstration test should lead to the necessary operating permits and understanding of operating requirements and operating costs. A demonstration test typically has a duration of 12 months.

4.2.3.3 Testing design. A testing plan should be carefully designed to accurately assess testing objectives and generate quantifiable and meaningful results. In general, testing design is dictated by testing goals, available budget, and the selected testing scale. The testing design should generate sufficient and accurate data through continuous and regular monitoring of key performance indicators to allow for clearly evaluating whether each testing objective is met. The following parameters and considerations need to be taken into account when developing a testing plan. Detailed discussion on other important aspects (resource planning, data management and interpretation, quality control, safety, etc.) to be considered when designing a testing plan are provided in Section 2.2.3. Process design criteria are discussed in detail in Section 4.3.

4.2.3.3.1 Process configuration and media type. If a treatment configuration has not been preselected, the testing plan might involve evaluating multiple system configurations in parallel to determine the best configuration for a given application. Parallel testing can also be considered with different biomass support media. GAC might provide advantages over other growth support media when the contaminant of interest can be removed through adsorption (e.g., VOCs). Testing should be performed that isolates and quantifies to the extent possible the degree of adsorptive versus biological removal and establishes adsorption exhaustion curves for any target contaminants. In some cases, GAC might better support biological growth and prevent inhibition of biological activity during accidental exposure to toxic levels of contaminants due to adsorption–desorption phenomena (Chang and Rittmann 1987).

4.2.3.3.2 Biological acclimation. It is critical to assess biological acclimation, which refers to the time required to establish a consistent, mature, and active microbial community on the support media before achieving steady-state targeted contaminant removal. Multiple parameters, including source water quality, media type, and the contaminant of interest, affect the duration required for biological acclimation. Complete biological acclimation could take two weeks (e.g., nitrate removal) to months (e.g., selenium removal). The testing duration should be long enough to complete biological acclimation and establish pseudo–steady-state conditions before each change in operating conditions. This will help characterize potential challenges that might occur because of increased biological activity, including fouling of sampling lines and analyzers, turbidity and contaminant breakthrough, and hydraulic challenges. During acclimation, recirculating the treated effluent to the upstream of the biological treatment process may help expedite biological acclimation.

4.2.3.3.3 *Contaminant removal kinetics and process optimization.* Once biological acclimation is complete, it is critical to quantify contaminant degradation, which is directly related to the electron donor dose and contact time. The process flow rate along with influent and effluent substrate and contaminant concentrations should be monitored regularly (e.g., inline monitoring for flow and more than three times per week for target contaminant). Testing often includes an evaluation of performance improvement strategies, including varying the substrate dose, supplementation of phosphorus, and varying the contact time.

4.2.3.3.4 *Biofilm control strategy.* Biomass growth is intentionally enhanced in an anoxic system. Therefore, higher biomass growth is observed in anoxic biological treatment systems and the biofilm control strategy is critical. Inadequate biofilm removal results in poor contaminant diffusion into the biofilm, clogging, and/or short-circuiting. Conversely, overly aggressive biofilm removal can lead to poor system performance due to inadequate bioactivity to meet treatment objectives. Although pilot testing is not typically used to generate full-scale design criteria for biofilm control strategies, it is important to determine the general effectiveness of biofilm control strategies during pilot-scale testing.

4.2.3.3.5 *Seasonal water quality variations.* Influent water quality (e.g., pH, temperature, alkalinity, DO, contaminant concentration, and orthophosphate concentration) can affect anoxic biological treatment. In general, because anoxic biological treatment is typically used for groundwater treatment, seasonal variability might not be as pronounced as with the surface water sources. However, if historical data indicate the possibility of seasonal variations (e.g., temperature variations), the testing plan should include provisions for quantifying these parameters and correlating changes in water quality with treatment performance.

4.2.3.3.6 *Challenge testing.* The testing plan should characterize the effects of any challenges that might be experienced at full scale, including chemical feed failure, system shutdowns, and changes in influent contaminant concentration.

4.2.3.3.7 *Assessing microbial parameters.* Anoxic biological treatment occurs through the activity of various microbial groups with specific functional genes/proteins (e.g., nitrate and nitrite reductases). The presence of these microbial groups can be assessed through the application of molecular techniques, such as polymerase chain reaction, targeting these functional genes. Characterizing the microbial community or their functional genes might be considered during testing, especially when the effects of multiple electron donors, nutrients, or oxidants are being evaluated. Additionally, establishing a baseline characterization of the microbial community is important, so that any changes in system performance can be potentially tracked to not only operational or water quality conditions but also to shifts in the microbial community.

4.2.3.3.8 *Posttreatment requirements.* The testing plan should also account for downstream treatment requirements, including reoxygenation, electron donor removal, filtration, disinfection, and pH adjustment. For autotrophic systems, recarbonation may also be required.

4.2.3.4 Testing outcomes. Testing results should be carefully reviewed to assess whether the testing objectives were met. If pilot testing was conducted, the outcome assessment should focus on confirming the establishment of design and operating information for full-scale implementation. If there are any shortfalls in the testing results, either additional testing should be planned, preferably with a testing scale that would be more representative of a full-scale application, or the design should be completed with larger safety factors to account for unknowns.

4.3 DESIGN

Full-scale anoxic biological treatment system design should be based on the design criteria determined through testing conducted at the appropriate testing scale (i.e., pilot scale or demonstration scale). Key design aspects, such as the type of biological process configuration, biofilm support media, contact time and/or hydraulic loading rate, nutrient feed dose, biomass control strategy, residual handling, and posttreatment requirements, are discussed subsequently.

4.3.1 Process Configurations

Anoxic biological treatment systems implemented for drinking water applications are biofilm-based processes that rely on microorganisms growing in a microbially produced extracellular polymeric matrix on a solid surface (e.g., granular media, membrane, or specialized biofilm support media). The most common anoxic biological treatment configurations are briefly described subsequently. The process configurations are not unique to anoxic biotreatment and can be used for aerobic biotreatment, as well. However, the use of a chemical electron donor is unique to anoxic biotreatment. Out of these configurations, only the membrane biofilm reactor (MBfR) is an autotrophic system. Most of these technologies are commercialized; selection is typically based on pilot testing performance comparisons, equipment cost estimates, and/or full-scale track records.

4.3.1.1 Fixed-bed bioreactor (FXB). FXBs are plug-flow reactors that use a stationary bed of granular media (e.g., sand, anthracite, or GAC) or proprietary biofilm carrier. The media bed can be packed in a pressure vessel or can be contained in an open basin or tank. Influent water, amended with electron donor and possibly nutrients, is either pumped or flows by gravity across the media. Over time, microbial growth and solids accumulation (if applicable) generate additional pressure drop across the media bed, requiring regular backwashing to remove excess biomass. The backwashing approach is similar to conventional filtration with air scour and low/high-rate fluidization, and it might also include oxidant addition. An example schematic of a common FXB bioreactor is shown in Figure 4-2.

4.3.1.2 Fluidized-bed bioreactor (FBR). An FBR configuration uses cylindrical concrete tanks containing granular media. The influent water is amended with external electron donor and is pumped in an upward mode. The granular media is continuously fluidized by recirculating a portion of the bioreactor effluent along with the influent

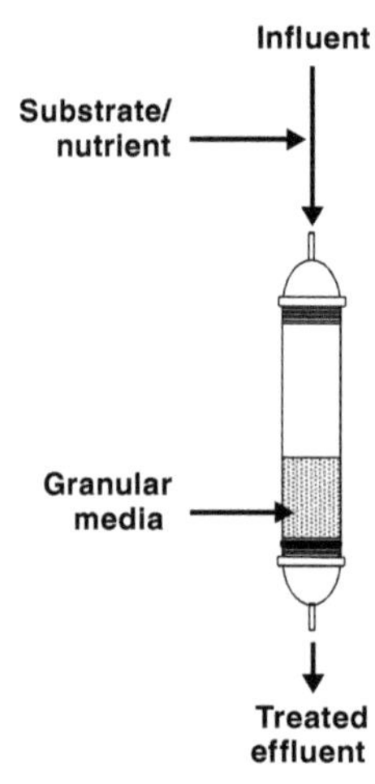

Figure 4-2 Simple schematic for a downflow fixed-bed bioreactor configuration

water. Typically, 25–30 percent bed expansion relative to the resting bed height is targeted during the startup, whereas bed expansion during the steady-state removal phase can be 40–60 percent. The high flow velocity and particle-to-particle abrasion help control biofilm accumulation; therefore, there is typically less downtime for biomass control in an FBR than in an FXB. Inline mechanical biomass/media separation devices can be used to prevent media loss from the reactor. Figure 4-3 shows an example schematic of a common FBR bioreactor.

4.3.1.3 Fixed-film suspended media bioreactor. A fixed-film suspended media bioreactor uses a plastic biofilm support medium (i.e., biofilm carrier) or a proprietary polymeric medium with encapsulated microorganisms. The media are kept in suspension using upflow hydraulics and a mechanical stirrer, and a screen may be used within the reactor to prevent media washout. In these systems, excess microbial growth is controlled through hydrodynamic shear, collisions among the media, and occasionally air sparging. Figure 4-4 shows an example schematic of a common fixed-film suspended media bioreactor.

4.3.1.4 MBfR. An MBfR uses modular hollow-fiber gas-transfer membranes that provide the biofilm support and the means to deliver H_2 gas as the electron donor (Martin and Nerenberg 2012; Rittmann 2018, 2007). Membranes are submerged in a reactor vessel fed with raw water. H_2 is fed to the lumen of the membranes, diffuses through the nonporous membrane walls, and is delivered directly into the biofilm attached to the outer side of the membrane. This helps maintain a high H_2 delivery efficiency. Although

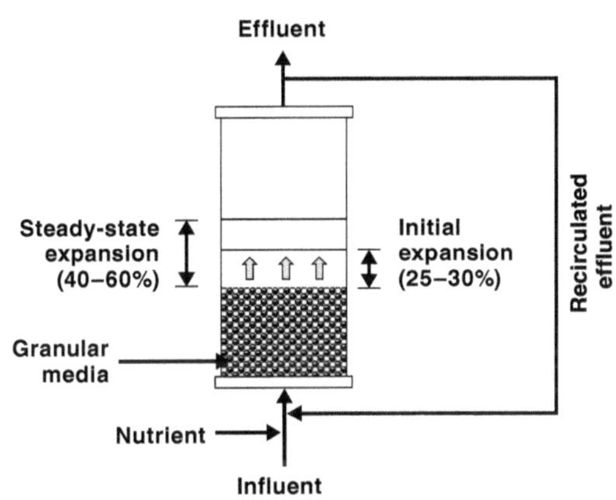

Figure 4-3 Simple schematic for a fluidized-bed bioreactor configuration

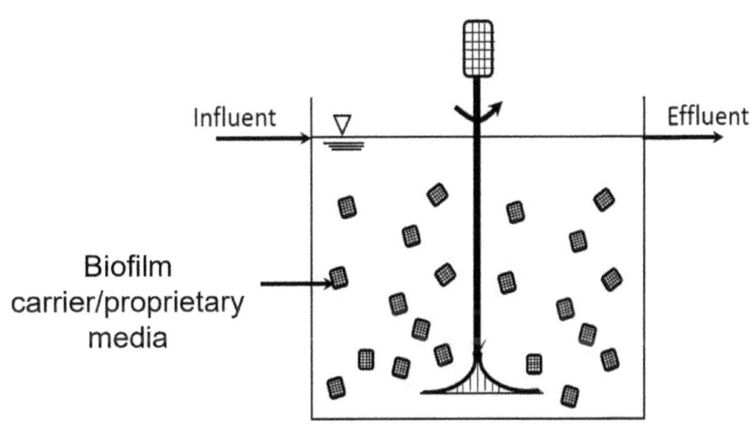

Figure 4-4 Simple schematic for a fixed-film suspended media bioreactor configuration

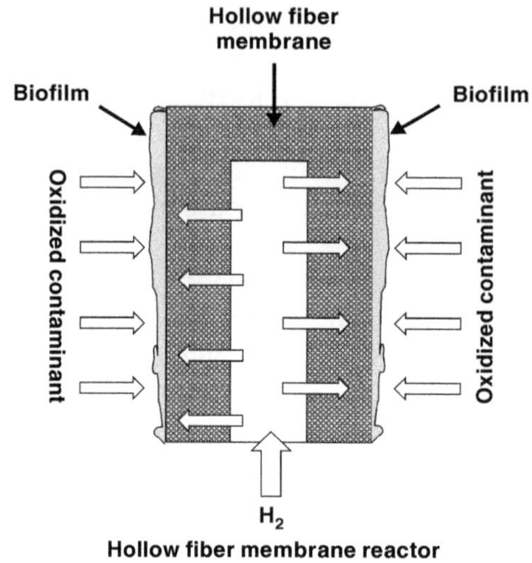

Figure 4-5 Simple schematic of membrane and biofilm in a membrane biofilm reactor

biomass accumulation is lower with autotrophic growth compared with heterotrophic growth, the membranes are occasionally backwashed to preclude the accumulation of excessive biomass. An example schematic of an MBfR membrane and its biofilm is shown in Figure 4-5.

4.3.2 Design Components

Irrespective of the process configuration, several critical design components should be carefully considered when designing a full-scale anoxic biological treatment system. These design components are discussed subsequently.

4.3.2.1 Media type. The media type used for anoxic biological treatment is selected on the basis of the reactor configuration. For example, MBfR configurations use membranes as the biofilm support medium, whereas FXB and FBR systems use granular media (e.g., GAC, sand, expanded clay, and anthracite). Special plastic biofilm carriers or proprietary media are typically used in fixed-film suspended media systems. Although all these media have been demonstrated to effectively support biological growth and contaminant removal, GAC may be preferred over the inert granular media when the target contaminant can be removed through adsorption and toxicity is a concern. With such contaminants, contaminant removal can be observed through adsorption on GAC before biological acclimation occurs. Contaminant adsorption on GAC can protect microorganisms by minimizing exposure to toxic concentrations if contaminant spikes occur. When the contaminant concentration in the bulk water becomes low, previously adsorbed contaminants can be released into the water, allowing the microorganisms to degrade contaminants at nontoxic concentrations.

4.3.2.2 Electron donor type. Anoxic biological treatment has been demonstrated with autotrophic and heterotrophic bacteria. As discussed in Section 4.1.3, heterotrophic systems are fed with organic electron donors, such as acetic acid. Autotrophic reactors, such as the MBfR, require an inorganic electron donor, such as H_2. Bicarbonate present in the raw water typically serves as the carbon source for biological growth in autotrophic systems. In the MBfR system, carbon dioxide may also be used to control pH in the system.

4.3.2.3 Electron donor dose. Detailed discussion on determining the substrate dose is presented in Section 4.1.3. Typically, the approximate electron donor dose is determined on the basis of influent DO and contaminant(s) concentration and the targeted effluent contaminant(s) concentration. Underdosing of substrate will likely result in effluent contaminant concentrations exceeding target levels, whereas overdosing of substrate can result in undesired reduction reactions (e.g., sulfate reduction to sulfide) and substrate breakthrough into the system effluent (if no downstream electron donor removal step is included). In principle, feed-forward and feedback control of chemical dosing should be used to avoid under- and overdosing the substrate. In the MBfR configuration, H_2 delivery is on-demand and requires that the delivery capacity be modestly in excess of the stoichiometric demand (Zhou et al. 2019, Tang et al. 2012).

4.3.2.4 Phosphorus dose. Bioavailable phosphorus (monitored as orthophosphate) is a critical nutrient for microbial growth and activity. When phosphorus is limited, bacteria can respond by producing excessive extracellular polymeric substances (EPS) (Liu et al. 2006), which can result in incomplete contaminant removal (due to low biomass) and/or hydraulic challenges (Mauclaire et al. 2004). In a pilot-scale study, phosphorus limitation resulted in incomplete nitrate removal (Figure 4-6). The pilot system received 30.5 mg/L acetic acid as the electron donor. When 0.35 mg/L orthophosphate as P was fed to the system, nitrate reduction improved, lowering the effluent concentration to below detection within two days. In another full-scale perchlorate-removing FXB bioreactor, phosphorus supplementation (100–150 μg/L orthophosphate as P) lowered terminal head loss from approximately 9 to less than 2 psi (Figure 4-7). The system received 18.7 mg/L ACA as the electron donor. In general, a 100:1 C:P molar ratio is a recommended starting point, but enzyme activity measurements (Keithley and Kirisits 2019) should be performed to confirm that phosphorus is truly limiting. Additionally, bench and/or pilot testing should be performed to confirm site-specific phosphorus dosing requirements.

4.3.2.5 Other chemical feed. Depending on the quality of the source water, other nutrients that serve as building blocks of microbial cells (e.g., N) might need to be supplemented to ensure biological growth and activity and to achieve consistent contaminant removal. Furthermore, most of the reduction reactions produce base (OH^-, Table 4-3), and pH may need to be adjusted for sustained contaminant removal. In general, sustained biological activity and contaminant removal are observed when the pH is

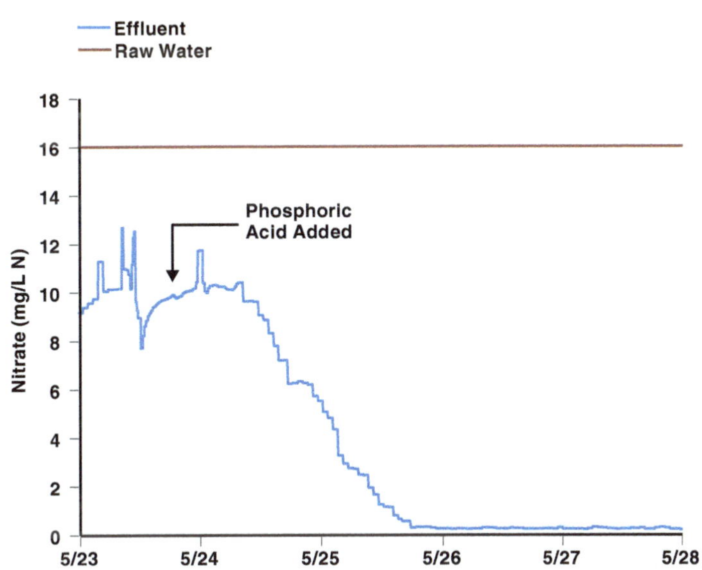

Source: Adapted from Brown et al. (2017)

Figure 4-6 Effect of phosphorus limitation on nitrate removal across a fixed-bed bioreactor

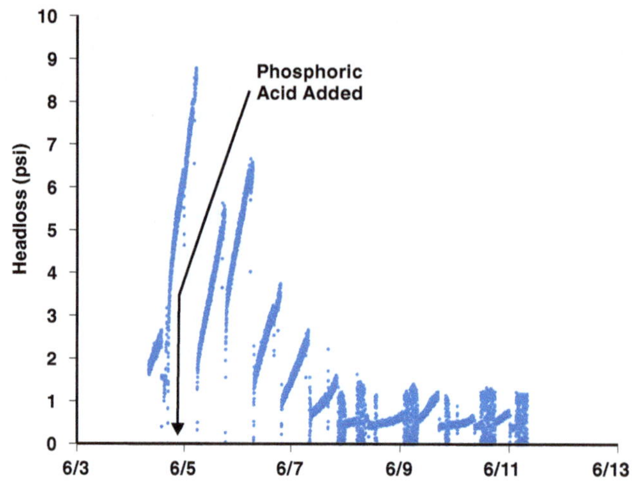

Source: Adapted from Brown et al. (2017)

Figure 4-7 Effect of phosphorus limitation on terminal head loss across a perchlorate-reducing fixed-bed bioreactor

maintained between 6.5 and 8 SU, though it is possible to have appreciable contaminant removal outside this range.

4.3.2.6 Contact time. Electron acceptor flux (i.e., *J*) or surface loading (i.e., *QSi*/*aV*) is the fundamental parameter for controlling contaminant removal kinetics, and the theoretical bioreactor volume can be determined as discussed in Section 4.2.2. Once *V* is established, it can be used to compute contact time, which is often empirically derived through pilot-scale testing. Testing for the determination of contact time (calculated as *V*/*Q* for most systems; for FXB systems, contact time is referred to as empty bed contact time [EBCT] and is calculated as the volume of the packed media bed/*Q*) and other design parameters should cover a wide range of seasonal changes, including temperature, contaminant loading, and water demand. Pilot testing should be considered to determine the required contact time for achieving targeted contaminant removal performance. In general, each application has a contact time threshold, beyond which additional contact time adds limited/negligible performance benefits.

4.3.2.7 Biomass control strategies. With the addition of an electron donor, biofilm accumulation can be significant, making biomass control methods essential. Biomass control prevents short-circuiting or excessive head loss without compromising overall biological activity and contaminant removal. Biomass control strategies include mechanical (e.g., fluidization and air scour) and chemical (i.e., oxidant addition) approaches. Biomass control strategies are established on the basis of the reactor configuration used, the equipment/system supplier specifications, and field testing.

4.3.2.8 Residuals handling. The main residual stream from anoxic biological treatment systems is the washwater produced during biomass control. If the washwater contains metals, the metals concentrations should be quantified to identify the available handling and disposal options. If the washwater does not contain metal or metalloid pollutants (e.g., Cr(III), Se°, U(IV), or arsenic) above locally regulated threshold levels, it may be discharged to the sewer or an onsite percolation basin contingent upon regulatory requirements. Waste volumes can be reduced through a settling- or floatation-based biomass separation step with recovered water recycled to the upstream of the bioreactor. If biomass separation is used, care should be taken to ensure that odors (e.g., sulfide) are not generated. Water production efficiency, which can be calculated as follows, can be as high as 95 and greater than 99 percent without and with a biomass separation step, respectively.

$$\text{Production Efficiency (\%)} = (V_W - V_T - V_B - V_{BW} + V_R)/V_W \times 100$$

where,

V_W = volume pumped from a well
V_T = volume treated but wasted
V_B = volume bypassed
V_{BW} = backwash volume
V_R = volume recovered through biomass separation and water recycling step

4.3.2.9 Posttreatment. Filtration and disinfection requirements downstream of an anoxic biological treatment process remain similar to surface water treatment processes. However, anoxic processes may also require a reoxygenation step because the redox conditions in the treated effluent can be considerably lower. Furthermore, if an organic electron donor is used, posttreatment should also remove any electron donor carryover. If hydrogen sulfide is generated during anoxic biological treatment, a sulfide removal step might also be required. Air stripping or oxidant (e.g., hydrogen peroxide or chlorine) treatment can be implemented for sulfide removal. For autotrophic systems (e.g., MBfR), recarbonation may also be required for pH adjustment.

4.4 OPERATION

Successful operation of an anoxic biological treatment process starts with the development of a detailed monitoring plan. The monitoring plan should clearly identify monitoring parameters, methods, sampling frequency and location, and sample collection and handling protocols. The plan should also detail safety protocols, data management, analysis, and reporting procedures. An O&M plan should also be developed that outlines expectations and protocols for the various phases of operation: startup and biological acclimation, optimization, steady-state operation, shutdown and restart, and trouble shooting.

4.4.1 Monitoring

Regular monitoring is critical for effectively assessing process performance and detecting and addressing any performance concerns. A monitoring plan should be developed with the key monitoring parameters and monitoring frequency. For accurate assessment of system performance, a combination of inline and grab sample analyses should be included in the plan. The use of inline monitoring tools/equipment is particularly important for anoxic biological treatment because contaminant removal and chemical feed systems need to be directly tied together through feed-forward and feedback algorithms to satisfy the contaminant removal and posttreatment requirements. Table 4-5 summarizes recommended monitoring parameters and their relevance.

A clear and organized data management plan should also be developed, including a schedule and well-defined protocol for inputting and analyzing data. Many of the parameters listed in Table 4-5 can be effectively monitored and recorded using supervisory control and data acquisition (SCADA) for real-time and long-term performance metrics. Grab sample results, generated either through in-house testing or testing at an external lab, should be compiled using software, such as Microsoft Excel or Microsoft Access. Data analysis tools, such as R and Python can be used for data management and regular generation of graphs, tables, trend plots, and statistical summary tables, which would help regularly assess treatment performance.

Table 4-5 Recommended monitoring parameters

Category	Parameter	Sampling Location	Sampling Method	Sampling Frequency	Performance Indication
Operating parameters	Flow/hydraulic loading rate	Inf	Inline	Continuous	Dictates contact time, which is directly related to biological contaminant degradation.
	Head loss/pressure drop	Across the system	Inline	Continuous	Allows the characterization of biofilm growth. In FXB configuration, clean-bed head loss signifies the efficacy of backwashing. The protocol and head loss accrual trend may help determine preferential channel formation. Pressure drop across a membrane signifies membrane fouling and the efficacy of membrane cleaning strategy.
	Backwash pressure	Downstream of the underdrain	Inline	Continuous	Indicates underdrain clogging in the FXB configuration if backwash pressure is increased.
	Water production efficiency*	-	Calculated	Continuous	Indicates water sent to distribution system compared with water pumped from the well.
Water quality	Temperature	Inf	Inline; grab	Continuous; daily	Affects biological growth and activity.
	pH	Inf, Eff	Inline; grab	Continuous; daily	Affects biological growth and activity. Biological reduction of oxidized contaminants can raise pH, requiring a pH adjustment.
	Alkalinity	Inf	Grab	Regular (three times per week)	Indicates buffering capacity of the water. A pH decrease or increase can be significant in water with low alkalinity. In autotrophic systems, alkalinity also reflects carbon availability for biogrowth.
	DO	Inf, Eff, downstream of reoxygenation step	Inline	Continuous	Affects substrate requirement, and near-complete DO consumption is required for anoxic biological treatment. Effluent DO monitoring allows rapid detection of performance anomaly (e.g., rising effluent DO indicates substrate feed failure). Monitoring DO downstream of reoxygenation step allows determining the reoxygenation efficacy.
	ORP	Inf, Eff	Inline	Continuous	Indicates electron acceptor utilization. Effluent ORP monitoring allows rapid detection of performance anomaly (e.g., rising effluent ORP indicates substrate feed failure).

(continued)

Table 4-5 Recommended monitoring parameters (*Continued*)

Category	Parameter	Sampling Location	Sampling Method	Sampling Frequency	Performance Indication
	Contaminant and intermediate product (e.g., nitrite) concentration	Inf, Eff	Inline, grab	Continuous; regular (e.g., two times per week)	Determines electron donor requirement. Monitoring contaminant and intermediate products (e.g., nitrite) concentrations is critical for regulating substrate feed.
	TOC/DOC	Inf, Eff	Grab	Regular (e.g., two times per week)	Indicates organic carbon, including the substrate in heterotrophic systems, present in the water.
	Turbidity	Inf, Eff	Inline	Continuous	Indicates biomass/solids present in treated water. Influent turbidity is an indicator of particle loading. Effluent turbidity monitoring helps assess the disinfection requirements.
	Electron donor concentration	Inf, Eff	Grab	Occasional (e.g., every month)	Affects biological growth and contaminant removal.
	Orthophosphate	Inf	Grab	Occasional (e.g., every month)	Affects biological growth and activity. Monitoring influent phosphorus is critical for tailoring nutrient dose.
	AOC	Eff	Grab	Occasional (e.g., every two to three months)	Indicates regrowth potential in distribution system.
	BDOC	Eff	Grab	Occasional (e.g., every two to three months)	Indicates regrowth potential in distribution system.
Microbial	HPC	Eff b pre- and post-disinfection	Grab	Regular (e.g., two times per week)	Indicates disinfection efficacy.
	Total coliform/ *E. coli*	Eff pre- and post-disinfection	Grab	Regular (e.g., two times per week)	Indicates disinfection efficacy.
	EPS	Growth medium	Grab	When hydraulic issues are observed	Indicates hydraulic performance. Hydraulic challenges are observed when excess EPS is produced.
	Biofilm morphology	Growth medium	Grab	When sustained hydraulic issues are experienced	Indicates microbial community and overall biomass health.
	Microbial community structure	Growth medium	Grab	When sustained hydraulic issues are experienced	Provides information on "who is present in the system?" Requires the use of molecular microbial methods; typically performed at specialty labs and universities.

(continued)

Table 4-5 Recommended monitoring parameters (*Continued*)

Category	Parameter	Sampling Location	Sampling Method	Sampling Frequency	Performance Indication
	ATP	Growth medium	Grab	During biological acclimation or system recovery	Indicates biological activity. Because biological growth and activity are intentionally enhanced in anoxic biological treatment, ATP monitoring is recommended only during the initial biological acclimation or during reacclimation.
Process integrity evaluation	Media characteristics	Growth medium	Grab	Regular (e.g., annual)	Allows assessment of particle size distribution and media loss in FXB and FBR configurations.
	Tracer testing	Across media bed in FXB configuration	Inline or grab	when abrupt changes in head loss or contaminant removal is observed	Allows determination of the presence of preferential channels.
	Chemical levels and feed flow	Upstream of the system	Inline	Continuous	Affects biological growth and activity and contaminant removal. Continuous monitoring chemical feed is critical for ensuring consistent treatment performance.
	H_2 demand versus actual delivery rate	Membrane in MBfR configuration	Inline	Regular (e.g., every three months)	Indicates membrane leakage or breakage in MBfR configuration. Serious leakage results in H_2 bubbling and can be dangerous.
	Inline analyzer maintenance	Inf, Eff	N/A	Regular (e.g., every week)	Allows achieving effective and consistent system performance.

Source: Adapted from Brown et al. (2017) *Blank cells in table indicate field not applicable.*

AOC—assimilation organic carbon, ATP—adenosine triphosphate, BDOC—biodegradable organic carbon, DO—dissolved oxygen, DOC—dissolved organic carbon, Eff—effluent, EPS—extracellular polymeric substance, FBR—fluidized-bed bioreactor, FXB—fixed-bed bioreactor, HPC—heterotrophic plate count, Inf—influent, MBfR—membrane biofilm reactor, ORP—oxidation-reduction potential, TOC—total organic carbon

*Production efficiency (%) = (volume pumped from a well – volume treated but wasted – volume bypassed – backwash volume + volume recovered through settling)/volume pumped from the well * 100).

Once the monitoring plan and O&M plan are developed, process operation and monitoring can be started. The operation phase can include the following subphases.

4.4.2 Initial Startup/Acclimation

Once the system is started, in most of the systems, microorganisms indigenous to the source water gradually establish on the growth support media (see Section 1.7). Depending on the target contaminant, it may take days (e.g., nitrate) to weeks (e.g., selenium) to observe appreciable contaminant reduction. Afterward, contaminant degradation improves exponentially until complete biological acclimation. The acclimation duration can be lowered by operating the system with a longer contact time (i.e., reduced hydraulic loading rate) or by recirculating a fraction of the treated effluent. It is critical to avoid substrate and nutrient limitation during biological acclimation. Once the system is fully biologically acclimated, the design operating conditions should be practiced.

4.4.3 Steady-State Operation

In general, after complete biological acclimation, steady-state system performance can be assumed if system performance is relatively stable (both in terms of contaminant removal and hydraulic performance) under relatively steady influent quality conditions. The steady-state phase should exclude process optimization and troubleshooting periods. A constant dose equal to or slightly higher than the stoichiometric requirement should be fed during the acclimation phase. Underdosing or overdosing can be controlled through the implementation of feed-forward and feedback algorithms programmed around calculated demand and inline DO and nitrate analyzers. Though operator intervention should be minimal during steady-state operation, it is critical to closely monitor the system to ensure early detection of any performance deviations so that corrective measures can be timely implemented. A well-developed monitoring plan and data management protocol (Section 4.4.1) should be used to characterize baseline performance and allow for identification of any subsequent performance deviations. Additionally, standard operating procedures should be developed and implemented for maintaining and calibrating chemical feed systems and analyzers.

4.4.4 Shutdown And Restart

Similar to any water treatment process, anoxic biological treatment systems perform the best under continuous, steady-state operation. However, intentional (e.g., for scheduled maintenance or due to low-flow conditions) and unintentional (e.g., emergency maintenance and repair) shutdowns are unavoidable. Shutdowns should be minimized to the extent possible. In the absence of continuous supply of fresh electron donor and electron acceptor, the biological health of any anoxic system might diminish, especially with extended shutdown. Although denitrification processes can withstand shutdowns up to 45 days with minimal effects and rapid recovery after the shutdown (Brown et al. 2016), shutdown greater than 48 hours might significantly affect other contaminants (e.g., hexavalent chromium) removing anoxic biological processes (Upadhyaya et al. 2019). Extended shutdowns could eventually result in unpleasant odors and can necessitate a reacclimation period upon restart. To address these issues, recirculation and/or treat-to-waste options can be considered when restarting the system. Because anoxic biological activity can be affected because of desiccation if stored dry during a shutdown, the biotreatment processes should be stored wet (i.e., filled with water). Dead biomass could pass through to the system effluent when the system is restarted after a shutdown, resulting in a turbidity spike and/or increased hydraulic pressure across a bioreactor. To minimize these issues, a biomass control step (e.g., backwashing) might be considered before restart. However, an overly aggressive biomass control step can lengthen the reacclimation period; site-specific testing should be implemented to establish an appropriate biomass control step before restart.

Table 4-6 Examples of optimization strategies for full-scale anoxic biological treatment systems

Strategy	Explanation
Electron donor dose	Electron donor dose optimization does not require additional investment of resources.
Phosphorus dose	Phosphorus dose optimization does not require additional investment of resources.
Contact time	With a reactor volume or bed volume, changing the flow rate allows evaluating different contact times. In an FXB configuration with sampling ports along the depth of the bed, various contact times can be evaluated with a single flow rate.
Biomass control strategies	Biomass control strategy options and optimization will be specific to the configuration used. Strategies may include fluid-induced-shear (e.g., backwashing), air-induced shear (e.g., air scour), or oxidant dosing (e.g., chlorine or hydrogen peroxide) for (direct biomass oxidation), chemical treatment.
Posttreatment requirements	For a new facility, the design is expected to account for the posttreatment needs listed in Section 4.2.3.9.

FXB—fixed-bed bioreactor

4.4.5 Optimization

Full-scale optimization should focus on verification of optimization strategies evaluated during pilot testing (Section 4.2.3). This includes the optimization of substrate and nutrient dose, contact time, biomass control strategy, and posttreatment processes. Full-scale process optimization should occur after complete biological acclimation with relatively steady system performance. It has been observed that the biomass control strategy developed for anoxic processes during pilot testing may not scale well. Therefore, the biomass control strategy should be re-evaluated and optimized at full scale. Long-term operation at full scale may also reveal optimization strategies that were not observed or anticipated during pilot testing (e.g., effects of electron donor carryover, chemical feed pump maintenance). Table 4-6 provides examples of full-scale anoxic biological treatment system optimization strategies and level of efforts required for the evaluation. The strategies are listed in recommended order of considerations based on relative effort.

4.4.6 Troubleshooting

Operational challenges most often observed with anoxic biological treatment systems are caused by (1) chemical feed system failures/malfunctions, (2) poor reactor hydraulics caused by inadequate biomass control strategies, and (3) biofouling of any downstream process and on-line instrumentation. Troubleshooting options are technology/process-specific and will likely be specified by the equipment/process manufacturer.

4.5 CASE STUDIES

The technical focus of the case studies presented in this section is on the biodenitrification process only. Posttreatment exists for each of these systems and is shown in the respective process flow diagram. Discussion of the posttreatment unit processes is beyond the scope of this MOP.

The case studies presented in this chapter represent a variety of applications regarding process design, plant capacity, and geographic location.

Table 4-7 Biodenitrification case studies

Facility, Location	Treatment	Capacity
Crescenta Valley Water District, La Crescenta, Calif.	Membrane biofilm reactor	150 gpm
City of Taylorville, Taylorville, Ill.	Two-stage, fixed-bed biological NRS Packaged plant (biottta®)	1,050 gpm
Sdot Yam Drinking Water Plant, Kibbutz Sdot Yam, Israel	MBBR	0.35 mgd
Sunny Slope Water Company, Pasadena, Calif.	MicroNiche Engineering (MNE)	350 gpm

MBBR—moving-bed biofilm reactor, MNE—MicroNiche Engineering, NRS—nitrate removal system

Table 4-8 Raw water quality

Parameter	Unit	Concentration/Level
Temperature	°C	15–20
pH	SU	6.1–7.3
Nitrate	mg/L N	8–11
Nitrite	mg/L N	ND
DO	Mg/L O_2	4–5
Phosphorus	mg/L PO_4-P	ND–0.06

DO—dissolved oxygen, ND—not detected

Descriptions of these facilities are presented in this chapter, such as process flow diagrams, raw water quality, treatment goals, and performance of each facility.

These facilities are listed in Table 4-7, in which location, treatment type, and capacity are outlined.

4.5.1 Case Study – CS01, MBfR

Crescenta Valley Water District, La Crescenta, Calif.
Author: David Friese

4.5.1.1 Facility description. Crescenta Valley Water District's Well 2 was taken out of service in 1976 due to nitrate concentrations over the MCL (Table 4-8). A 150-gpm MBfR (APTwater's ARoNite) system was designed and constructed in 2018 to treat the groundwater contaminated with 11-12 mg NO_3-N/L. The treatment train (Figure 4-8 and Figure 4-9) consists of an ARoNite reduction zone, an aeration and coagulation chamber, and a mono-media filtration unit (size 0.4–1.4 mm, uniformity coefficient of 1.7) in series. The treatment system is highly automated with remote monitoring and control and requires site visits lasting about 1 hour two or three times per week. The ARoNite reduction zone consists of 48 hollow-fiber membrane modules (Figure 4-10). The process treats a portion of the contaminated groundwater to less than 2 mg-N/L, resulting in less than 7 mg-N/L after blending with untreated groundwater. Since this was the first of its kind installation, the design was purposely conservative compared with optimized systems; design parameters are in Table 4-9. A demonstration test was conducted for 21 days, which led to a conditional operational permit from the California Division of Drinking Water in June 2019. The system has been in production since June 2019.

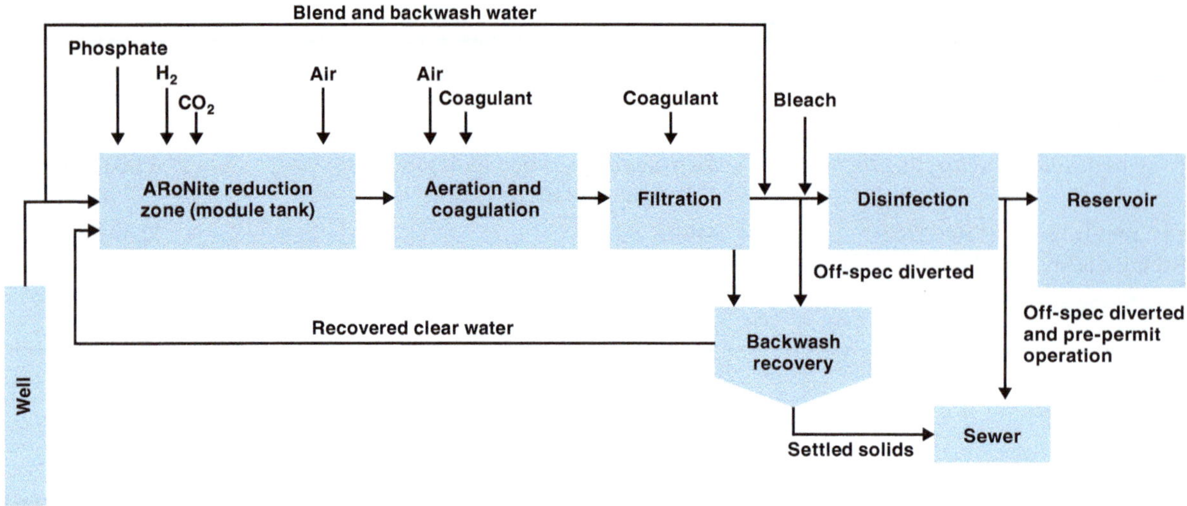

Figure 4-8 Process flow diagram for the membrane biofilm reactor system

Figure 4-9 The membrane biofilm reactor setup

4.5.1.2 Facility treatment goals. The primary goal is to treat nitrate to less than 7 mg/L N in the blended water. In addition, filtration to less than 0.3 NTU and disinfection to 4-log virus inactivation are state-required.

4.5.1.3 Operation and performance. The MBfR treats 60 gpm out of the total well flow of 150 gpm. Hydrogen gas, produced on site and on demand, is used as the electron donor. The hydrogen is generated from deionized water in an electrolysis unit, and its on-demand delivery precisely controls the nitrate reduction in the module, lowering nitrate concentrations in the treated effluent to less than 2 mg/L N (Figure 4-11). Orthophosphate (0.1–0.2 mg P/L) is dosed into the reduction zone for enhanced biological growth. pH control is achieved by feeding CO_2, as needed. The effluent of the reduction zone passes through the aeration and coagulation chamber, in which the water is reoxygenated and aluminum chlorohydrate coagulant is added to aid turbidity removal in the downstream filters. The filtration unit effectively lowers turbidity in the treated effluent to 0.05–0.12 NTU with filter run times ranging from 100 to 240 hours at a filter loading of

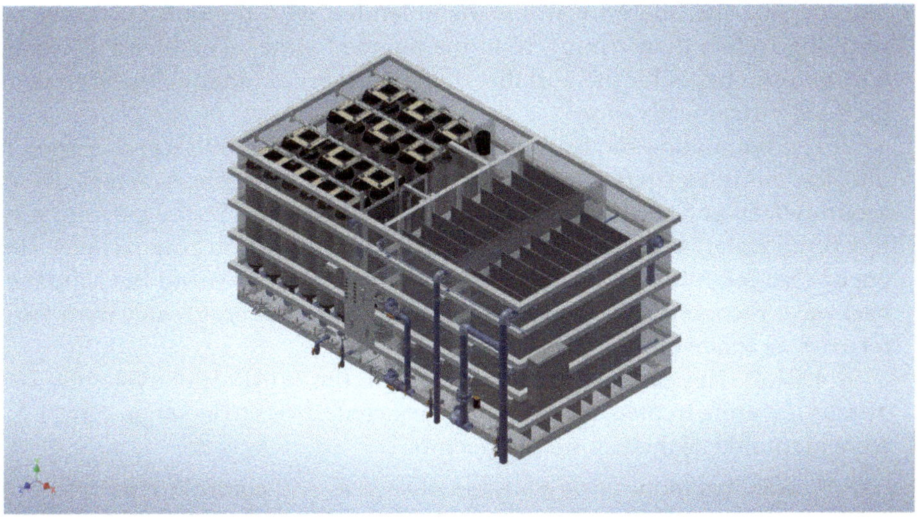

Figure 4-10 A 3-D model of the membrane biofilm reactor system

Table 4-9 Biological treatment design criteria

Parameter	Unit	Case Study Design Criteria	Optimized Systems
Nitrate removal flux	g N/m^2-d	0.6	1.5–4
H_2 delivery Flux	g H_2/m^2-d	110–130% of the total acceptor demand*	110% of the total acceptor demand*
Reduction zone HRT (resulting from removal flux, module size and arrangement)	min	60	10–15
Aeration (fast mix) HRT	min	2	0.5
Coagulation (slow mix) HRT	min	15	10
Aerobic media filtration loading	gpm/ft^2	3	5–15

HRT—hydraulic residence time

*The total acceptor demand is computed from stoichiometry for the reductions of NO_3^-, NO_2^-, plus O_2. The H_2 pressure in the lumen must be great enough to provide the needed H_2-delivery flux.

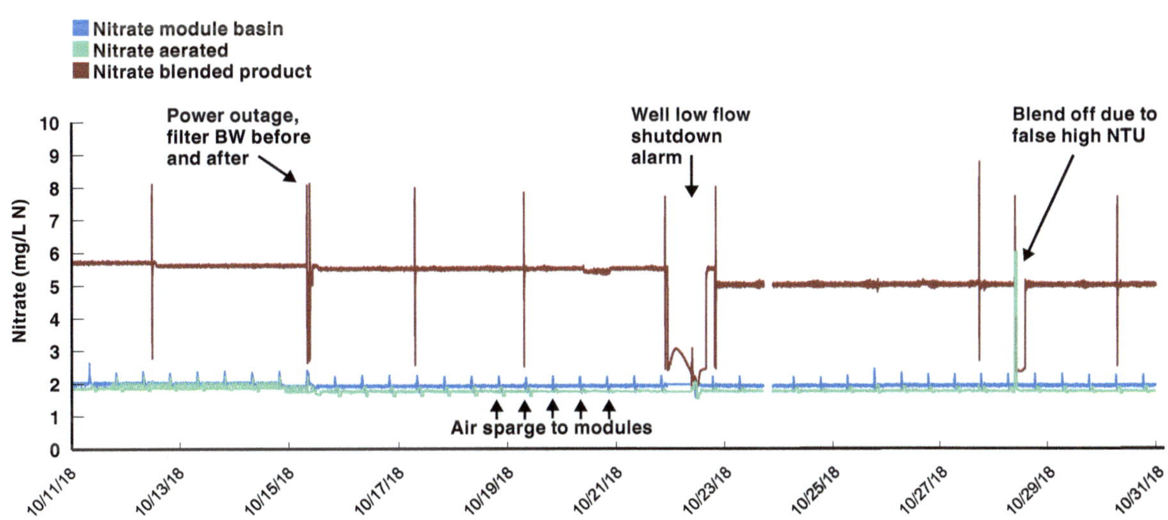

BW—backwash, NTU—nephelometric turbidity units

Figure 4-11 Example of consistent nitrate removal in the membrane biofilm reactor

3–4 gpm/ft^2. The filtered water is then blended with the untreated well water (90 gpm), resulting in less than 7 mg/L N in the blended water. The blended water is chlorinated with sodium hypochlorite and directed to a reservoir after chlorine contact time required for 4-log virus inactivation.

The membrane modules are cleaned by using a 30-second compressed air sparge into the modules every 12 hours. The filters are then backwashed using the untreated groundwater, and the backwash wastewater is sent to a backwash recovery chamber. The recovered water is recycled back to the ARoNite reduction zone, whereas the settled solids are discharged into a sewer. With the long filter run times and the solids separation in the backwash recovery and recycle of decanted clear water, greater than 99.8 percent water recovery is achieved.

4.5.1.4 Key success factors. Overall, the MBfR pilot system effectively removed nitrate, lowering treated effluent nitrate concentrations to less than 7 mg/L N. The following were identified as the key success factors:

- effective monitoring with remote access and control to the system and
- onsite hydrogen generation and on-demand delivery, which minimize costs and eliminate shipping and storage.

4.5.2 Case Study – CS02, Two-Stage, Fixed-Bed Biological Nitrate Removal System (NRS) Packaged Plant (biottta®)

City of Taylorville, Ill.
Authors: Christopher E. Clark P.E. and Martin A. Lawrence

Nitrate in the groundwater in the City of Taylorville (City), a small community of approximately 11,000 residents located southeast Springfield, Ill., has exceeded the MCL of 10 mg/L N over time. Due to limited alternative water sources, the City decided to treat the existing groundwater wells. An initial feasibility study revealed that ion exchange (IX) would generate concentrated waste streams with nitrate and chloride concentrations that were higher than the wastewater facility could process. Since biological treatment has been demonstrated to effectively remove nitrate, a two-stage, FXB biotreatment option was considered as the potential treatment alternative to the IX process.

A two-stage, FXB biological treatment system, consisting of two pressure vessels in series, was selected for nitrate removal from the water. In this system, acetic acid and phosphoric acid are fed to the raw water. Naturally occurring microorganisms remove nitrate in the Stage 1 (bioreactor). The bioreactor effluent is supplemented with hydrogen peroxide (for reoxygenation) and a filter-aid polymer (for turbidity removal). The Stage 2 (aerobic biofilter) polishes the bioreactor effluent by removing turbidity and any organic carbon carryover from the bioreactor.

A three-month pilot study (2,600 gallons per day [gpd]) was conducted, demonstrating nitrate removal to less than 0.06 mg/L N and system resiliency with 4–8 min EBCT. The pilot testing developed design and operating criteria for a full-scale FXB system. The pilot-scale testing demonstrated that the City would be able to treat half of the well flow and meet the treatment goal of 7 mg/L N in the blended water.

After thorough assessment of the pilot-scale results, the Illinois Environmental Protection Agency (IEPA) permitted the construction of a 5.2-mgd combined surface water and groundwater treatment full-scale facility in January of 2017.

4.5.2.1 Facility description. A full-scale treatment plant (Figure 4-12) was designed and built to treat 1,511 gallons per minute (gpm) surface water and 2,100 gpm groundwater (totaling 3,611 gpm). The treatment train consists of the two-stage, FXB biological treatment system for nitrate removal from the groundwater. The groundwater treatment system consists of two treatment trains (Trains A and B with lead/lag vessels)

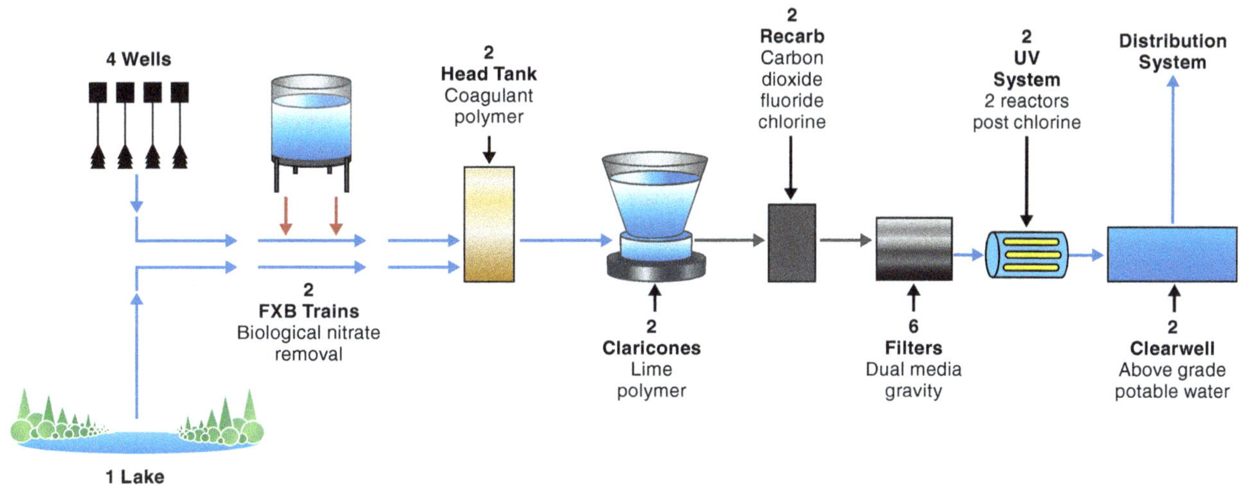

Figure 4-12 Process flow diagram for City of Taylorville, Ill., Water Treatment Facility

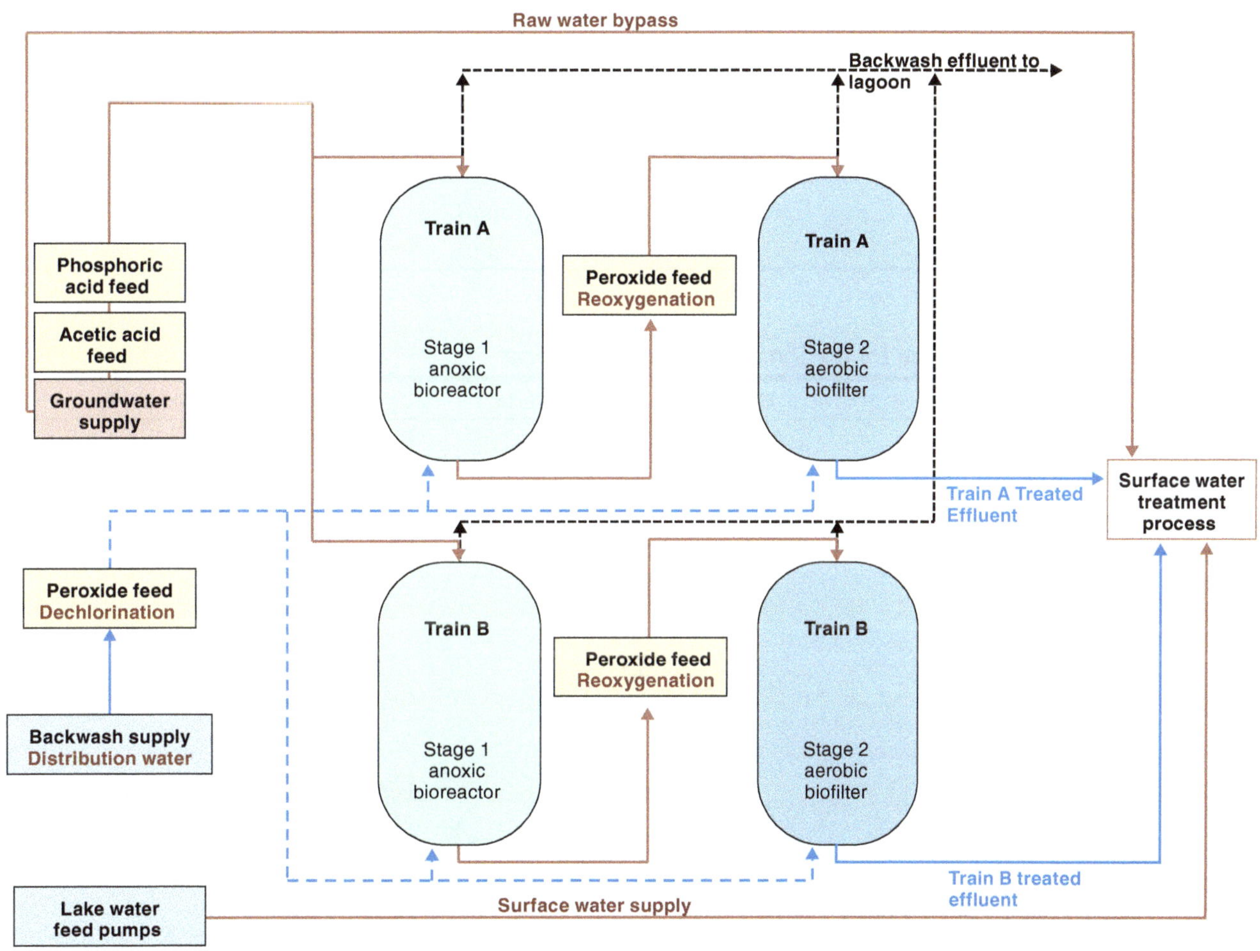

Figure 4-13 Process flow diagram

with chemical feed systems (Figure 4-13). The groundwater treatment train is followed by surface water treatment train, which includes coagulation, clarification, recarbonation, filtration, and ultraviolet disinfection. Due to the presence of the downstream processes, filter-aid polymer was not fed to the bioreactor effluent.

The FXB system treats a maximum of 1,050 gpm groundwater, removing nitrate from 13 mg/L to less than 1 mg/L N, and the remaining well flow is bypassed around the FXB, which results in 7.0 mg/L N nitrate or less in the blended water. The treated and blended groundwater is then blended with the surface water and directed to the surface water treatment processes.

Table 4-10 presents groundwater quality. The raw water contains 8–13 mg/L N nitrate with elevated salinity. Table 4-11 presents the design and operating criteria for the FXB system.

4.5.2.2 Facility biological treatment goals. The treatment goals for the FXB, two-stage biological treatment system include

- sustained removal of groundwater nitrate to less than 1 mg/L N and maintaining less than 7 mg/L as N in the blended water; and
- maintaining concentration of 2.5 mg/L DO or more in the FXB effluent.

Table 4-10 Groundwater quality

Parameter	Location	Unit	Concentration (Average±S.D.)
Temperature	Combined influent (i.e., blended water from the three wells)	°C	8.0–13.0
pH		SU	6.9–7.2
Nitrate		mg/L NO_3-N	8.0–13.0
Phosphorus		mg/L PO_4-P	0.15–0.16
DO		mg/L	2.5–2.7
Chloride		mg/L	122

DO—dissolved oxygen

Table 4-11 Fixed-bed bioreactor design and operating criteria

Parameter	Design/Operating Criteria	Remark
Treatment train (count)	2	Parallel treatment trains
Pressure vessels/train (count)	2	Lead (bioreactor – Stage 1)/lag (biofilter – Stage 2)
Flow rate/train (gpm)	525	
Loading rate (gpm/ft^2)	4.6	
Stage 1 EBCT (min)	8.1	
Stage 2 EBCT 2 (min)	6.5	
Chemical addition		
Mass ratio of acetic acid to nitrate N	5.0	
Typical acetic acid dose (mg/L)	43.0	Fed as electron donor and C source for microbial growth
Typical phosphoric acid dose (mg/L H_3PO_4)	1.9	Fed as micronutrient to support microbial growth
Typical hydrogen peroxide dose (mg/L)	7.0	For reoxygenation, targeting ≥2.5 mg/L DO in the biofilter effluent
Backwash water	Hydrogen peroxide	For dechlorination and biomass control
EBCT (min)		
Stage 1	8.1	
Stage 2	6.5	

DO—dissolved oxygen, EBCT—empty bed contact time. *Blank cells in table indicate field not applicable.*

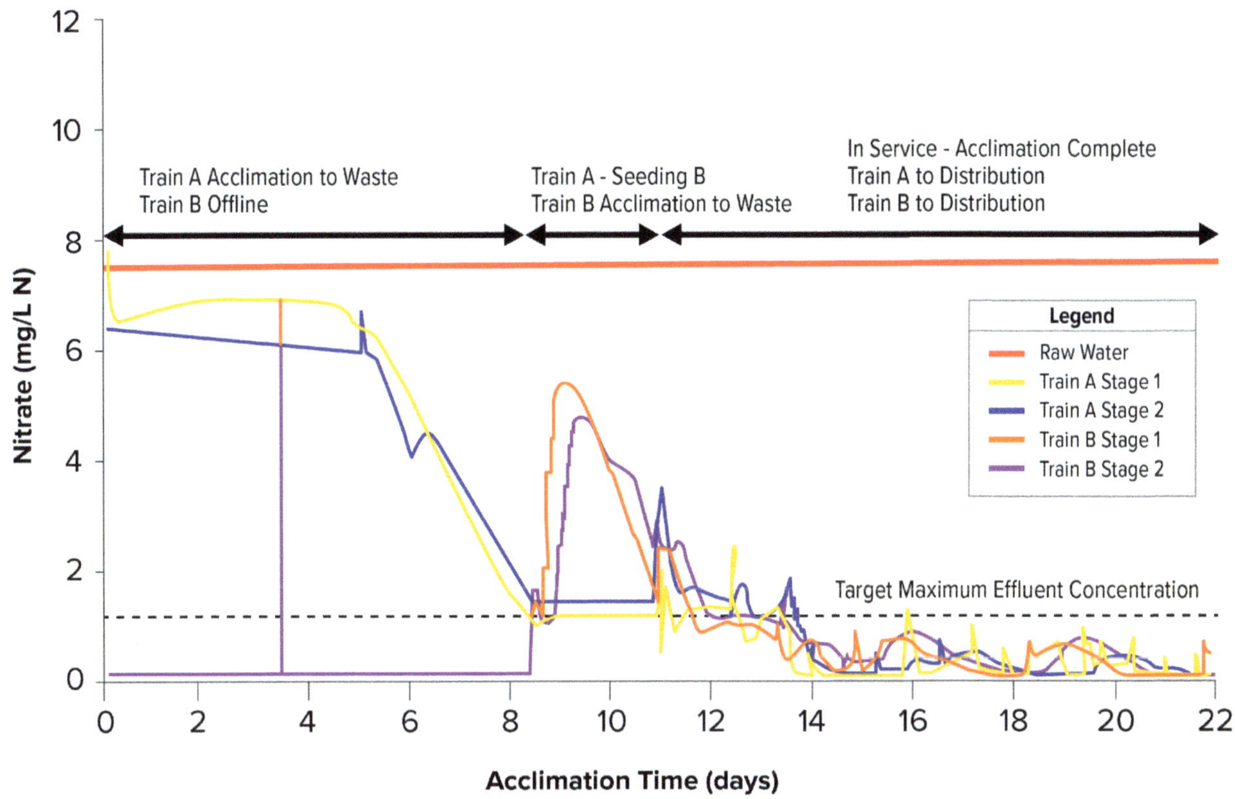

Figure 4-14 Effluent nitrate concentrations in Trains A and B during biological acclimation

4.5.2.3 Operation and performance.

4.5.2.3.1 Train A acclimation. Full-scale operations began on Dec. 9, 2019, with continuous operation of only Train A, treating 525 gpm (i.e., 25 percent of the well flow). Significantly faster biological acclimation was observed in Train A (i.e., within 8.5 days; Figure 4-14) compared with the pilot system (14 days). The faster acclimation resulted in 39 percent waste volume reduction (5.4 million gallons wasted) compared with the pilot system. The Train A acclimation phase is summarized as follows:

- Flow rate was 525 gpm; EBCT was 8.1 min.
- Excess acetic acid and phosphoric acid were injected (i.e., 10–15 percent more than stoichiometric requirements) to avoid substrate and nutrient limitation.
- Stage 2 hydrogen peroxide feed started after near-complete acclimation.
- Train A biofilter effluent was routed through Train B for seeding and removing acetic acid and phosphoric acids carried over from Train A due to excess feeding.
- The bioreactor and biofilter were not backwashed during the acclimation.
- Treated effluent was directed to lagoons.

4.5.2.3.1 Train B acclimation. Biological acclimation of Train B started immediately after achieving complete acclimation in Train A. The train was continuously fed with the well water (525 gpm). Because of the seeding during Train A acclimation, Train B acclimation was complete within 2.5 days (Figure 4-14) with only 1.9 million gallons wasted, which resulted in 86% and 71% waste volume reduction compared with the pilot system and Train A, respectively. The Train B acclimation phase remained similar to that in Train A, except that Train B effluent was not routed to Train A.

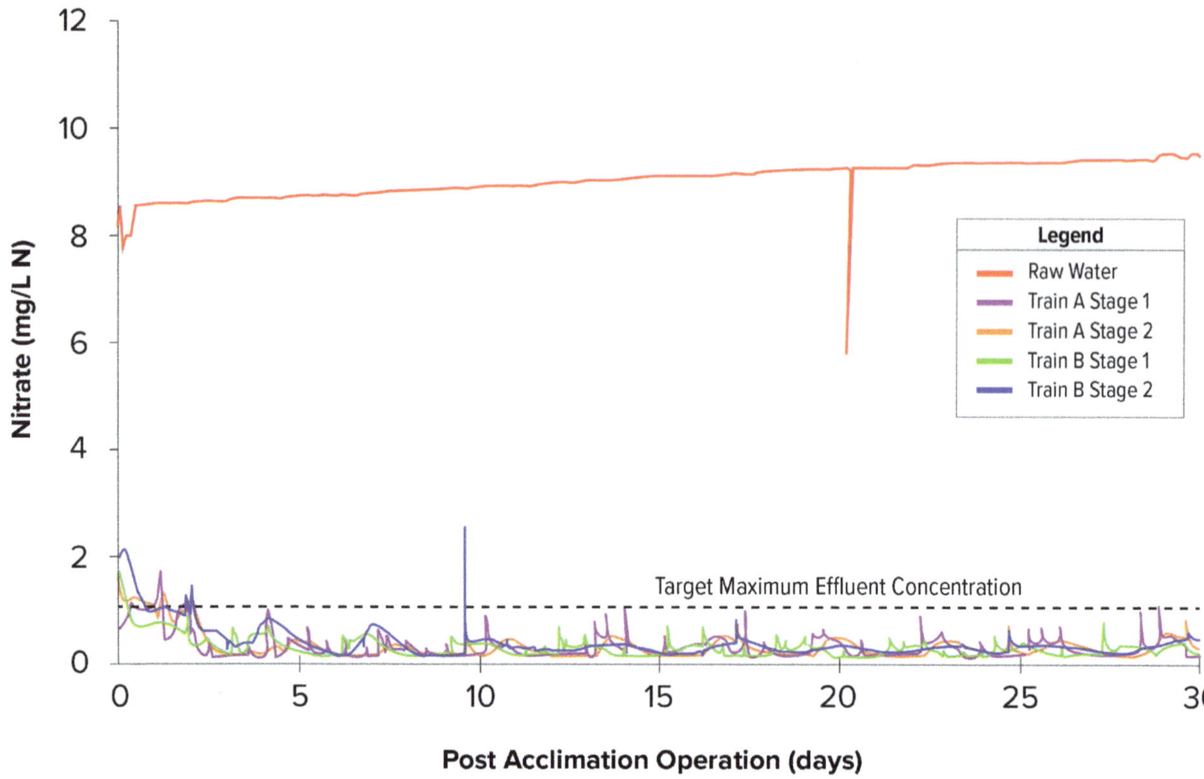

Figure 4-15 Nitrate concentrations in the raw water, and State 1 and Stage 2 effluent during the steady-state operation period

4.5.2.3.1 Steady-state operation with potable discharge. Full-scale operation began immediately after complete acclimation of B. The plant has been operating 24/7 with 0.3 mg/L NO_3-N in the biofilter effluent (Figure 4-15) despite the continuously rising raw water nitrate concentrations. Nitrate in the blended groundwater (i.e., FXB treated plus FXB bypassed) consistently remains less than 7.0 mg/L NO_3-N. The steady-state operation phase is summarized as follows:

- Trains A and B had a flow rate of 525 gpm and EBCT of 8.1 min.
- Design doses of acetic (43 mg/L) and phosphoric (1.9 mg/L) acids were used.
- Design dose of Stage 2 hydrogen peroxide (7.0 mg/L) was used, maintaining 2.5 mg/L DO or more in the FXB effluent.
- FXB effluent consistently contained the target nitrate concentration (i.e., less than 1.0 mg/L NO_3-N).
- FXB groundwater bypass was maintained and monitored by Plant SCADA.
- Periodic "gas release" (i.e., every 12 hours) was performed in Stage 1 to eliminate head loss accrual due to N_2 gas buildup in the media bed.
- Blended groundwater was further blended with surface water and directed to downstream surface treatment processes.
- System performance was evaluated with peak and 50% well flows within 30 days after complete biological acclimation. The changes in the feed flow rate did not affect system performance.

4.5.2.4 Key success factors and lessons learned. In summary, directing the treated effluent of Train A to the influent of the Train B when acclimating Train A helped expedite the acclimation in Train B. Sustained nitrate removal was achieved, maintaining

N concentration below the target in the blended treated effluent. The following were the major success factors and lessons learned during this study:

- 24 years of testing (Brown et al. 2002) and scaleup plus the benefit of learning from two prior full-scale biottta installations, all integrated into the Taylorville facility;
- inline water quality and hydraulic monitoring integrated into automated operation to improve chemical dosing and real-time performance feedback; and
- routing of effluent from one train through parallel trains during startup to minimize the biological acclimation period.

4.5.3 Case Study – CS03, Moving-Bed Denitrification (MBDEN)

Sdot Yam Drinking Water Plant, Israel
Author: Keren Nof

A production well, supplying both irrigation and drinking water to a small community in Israel, was shut down as nitrate concentrations exceeded the Israeli regulated drinking water level of 16 mg/L NO_3-N. Following a site analysis, biological treatment was selected as the most practical and cost-effective treatment option, and a full-scale MBDEN treatment plant was implemented for nitrate removal from the contaminated well in 2013.

4.5.3.1 Facility description. The full-scale treatment plant is located in Sdot Yam, Israel, with a total footprint of 6,458 ft^2. This area is characterized by significantly variable seasonal water demands. The plant has a total average design capacity of 0.21 mgd and a maximum design capacity of 0.35 mgd. The plant has been supplying irrigation and drinking water to more than 1,000 people since 2014. Well water quality characteristics are presented in Table 4-12.

To provide an economical and sustainable solution for the small community, a "treat and blend" approach was designed (Figure 4-16). The contaminated groundwater is pumped through a hydrocyclone for solids separation and then divided into two streams: "treated" and "untreated." The "treated" stream passes through the biological treatment unit, which consists of two-stage moving-bed biofilm bioreactor (MBBR) process (Figure 4-17). The effluent of the biological process is directed to the UF unit and then to the GAC filter for filtration and additional polishing before disinfection. The combination of ultrafiltration (UF) and GAC was an original safety requirement by the Israeli authorities during plant startup for TOC and odor concerns. The treated stream is then blended with the "untreated" stream before disinfection, resulting in a supplied drinking water containing nitrate below the regulated concentration (i.e., 16 mg/L NO_3-N). Typical blend ratios (i.e., treated to untreated water) range from 1:2 to 1:3 during winter months and 1:2 to 3:8 during summer months.

The biological treatment unit consists of a two-stage MBBR process: DEOX and ANOX, which are both filled with Aqwise Biofilm Carriers made of high-density polyethylene (HDPE). Biomass in the reactor grows as a biofilm on the carriers, which have an effective surface area of 650 m^2/m^3 and a highly open external design to promote oxygen and nutrient transfer. The carriers are kept in suspension using mechanical mixers,

Table 4-12 Raw water quality

Parameter	Unit	Concentration/Level
Temperature	°C	20–22
pH	SU	7.0–8.0
Nitrate	mg/L NO_3-N	17–23
DO	mg/L O_2	5.0–8.0

DO—dissolved oxygen

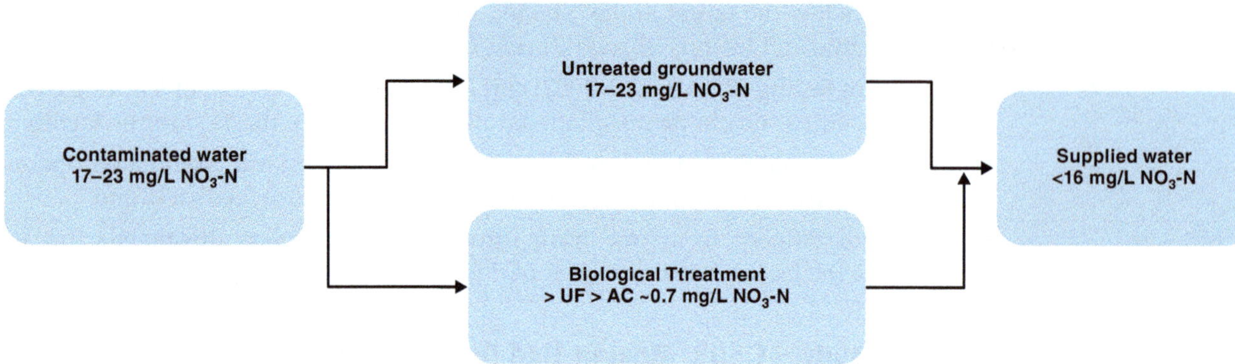

AC—activated carbon, UF—ultrafiltration

Figure 4-16 Treat and blend approach

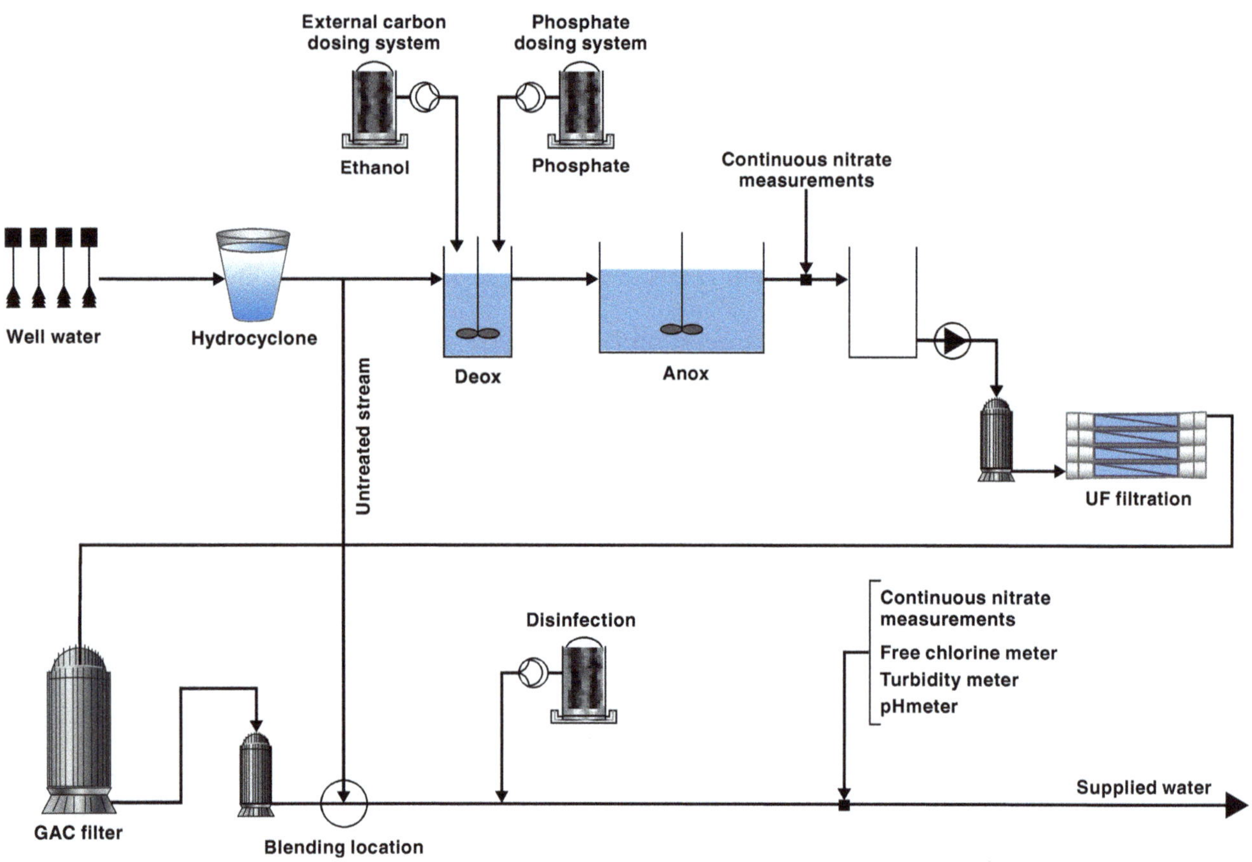

GAC—granular activated carbon, UF—ultrafiltration

Figure 4-17 Process flow diagram for Sdot Yam Drinking Water Plant

which support constant introduction of substrate and oxygen to the biomass. The mixing causes excess biomass to slough off from the carrier and continue to the UF, preventing system clogging and eliminating carrier cleaning and backwash. The first stage, DEOX, reduces oxygen and establishes initial denitrification. Ethanol, used as an electron donor/carbon source, and phosphorus (as orthophosphate) are added to this stage for enhancing the growth and activity of heterotrophic bacteria. In the second (ANOX) stage, further denitrification occurs, lowering nitrate to less than 0.7 mg/L NO_3-N. Table 4-13 presents the design criteria and water treatment goals for the system.

Table 4-13 Design criteria and water treatment goals for the Sdot Yam Treatment Plant

Parameter	Criteria
HRT (hours)	
Stage 1 (DEOX)	0.5–1.0
Stage 2 (ANOX)	1.5–2.0
pH	6–8
Nitrate (mg/L)	
Treated water	~0.7
Blended water	<16
DO (mg/L)	
Stage 1 (DEOX) effluent	0.5–1.0
Stage 2 (ANOX) effluent	<0.5

ANOX—, DEOX—

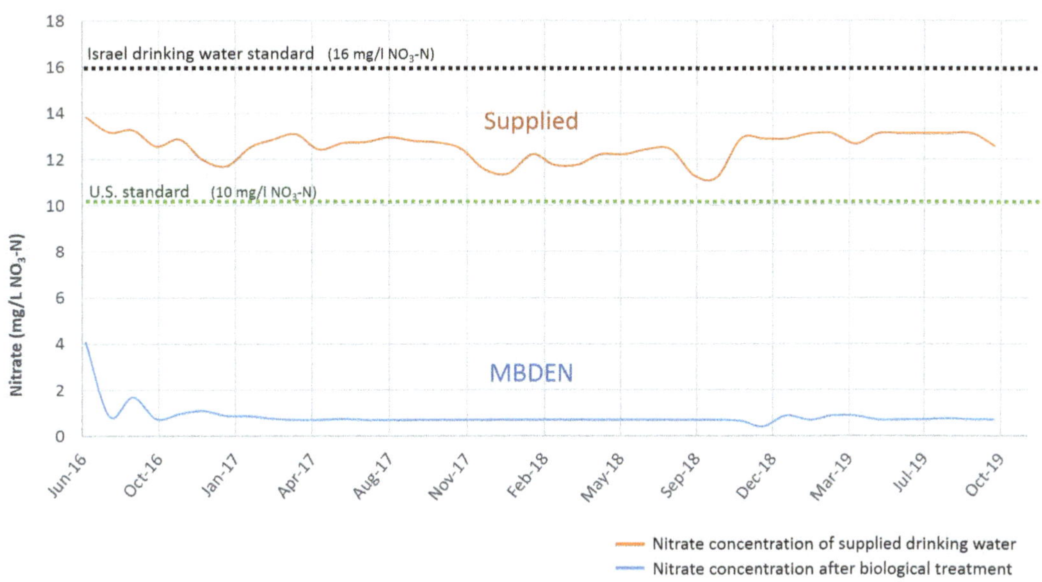

MBDEN—moving-bed denitrification

Figure 4-18 Nitrate concentrations in the MBDEN-treated water and supplied water

4.5.3.2 Operation and performance. Figure 4-18 shows the MBDEN-treated and blended nitrate concentrations. The MBDEN efficiently and consistently lowered the nitrate to below 0.7 mg/L NO_3-N, and the blended (i.e., supplied) water was consistently less than 16 mg/L NO_3-N. The MBDEN system is capable of consistently reducing nitrate even during significant seasonal flow fluctuations observed in this location (Figure 4-19).

Microbial parameters are regularly monitored in the disinfected water. Since operation began, *Pseudomonas*, *Streptococci* faecalis, fecal coliform, and total coliform have not been detected in the distributed water, while HPCs remained less than 20 CFU/mL. Turbidity in the supplied drinking water remained below 0.2 NTU, which was well below the drinking water standard in Israel (i.e., 1 NTU).

4.5.3.3 Key success factors and lessons learned. The system was designed with the combination of the UF and GAC filter as per the government's requirement. However, the plant performance has proved that a single-stage filtration downstream of the biological

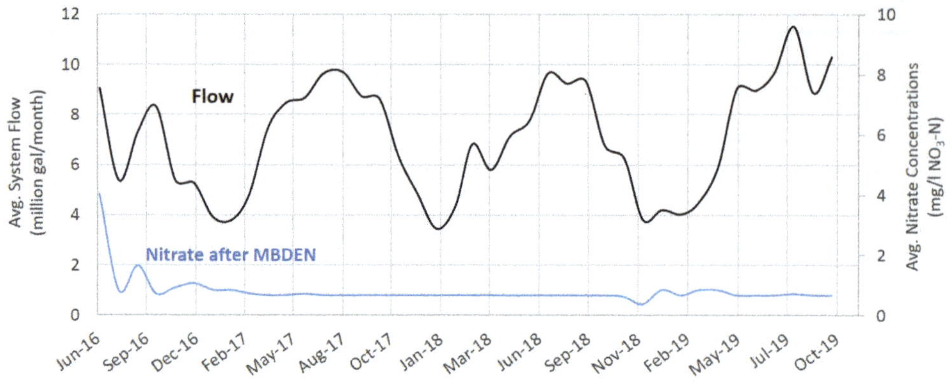

MBDEN—moving-bed denitrification

Figure 4-19 MBDEN nitrate removal performance during seasonal flow fluctuations

treatment will be enough. This will significantly lower capital and O&M costs. Key lessons, including the following, have been learned through process optimization:

- Feeding ethanol directly below the water surface in the DEOX rather than from the top of the MBBR allows avoiding ethanol loss.
- Review of historical operational and performance data (i.e., from the last five years) allows optimizing carbon and nutrients doses and avoiding chemical carry-over downstream of the MBBR unit. This allows lowering the O&M costs.
- Mixing speed can be optimized for maintaining good carrier mixing while sloughing of excess biomass.

4.5.4 MicroNiche Engineering System for Nitrate Removal

Sunny Slope Water Company *Pasadena, Calif.*
Authors: Ali Dorri and Ameen Razavi

4.5.4.1 Background. Sunny Slope Water Company (SSWC; Pasadena, Calif.), provides water to over 30,000 households, primarily from its own wells in the Raymond Basin and the San Gabriel Main Basin. Nitrate contamination that naturally exceeded limits for drinking water in two of their wells forced SSWC to purchase Tier 2 replenishment water from the San Gabriel Main Basin, which is at a higher price and relies on available allocation from the member agency. That water was used to dilute well water nitrate to safe levels for drinking water. Conventional options were considered to remove the nitrate, including reverse osmosis (RO) and various IX processes. The RO option was ruled out due to the high energy cost and low water recovery. With IX, there are high capital costs and high operating costs for electricity, chemicals, and disposal of concentrated waste brine. Both technologies proved to be too expensive for the facility's needs on a life cycle basis. As a result, SSWC chose MicroNiche Engineering (MNE) for nitrate treatment.

4.5.4.2 Facility description. The MNE nitrate treatment system, placed on a 25-ft by 45-ft concrete pad, is based on the use of biocatalyst composites developed through MNE platform technology. The MNE biocatalysts contain a high density of a specifically selected microorganism that converts nitrate into nitrogen gas.

The MNE nitrate treatment system (Figure 4-20) was designed as a stand-alone process that would produce water meeting all drinking water standards. The treatment system treats water from Sunny Slope's GAC filters. The GAC filters are used to remove VOCs. The treatment train (Figure 4-21) consists of a feed tank for flow equalization, a nutrient dosing system, two bioreactor tanks (in parallel) with biocatalyst for nitrate

Figure 4-20 Photograph of Microvi nitrate treatment system

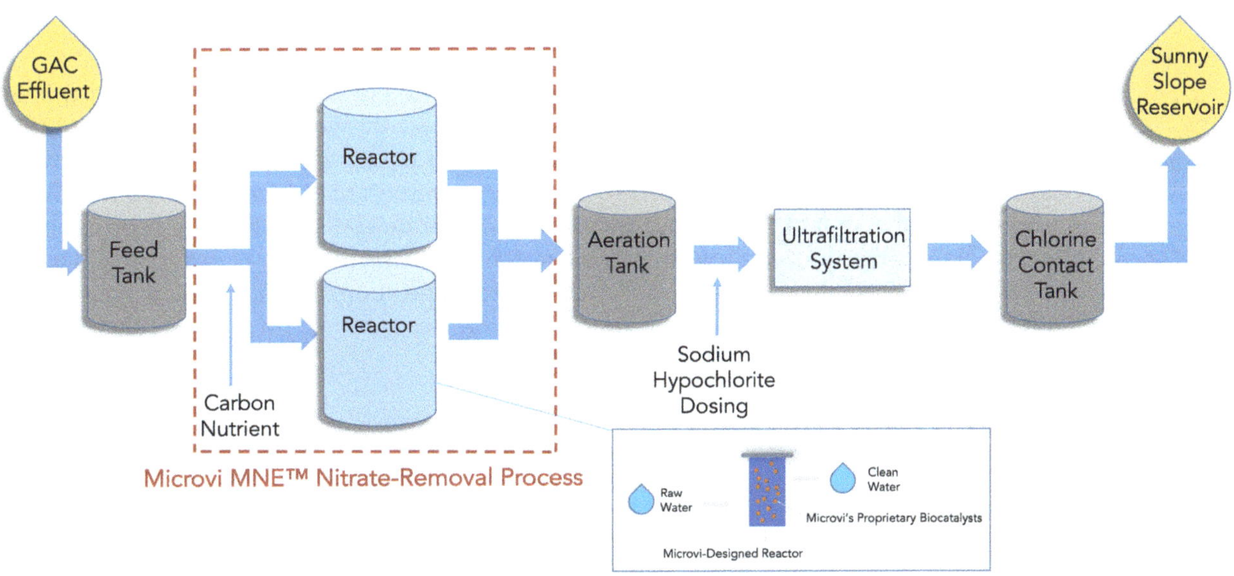

For nitrate treatment

GAC—granular-activated carbon

Figure 4-21 Process flow diagram for the MicroNiche Engineering system

removal, an aeration tank, a UF system for turbidity removal, and a chlorine contact tank (CCT) for disinfection.

During operation, the flow path is as follows:

- Water from SSWC's GAC filters enters the MNE system through a flow control valve and fills a feed tank.
- Water from the feed tank is pumped into the bioreactor tanks. Flow is dosed with an electron donor and carbon source (acetic acid), and a very small dose of phosphoric acid as it is being pumped into the bioreactor tanks.
- In the bioreactor tanks, the biocatalysts are continuously mixed and use the nutrient to reduce nitrate to nitrogen gas.
- Flow from each bioreactor tank gravitates to the aeration tank where jet mixers add dissolved oxygen (DO).
- Flow is pumped from the aeration tank through the UF system where turbidity is removed. Sodium hypochlorite is dosed upstream of the UF system to provide disinfection dosing. Aluminum chlorohydrate (ACH) is also dosed upstream of the UF system to help improve performance of the UF membranes.
- Residual pressure from the UF feed pump directs flow through the CCT.
- The CCT provides residence time to meet the disinfection CT (chlorine concentration times time) requirement.
- Water gravitating out of the CCT is blended with SSWC's GAC filter effluent.

Table 4-14 and Table 4-15 provide the raw water quality and operating parameters for the MNE Nitrate treatment system for SSWC.

4.5.4.2.1 Inline monitoring. The nitrate treatment system had a robust online monitoring and nutrient dosing system that was critical to the operation of the system. Nitrate concentration is monitored at four locations (the feed tank outlet, bioreactor outlet[s], and the CCT outlet) by an automated sample sequencer. Measured concentrations are logged by the Programmable Logic Controller (PLC) and displayed by the Human Machine Interface (HMI).

Flow to the sample sequencer is provided by a dedicated peristaltic pump. During sampling, automatic valves direct flow from the desired sampling location to the nitrate analyzer. An operator-adjustable time interval is allowed to elapse to ensure that the sample line has been flushed and that the sample is representative of the process water at the sample location. The control system will only record results for the nitrate analyzer after this preset flushing time has elapsed so that only representative values are used to monitor and control the system. Results for the feedwater nitrate concentration are returned once every 20 min. The nutrient dose was paced according to the feed flow rate measured by the flowmeter on each Bioreactor Feed Pump, DO concentration, and nitrate concentration of the water entering the feed tank. Nutrient addition was trimmed on the basis of the nitrate concentration of each bioreactor tank effluent.

Table 4-14 Raw water quality

Parameter	Units	Value
Raw water nitrate (as N)*	mg/L	8–12
Temperature	C	15–22
pH	SU	7.0–7.3
DO	mg/L	6.0

DO—dissolved oxygen

Notes: The system treats water from two wells. One of the wells has lower nitrate concentration (ranging from 5 to 6.5 mg/L as N), which results in the wide variation observed in the raw water nitrate.

Table 4-15 Operating parameters for Sunny Slope Water Company

Parameter	Units	Value
Well flow rate	gpm	950
Nitrate treatment flow rate	gpm	150–350
HRT	min	20
Daily run time	h	8–24
Mass ratio or acetic acid C to DO		0.38
Mass ratio of acetic acid C to nitrate N		1.4
Mass ratio of acetic acid C to PO4-P		50
UF backwash frequency	h	6–8
Water recovery	%	98.5
CIP frequency	times/year	2

CIP—clean-in-place, DO—dissolved oxygen, HRT—hydraulic residence time, UF—ultrafiltration

The speed of the pump was mediated by a proportional–integral–derivative (PID) control loop that will control the flow to the flow paced or set point value. The PID loop would reference the flow indicator and transmitter at each bioreactor inlet, the nitrate concentration at the feed tank inlet, and the nitrate concentration at each bioreactor outlet. The PID also had low- and high-speed limits so the pump does not over or under dose past a certain point in the event the online monitoring system failed or was reporting incorrect values.

4.5.4.3 Facility biological treatment goals. SSWC has the following goals for the Microvi MNE Nitrate Treatment System:

- effluent nitrate from the treatment system to be less than 3 mg/L N;
- achievement of more than 98% water recovery through the entire treatment system; and
- effluent water to meet drinking water standards for nitrate, turbidity, total trihalomethanes, haloacetic acids, corrosivity, heterotrophic plate count, and total coliform.

4.5.4.4 Operation and performance. A key performance and optimization metric for the Sunny Slope Nitrate Treatment System was to be able to scale and replicate effluent nitrate results (less than 3 mg/L N) seen at lab scale and pilot systems. Developing a strong understanding of long-term nitrate treatment performance at scale is critical to the future Microvi MNE Nitrate Treatment Systems. As observed in Figure 4-22, nitrate was consistently lowered to less than 3 mg/L N from the raw water concentration of 4–10 mg/L N.

In addition to demonstrating nitrate removal performance, the postfiltration process was optimized. A UF system was used downstream of the biological reactors as a postpolishing filtration step. Initial operation of the UF system demonstrated difficulty in effectively recovering transmembrane pressure (TMP) after standard backwashing, requiring more frequent chemically enhanced backwashing (CEB). Introduction of a coagulant (ACH) dramatically improved TMP recovery performance and minimized the amount of CEB required to operate UF effectively (Figure 4-23).

4.5.4.5 Key success factors and lessons learned.

- Accurate and robust online monitoring using a robust suite of Hach instrumentation. Instrumentation maintenance was provided by Hach every six months and cross samples were sent to outside lab once per month to verify results produced by online instrumentation.

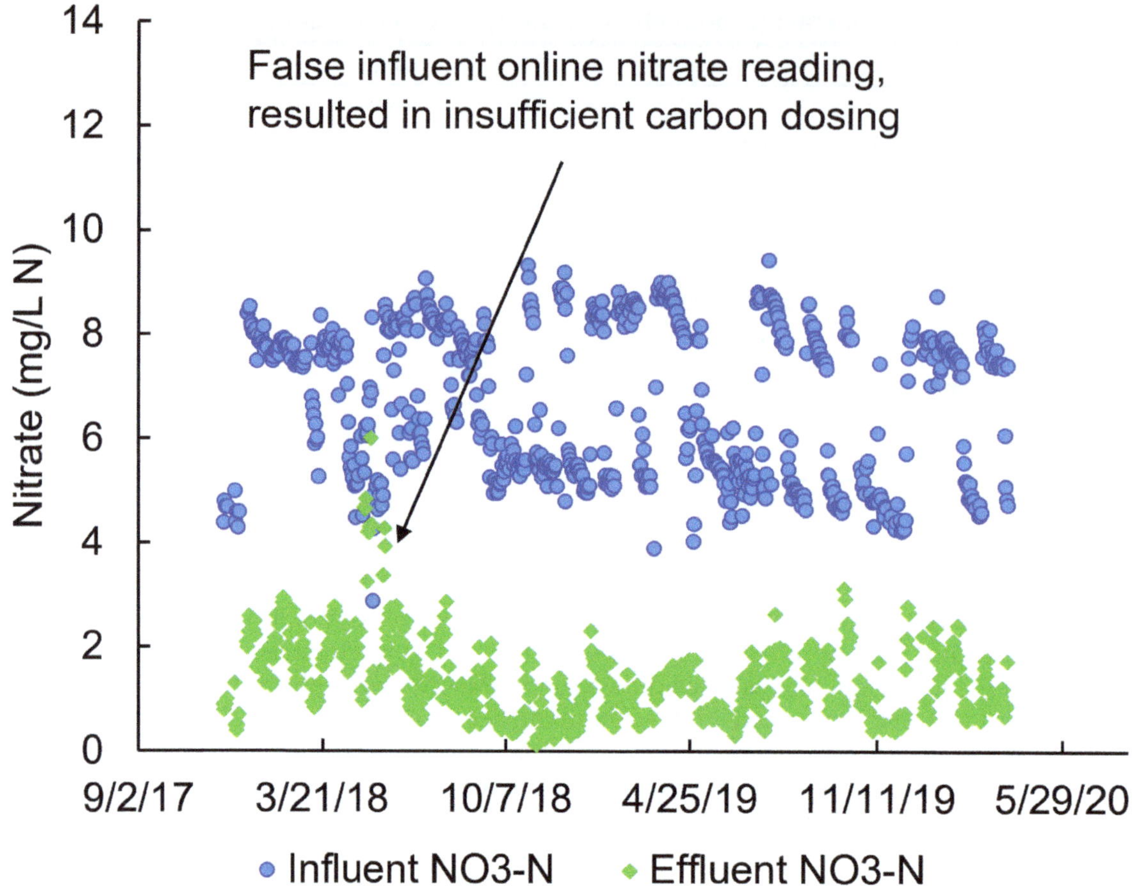

Figure 4-22 Long-term nitrate performance

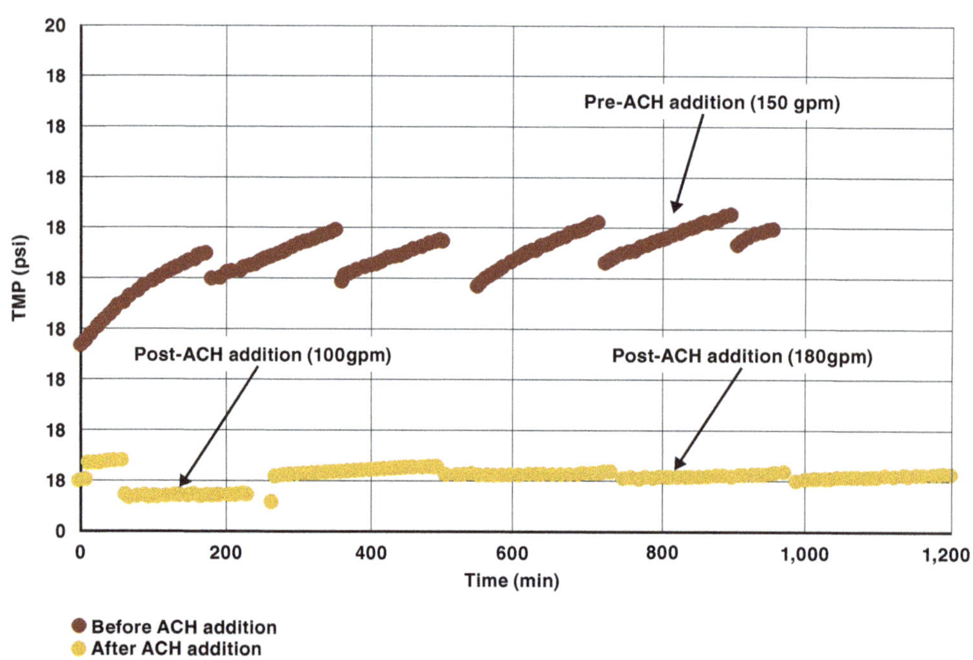

ACH—aluminum chlorohydrate, TMP—transmembrane pressure

Figure 4-23 Ultrafiltration performance with ACH dose

- Utilization of multichannel sample sequencing to allow for data from various points within the process. The sample sequencer allowed for the nutrient dose to be paced on the basis of the feed flow rate measured by the flowmeter on each Bioreactor Feed Pump, DO concentration, and nitrate concentration of the water entering the feed tank. Nutrient addition is trimmed on the basis of the nitrate concentration of each bioreactor tank effluent.
- Implementation of an absolute barrier postfiltration system (e.g., UF); significantly de-risking turbidity breakthrough.

4.6 REFERENCES

Ali, D. 2017. "A New and Innovative Approach to Drinking Water Nitrate Removal [Webinar]." Arizona Water Association Water Treatment Committee webinar series.

Bouwer, E.J. and J.P. Wright. 1988. "Transformations of Trace Halogenated Aliphatics in Anoxic Biofilm Columns." *Journal of Contaminant Hydrology*. 2(2):155-169. https://doi.org/10.1016/0169-7722(88)90005-8.

Brown, J., P. Carlo, C. Cleveland, G. Upadhyaya, J. Riley, and P.S. Wu. 2016. "Two-Stage, Fixed-Bed Biodenitrification at the City of Delano: Pilot Testing to Full-scale Implementation." AWWA International Biological Treatment Symposium, Long Beach, Calif.

Brown, J., R.S. Summers, E. Dickenson, and G. Upadhyaya. 2017. "Biological Removal of Hazardous Pollutants in Drinking Water Treatment." Hazardous Pollutants in Biological Treatment Systems: Fundamentals and A Guide to Experimental Research. 313-358.

Brown, J., J. Riley, C. Cleveland, P. Carlo, G. Upadhyaya, B. Silva, and D. Mount. 2015. "Two-Stage Fixed-Bed Biological Treatment: Unlocking Contaminated Groundwater Sources." AWWA Annual Conf. and Exposition, Anaheim, Calif.

Brown, J.C., V.L. Snoeyink, and M.J. Kirisits. 2002. "Abiotic and Biotic Perchlorate Removal in an Activated Carbon Filter." *Journal AWWA*. 94(2):70-79. https://doi.org/10.1002/j.1551-8833.2002.tb09408.x.

Carter, S.R. and W.J. Jewell. 1993. "Biotransformation of Tetrachloroethylene by Anaerobic Attached-Films at Low Temperatures." *Water Research*. 27(4):607-615. https://doi.org/10.1016/0043-1354(93)90170-M.

Chang, H. and B.E. Rittman. 1987. "Mathematical Modeling of Biofilm on Activated Carbon." *Environmental Science and Technology*. 21(3): 273-280.

De Bruin, W.P., M.J.J. Kotterman, M.A. Posthumus, G. Schraa, and A.J.B. Zehnder. 1992. "Complete Biological Reductive Transformation of Tetrachloroethene to Ethane." *Applied and Environmental Microbiology*. 58(6):1996-2000.

Dugan, N. R., D.J. Williams, M. Meyer, R.R. Schneider, T.F. Speth, and D.H. Metz. "The Impact of Temperature on the Performance of Anaerobic Biological Treatment of Perchlorate in Drinking Water." *Water Research*. 43(7): 1867-1878.

Fathepure, B.Z. and T.M. Vogel. 1991. "Complete Degradation of Polychlorinated Hydrocarbons by a Two-Stage Biofilm Reactor." *Applied and Environmental Microbiology*. 57(12):3418-3422.

Gudal, D. 2016. "Status of Biological Nitrate Removal in Drinking Water in Huntsville, Texas." AWWA International Biological Treatment Symposium, Long Beach, Calif.

Hatzinger, P. B., M. R. Greene, S. Frisch, A. P. Togna, J. Manning, and W. J. Guarini. 2000. "Biological treatment of perchlorate-contaminated groundwater using fluidized bed reactors." International Conference of Remediation of Chlorinated and Recalcitrant Compounds.

Hauck, R. and W. Hegemann. 2000. "Anaerobic Degradation of 1,2-Dichloropropane in Batch and Continuous Culture." *Water Science & Technology*. 41:7-13.

Hwu, C.S. and C.J. Lu. 2008. "Continuous Dechlorination of Tetrachloroethene in an Upflow Anaerobic Sludge Blanket Reactor." *Biotechnology Letters*. 30(9):1589-1593. https://doi.org/10.1007/s10529-008-9738-x.

Keithley, S.E. and M.J. Kirisits. 2019. "Enzyme-Identified Phosphorus Limitation Linked to More Rapid Headloss Accumulation in Drinking Water Biofilters." *Environmental Science and Technology*. 53(4):2027-2035. https://doi.org/10.1021/acs.est.8b04573.

Li, X., G. Upadhyaya, W. Yuen, J. Brown, E. Morgenroth, and L. Raskin. 2010. "Changes in the Structure and Function of Microbial Communities in Drinking Water Treatment Bioreactors upon Addition of Phosphorus" *Applied and Environmental Microbiology*. 76(22): 7473-7481.

Liu, J.R., C.T. Liu, E.A. Edwards, and S.N. Liss. 2006. "Effect of Phosphorus Limitation on Microbial Floc Structure and Gene Expression in Activated Sludge." *Water Science and Technology*. 54(1):247-255. https://doi.org/10.2166/wst.2006.393.

Logan, B.E., B. Min, K. Kim, J. Miller, D. Lapoint, J. Batista, J. Liu, J. P. Evans, A. Chu, S. Price. 2004. "Bioreactor Systems for Treating Perchlorate-Contaminated Water: Benchand Pilot-Scale Investigations" AWWA Research Foundation.

Mccarty, P. L. and T. E. Meyer. 2005. "Numerical Model for Biological Fluidized-Bed Reactor Treatment of Groundwater." *Environmental Science and Technology*. 39(3): 850-858.

Mauclaire, L., A. Schürmann, M. Thullner, S. Gammeter, and J. Zeyer. 2004. "Sand Filtration in a Water Treatment Plant: Biological Parameters Responsible for Clogging." *Journal of Water Supply: Research and Technology–AQUA*. 53(2):93-108.

Min, B., P.J. Evans, A.K. Chu, and B.E. Logan. 2004. "Perchlorate Removal in Sand and Plastic Media Bioreactors." *Water Research*. 38:47-60.

Najm, I., N.P. Brown, B. Gallagher, E. Seo, K. Gramith. 2014. "Minimizing Waste Backwash Water From a Biological Denitrification Treatment System." Water Research Foundation.

Ontiveros-Valencia, A., C.R. Penton, K-B. Rosa, and B.E. Rittman. 2016. "Hydrogen-fed Biofilm Reactors Reducing Selenate and Sulfate: Community Structure and

Capture of Elemental Selenium Within the Biofilm." *Biotechnology and Bioengineering*. 113(8):1736-1744.

Polk, J.C. Murray, C. Onewokae, D.E. Tolbert, A.P. Togna, W.J. Guarini, S. Frisch, and M.D. Vecchio. 2001. "Case Study of Ex-Situ Biological Treatment of Perchlorate-Contaminated Groundwater." 4th Tri-Services Environmental Technology Symposium. San Diego, California. June 18-20.

Richard Y.R. 1989. "Operating Experiences of Full-Scale Biological and Ion-Exchange Denitrification Plants in France." *Water and Environment Journal*. 3(2): 154-157.

Rittmann, B.E. and P.M. Huck. 1989. "Biological Treatment of Public Water Supplies." *Critical Reviews in Environmental Control*. 19(2):119-184. https://doi.org/10.1080/10643388909388362.

Rittmann, B.E. and P.L. McCarty. 2020 (2nd ed.). *Environmental Biotechnology: Principles and Applications*. New York: McGraw-Hill.

Schollhorn, A., C. Savary, G. Stucki, and K.W. Hanselmann. 1997. "Comparison of Different Substrates for the Fast Reductive Dechlorination of Trichloroethene Under Groundwater Conditions." *Water Research*. 31(6):1275-1282.

Silverstein, J.S. and G.L. Carlson. 1999. *Biological Denitrification of Drinking Water for Rural Communities*.

Stucki, G., C.W. Yu, T. Baumgartner, and J.F. Gonzalez-Valero. 1995. "Microbial Atrazine Mineralisation Under Carbon Limited and Denitrifying Conditions." *Water Research*. 29(1):291-296. https://doi.org/10.1016/0043-1354(94)E0101-B.

Tang, Y., Z.-E. Michal, C. Zhou, J. H. Shin, C.H. Ahn, K. Meyer, D. Candelaria, D. Friese, R. Overstreet, R. Scott, and B.E. Rittman. 2010. "Bioreduction of Nitrate in Groundwater Using a Pilot-Scale Hydrogen-Based Membrane Biofilm Reactor." *Frontiers of Environmental Science and Engineering in China*. 4(3): 280-285.

Tang, Y., C. Zhou, S.W. Van Ginkel, A. Ontiveros-Valencia, J. Shin, and B.E. Rittmann. 2012. "Hydrogen Permeability of the Hollow Fibers Used in H2-Based Membrane Biofilm Reactors." *Journal of Membrane Science*. 407-408:176-183. https://doi.org/10.1016/j.memsci.2012.03.040.

Tchobanoglous, G., F.L. Burton, and H.D. Stensel. 2002. *Wastewater Engineering: Treatment and Reuse*. New York: McGraw-Hill.

Upadhyaya, G., J. Jacksho, T.M. Clancy, S.P. Hyun, J. Brown, K.F. Hayes, and L. Raskin. 2010. "Simultaneous Removal of Nitrate and Arsenic from Drinking Water Sources Utilizing a Fixed-Bed Bioreactor System." *Water Research*. 44(17): 4958-4969.

Upadhyaya, G., T. Clancy, J. Brown, K.F. Hayes, and L. Raskin. 2012. "Optimization of Arsenic Removal Water Treatment System Through Characterization of Terminal Electron Accepting Processes." *Environmental Science and Technology*. 46(21): 11702-11709.

Upadhyaya, G. and J. Brown. 2018. "Fixed-bed Biologically Active Carbon Bioreactors Effectively Remove Selenium from Drinking Water Sources." Water Quality Technology Conference, Toronto, Ontario, Canada. Nov 11-15.

Upadhyaya, G. and J. Brown. 2016. "Biologically Mediated Simultaneous Removal of VOCs and Inorganic Pollutants from Drinking Water Sources." AWWA Annual Conf. and Exposition, Chicago, Ill.

Upadhyaya, G., B. Dunagan, C. Mattingly, and J. Brown. 2019. "Resilient Biological Hexavalent Chromium Removal with a Two-Stage, Fixed-Bed Biotreatment System." *AWWA Water Science*. 1(4):e1151. https://doi.org/10.1002/aws2.1151.

Upadhyaya, G., N. Kotlarz, K.F. Hayes, L. Raskin, and J. Brown. 2013. "Efficient Removal of Multiple Pollutants Using Two-Stage Biologically Active Carbon Bioreactors." AWWA Water Quality Technology Conference, Long Beach, Calif.

Upadhyaya, G., W. Raymond, T. Guarin, P.S. Wu, J. Witter, and J. Brown. 2022. "Efficient Removal of Nitrate and Arsenic Using a Two-Stage Biotic-Abiotic Treatment System." *AWWA Water Science*. 4(3):e1292.

Webster T. S. and P. Togna. 2009. "Demonstration of a Full-Scale Fluidized Bed Bioreactor for the Treatment of Perchlorate at Low Concentrations in Groundwater." Environmental Security Technology Certification Program.

Zhao, H.P., A.O. Valecia, Y. Tang. Bi-O. Kim, S. Vanginkel, D. Friese, R. Overstreet, J. Smith, P. Evans, R.K. Brown, and B. Rittman. 2014. "Removal of Multiple Electron Acceptors by Pilot-Scale, Two-Stage Membrane Biofilm Reactors." *Water Research*. 54:115-122.

Zhou, C., A. Ontiveros-Valencia, L. Cornette de Saint Cyr, A.S. Zevin, S.E. Carey, R. Krajmalnik-Brown, and B.E. Rittmann. 2014. "Uranium Removal and Microbial Community in a H_2-Based Membrane Biofilm Reactor." *Water Research*. 64:255-264. https://doi.org/10.1016/j.watres.2014.07.013.

Zhou, C., A. Ontiveros-Valencia, Y. Tang, D. Friese, R. Krajmalnik-Brown, and B.E. Rittman. 2019. "Hydrogenotrophic Microbial Reduction of Oxyanions with the Membrane Biofilm Reactor." *Frontiers in Microbiology*. 9:1-14. https://doi.org/10.3389/fmicb.2018.03268.

Chapter 5

Aerobic Biofiltration for Reuse

Jen Hooper, Christina Alito, Jennifer Nyfennegger, Ramola Vaidya, Leigh Terry, and Lynn Stephens

In recent years, there has been increasing interest and use of biofiltration as a sustainable treatment process for producing high-quality treated water in water reuse applications when coupled with strong oxidants such as ozone, as part of a multibarrier approach for the removal of trace chemical constituents, biodegradable organic matter, and pathogens. These water quality benefits have been observed in full-scale drinking water treatment applications, and application in a reuse context has been fairly recent (within the past decade). Biofiltration, when used downstream of oxidation and/or advanced oxidation processes (AOPs), has been a critical component to mitigate unique water quality challenges associated with treated wastewater. Further, biofiltration coupled with oxidation has been demonstrated to produce stable potable water quality without the use of reverse osmosis (RO). The applications and benefits of biofiltration for surface water have been discussed in Chapter 2. This chapter focuses on biofiltration in potable reuse applications and provides detail on the removal of contaminants specific to potable reuse as well as regulatory considerations, testing, and operational guidance. This chapter also includes case studies of three potable reuse projects currently implementing biofiltration in their treatment process.

5.1 APPLICATION OVERVIEW

With increasing demands to produce high-quality water because of growing populations and shifting climactic patterns, utilities have considered water reuse and potable water reuse to expand and diversify their water supply portfolio. Indirect potable reuse (IPR) and direct potable reuse (DPR) are sought after in the United States to meet supply needs. Biofiltration for drinking water has been used since the early 1900s. Starting in approximately 2014, IPR and DPR studies began evaluating biofiltration coupled with oxidation in their treatment processes to reduce costs associated with membrane operation and provide biologically stable water quality. Biofiltration is currently being implemented in nonpotable and IPR applications to meet filtration objectives and improve the removal of dissolved organic carbon (DOC) compounds, including disinfection by-product (DBP) precursors, taste and odor (T&O) compounds, trace contaminants (such as pharmaceuticals and personal care products [PPCPs]), and inorganic compounds (such as metals and inorganic salts). Although several studies have evaluated biofiltration in a DPR scenario, none are currently operating full-scale in the United States. There is one full-scale 5.5-million-gallon-per-day (mgd) facility that has operating biofiltration for DPR in an intermittent basis since 1968 in Windhoek, Namibia (USEPA 2012, 2002).

5.1.1 Benefits

As described in Chapter 2, biofiltration downstream of ozonation (ozone-biofiltration) can stabilize the water quality by removing ozonation by-products and improving the removal of emerging contaminants. Ozone, biofiltration, granular activated carbon (GAC) adsorption, and coagulation/sedimentation have also been used in potable reuse treatment trains because of the multiple barriers they provide for a multitude of water quality constituents. In particular, the combination of ozone-biofiltration can offer advantages over RO-based treatment trains. For instance, ozone-biofiltration does not produce a concentrated brine waste stream requiring capital-intensive disposal for inland facilities. In addition, ozone-biofiltration treatment trains offer lower capital costs, are more energy efficient, and have higher water recovery (Noibi et al. 2020, Gerrity et al. 2014, Schimmoller and Kealy 2014). Ozone-biofiltration can also be used upstream of full advanced treatment (FAT) processes, which include microfiltration/ultrafiltration (MF/UF), RO, ultraviolet (UV) light disinfection, and AOPs to improve operational performance (i.e., reduce membrane fouling), improve RO brine quality, provide added robustness, and act as a critical barrier for intermittent inputs to the sewer system that might pass through a typical FAT train (e.g., acetone). Biofiltration serves a critical role in both non-RO and RO treatment trains, particularly when coupled with upstream ozonation (or preozonation).

5.1.2 Contaminant Removal

Domestic wastewater, industrial wastewater, and stormwater runoff have substantially different chemical compositions, so the type of water reuse and associated relative contributions from each source will affect the overall composition of water requiring advanced treatment for reuse. Specific constituents include inorganic compounds (e.g., metals, salts, nutrients), bulk organic compounds (i.e., natural organic matter [NOM] and effluent organic matter [EfOM] measured as total organic carbon [TOC] or chemical oxidant demand [COD]), anthropogenic trace organic constituents (e.g., pesticides, PPCPs), and constituents formed during wastewater treatment (e.g., DBPs). The chemical constituents in wastewater include a combination of naturally occurring compounds from the environment (e.g., recalcitrant NOM), compounds associated with human or industrial waste, and anthropogenic compounds associated with human activity within the wastewater collection system. Table 5-1 describes contaminants commonly removed by ozone-biofiltration. The following section summarizes the current understanding of biofiltration contaminant removal capabilities in reuse applications.

5.1.2.1 Pathogens. Examples of pathogens include bacteria (e.g. Escherichia coli), viruses, protozoa (e.g., *Giardia, Cryptosporidium*), and helminths (e.g., worms) (Crittenden et al. 2012). Pathogen reduction recommendations for reuse trains are discussed in Section 5.1.4. Pathogen log removal by ozone-biofiltration is summarized in Table 5-1.

Table 5-1 Contaminants commonly removed by ozone-biofiltration in reuse applications

Contaminants	Source in Wastewater	Typical Removal by Ozone-Biofiltration	References
Pathogens	Human pathogens and industrial waste from food production and animal processing waste.	Viruses 4–6 log removal. *Giardia* 3–5 log removal. *Cryptosporidium* 2–4 log removal. Total coliform 3- to greater than 6-log removal.	Trussell et al. (2013)
Solids and particulates	Natural and anthropogenic activities.	Turbidity <0.1 NTU.	Hooper et al. (2020)
Bulk organic carbon compounds	NOM, industrial and domestic wastes.	16–26% TOC removal.	Sundaram et al. (2020), Gifford et al. (2018)
Trace contaminants	PPCPs, pesticides, flame retardants, etc.	40–80% removal of trace contaminants (highly dependent on contaminant type, ozone dose, biofiltration media type, and EBCT).	Sundaram et al. (2020), Vaidya et al. (2020), Sun et al. (2018)
Nitrogenous DBPs	Industrial wastes, formed during wastewater treatment from organic nitrogen.	Up to 90% NDMA removal can be achieved in biofilters and is highly dependent on EBCT.	Bacaro et al. (2019)
Inorganic compounds	Metals, inorganic salts, mineral acids, etc., discharged from industrial wastes.	Complete (<0.05 mg/L) manganese removal achieved after media acclimation and after achieving complete nitrification in biofilters. Oxidation and biofiltration are not effective at removing TDS. If recycled water TDS concentrations are a concern, then treatment with IX resin, RO, or EDR will be needed.	Hogard et al. (2021), Brandhuber et al. (2013), Kohl and Medlar (2006)

DBP—disinfection by-product, EBCT—empty bed contact time, EDR—electrodialysis reversal, IX—ion exchange, NDMA—N-nitrosodimethylamine, NOM—natural organic matter, PPCP—pharmaceutical and personal care product, RO—reverse osmosis, TDS—total dissolved solid

5.1.2.2 Solids and particulates. Solids removal related to turbidity and particulate removal are discussed in Chapter 2. Turbidity and solids removal may also be achieved in reuse scenarios, and the removal capabilities are similar to surface water treatment removals when coagulation, sedimentation, and media configurations are the same.

5.1.2.3 Bulk organic carbon compounds. Wastewater or water reclamation facility effluent can contain high organic carbon concentrations (refer to Table 5-2), which must be addressed before subsequent use in a potable or nonpotable reuse context. The concentration of organic carbon in a water reclamation facility effluent, referred to as EfOM, could be greater or less than typical surface water sources used to supply drinking water treatment plants (WTPs), depending on the upstream wastewater treatment processes (NWRI 2015). EfOM is highly distinguishable from surface water NOM, in that it typically has higher aliphatic, proteinaceous, carbohydrate, and halogen content relative to surface water NOM (Pernet-Coudrier et al. 2008, Sirivedhin and Gray 2005). Reuse trains with ozone/biofiltration produce low organic levels ranging from 2 to 5 mg/L, but RO can achieve even lower TOC concentrations.

There is an abundance of research evaluating biofiltration for the removal of NOM in potable surface water applications. NOM removal efficiencies are influenced by several factors, such as the preoxidant type and dose (or oxidant to TOC ratio), water temperature, empty bed contact time (EBCT), biofilter media type, and the quality (e.g., biodegradability) of organic carbon present; the same principles apply for removal capabilities in reuse scenarios and can be related to TOC, assimilable organic carbon (AOC), and biodegradable organic carbon (BDOC) removal (Terry 2018). TOC removal achieved by coupled ozone with biofiltration treatment of wastewater effluent will depend on ozone and biofiltration

Table 5-2 Typical TOC concentrations (NWRI 2015, Crittenden et al. 2012)

Water Type	Typical TOC *mg/L*
Surface water	1–2
Raw wastewater	100–300
Wastewater effluent (conventional activated sludge with filtration)	15–30
Wastewater effluent + water reclamation facility (MF and RO)	0.1–1
Wastewater effluent + water reclamation facility (ozone/biofiltration)	2–5

MF—microfiltration, RO—reverse osmosis, TOC—toxic organic carbon

performance as well as that of upstream wastewater treatment. Typically, 20–30 percent of TOC removal can be achieved by biofiltration following ozonation; ozone alone does not remove TOC but does increase the biodegradability of the organic carbon present. A polishing step, such as GAC adsorption, might be needed postbiofiltration to achieve TOC concentrations of 2–5 mg/L, as shown in Table 5-2.

5.1.2.4 Trace chemical constituents. Another important factor when considering biofiltration treatment for water reuse applications is the presence of trace chemical constituents, such as PPCPs, pesticides, endocrine disrupting compounds (EDCs), solvents such as trichloroethelene, and solvents such as 1,4-dioxane, which may be present in wastewater effluent.

Trace chemical constituents are typically detected in treated wastewaters at 1 μg/L or lower, which may be concerning, depending on the chemical's toxicity (Kumar and Xagoraraki 2010, Miege et al. 2009). In one study, organic wastewater contaminants were detected in 80 percent of 139 streams sampled, and although measured concentrations were generally low (<1 μg/L), about 5 percent of compounds were greater than 1 μg/L (Kolpin et al. 2002). Consequently, an important topic to evaluate when considering biofiltration for reuse water treatment is its ability to remove trace chemical constituents. Currently there is no evidence that trace levels of PPCPs or EDCs have measurable human health risk; however, development of regulations specific to several classes of trace contaminants, such as per- and polyfluoroalkyl substances (PFAS), are ongoing.

Studies have shown varying removal rates of trace chemical constituents associated with biofiltration. Biofiltration removes biodegradable contaminants by breaking down these constituents into smaller molecules (daughter products) or through complete mineralization to carbon dioxide (CO_2). Some constituents may also be adsorbed to the filter media, in the case of GAC, or absorbed into the biofilm matrix (McDonald et al. 2015). Trace chemical constituents are used by microorganisms as a secondary substrate through co-metabolism, which is highly dependent on the quality and quantity of primary substrate (e.g., bulk organic carbon present), temperature, trace constituent chemical structure, and overall biodegradability, as well as operational parameters, such as EBCT, and ozone demand/dose (Halle et al. 2015, Lee et al. 2012, Zearley and Summers 2012).

Treated water aesthetics are important for both potable- and nonpotable-water reuse objectives, thus removal and mitigation of the formation of odorous compounds should be included with a monitoring strategy (Lundgren et al. 1988).

5.1.2.5 Nitrogenous DBPs. Nitrosamines are a type of emerging DBPs associated with wastewater-derived constituents, given the presence of N-nitrosodimethylamine (NDMA) precursors such as dissolved organic nitrogen in wastewater effluent and use of nitrogenous coagulants such as poly diallyldimethylammonium chloride (polyDADMAC) during wastewater treatment. Nitrosamines pose health concerns, even at the very low (ng/L) level. The US Environmental Protection Agency (USEPA) has been monitoring the effects of nitrosamines and included five nitrogenous species on the Contaminant Candidate List 4. NDMA is the primary nitrosamine identified in drinking water and reuse water. Among other nitrosamines, N-nitrosomorpholine has been detected in wastewaters, surface waters, and drinking WTP effluents. Additionally, California currently has a notification level of 10 ng/L for three nitrosamines (N-nitrosodiethylamine, NDMA, N-nitrosodipropylamine), which multiple potable reuse projects use as a guideline value (California Department of Public Health 2013).

NDMA formation is highly dependent on the type and nature of precursors present. Some example NDMA precursors include dimethyl sulfide, dimethylhydrazine, and polyDADMAC. Although chlorine dioxide can form NDMA, preoxidation with chlorine dioxide and even chlorine can mitigate the formation of NDMA. Similarly, although ozone can form NDMA, it has been shown to remove monochloramine-related NDMA precursors (Vaidya et al. 2021). The precursors necessary for NDMA formation by ozone are different than those required for NDMA formation by chloramines (Marti et al. 2015).

Biofiltration has been shown to remove NDMA and NDMA precursors in reuse applications. Bacaro et al. (2019) found that NDMA removal in biofilters was dependent on EBCT and achieved greater than 90 percent NDMA removal for 10 min of EBCT, whereas an EBCT of 2 min achieved only 30 to 40 percent NDMA removal. In another study, preozonation resulted in 20 ng/L of NDMA formation that was removed to below 5 ng/L by downstream biofiltration after media acclimation (Sun et al. 2018). NDMA precursor removal was documented in a study in which a 10-min EBCT biofilter reduced NDMA formation potential by 74 percent (biofilter effluent NDMA <10 ng/L) after forming 30 to 35 ng/L of NDMA by ozone (Vaidya et al. 2021).

Other nitrogenous DBPs, such as haloacetonitriles, halonitromethanes, and haloacetomides, are being detected in reuse water and need further attention. Biofiltration has been shown to remove dissolved organic nitrogen, which is a precursor for these nitrogenous DBPs (Zhang et al. 2016).

5.1.2.6 Inorganic compounds. The USEPA has established a secondary maximum contaminant level (SMCL) of 0.05 mg/L for manganese. Manganese in raw wastewater ranges from 1 to 2.5 mg/L; however, wastewater treatment using activated sludge followed by downstream ozone/biofiltration can reduce the manganese concentration to 0.3 mg/L or less for an influent wastewater concentration of 1–2.5 mg/L (NWRI 2015, Brandhuber et al. 2013, Kohl and Medlar 2006).

Manganese can be removed in biofilters by multiple mechanisms. A detailed discussion of manganese removal is in Section 2.1.3. In reuse applications, consideration of manganese concentrations should account for blending from advanced treatment and raw drinking water sources to achieve manganese concentrations below the SMCL.

Biofiltration can remove other inorganic compounds such as iron and ammonia. This is important for reuse water because these compounds are typically present in wastewater effluent and may have implications for biological regrowth in downstream processes if not mediated.

5.1.3 Removal Mechanisms

As discussed in the Chapter 2, biofiltration removal mechanisms include adsorption, biodegradation/transformation, and physical separation (i.e., particle removal). GAC is a robust biofilter support medium compared with other granular media, such as anthracite or sand, making it a preferred media for ozone/biofiltration applications for potable reuse; thus, subsequent discussion assumes the use of GAC media for reuse applications.

5.1.4 Regulatory Considerations

There are currently no national regulatory requirements for implementing biofiltration for potable reuse treatment. Figure 5-1 illustrates the urban water cycle and how drinking water is regulated by the Safe Drinking Water Act (SDWA) and state regulations (often adopting SDWA requirements). Wastewater discharges to public waters are regulated by the Clean Water Act and National Pollutant Discharge Elimination System permits. Potable reuse streams for IPR and DPR applications are on the basis of state regulatory requirements without a national standard in place (Figure 5-1).

To understand the drivers for biofiltration in reuse applications, it is important to review the regulatory framework or guidance developed for potable reuse systems because they affect the design and operation of biofiltration systems. Despite no national requirements, several states have taken initiative to create regulations for IPR or DPR treatment schemes. Figure 5-2 depicts current water reuse frameworks and regulations nationwide. Colorado, New Mexico, Arizona, and California have developed state-specific DPR regulatory frameworks. Each of these regulatory frameworks is unique to their respective states. North Carolina, Texas, and Washington allow for DPR approval on a case-by-case basis. Oklahoma is studying DPR.

States are taking different approaches in granting pathogen log removal credits. Some states, such as California, grant pathogen log removal credit from raw wastewater to finished drinking water, whereas others, such as Texas, grant pathogen log removal credits from second or tertiary effluent to finished drinking water. Table 5-3 presents pathogen reduction recommendations/requirements by various entities.

In the *Framework for Direct Potable Reuse* (NWRI 2015), biofiltration is not shown to achieve any removal credits for viruses, *Cryptosporidium* and *Giardia*. This contrasts with drinking water regulations wherein filtration and biofiltration are granted credits for pathogen removal. However, biofiltration is typically used in conjunction with ozonation, and ozonation can be granted 5-log removal of viruses, 0-log removal of *Cryptosporidium*, and 3-log removal of *Giardia* (NWRI 2015) for advanced treated water.

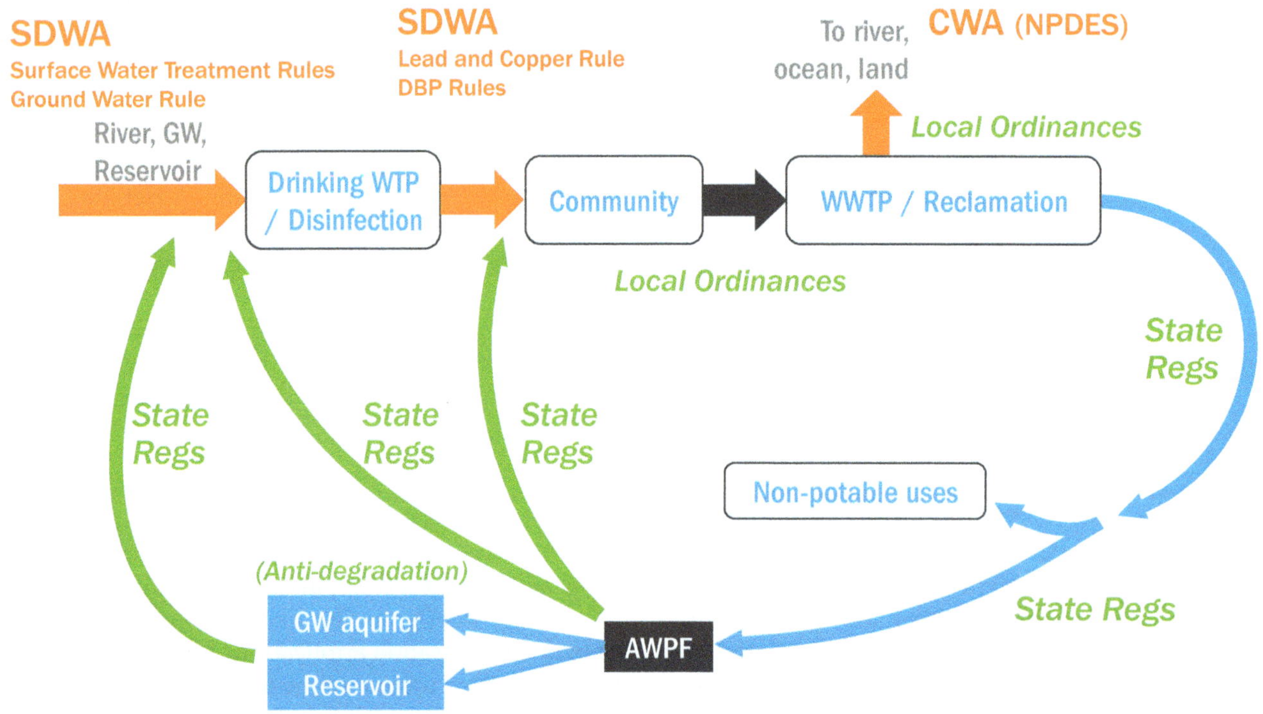

Source: Holmer (2018)

AWPF—advanced water purification facility, CWA—Clean Water Act, GW—groundwater, LT2ESWTR—Long Term 2 Enhanced Surface Water Treatment Rule, NPDES—National Pollutant Discharge Elimination System, SDWA—Safe Drinking Water Act, WTP—water treatment plant, WWTP—wastewater treatment plant

Figure 5-1 Urban water cycle and regulatory approach

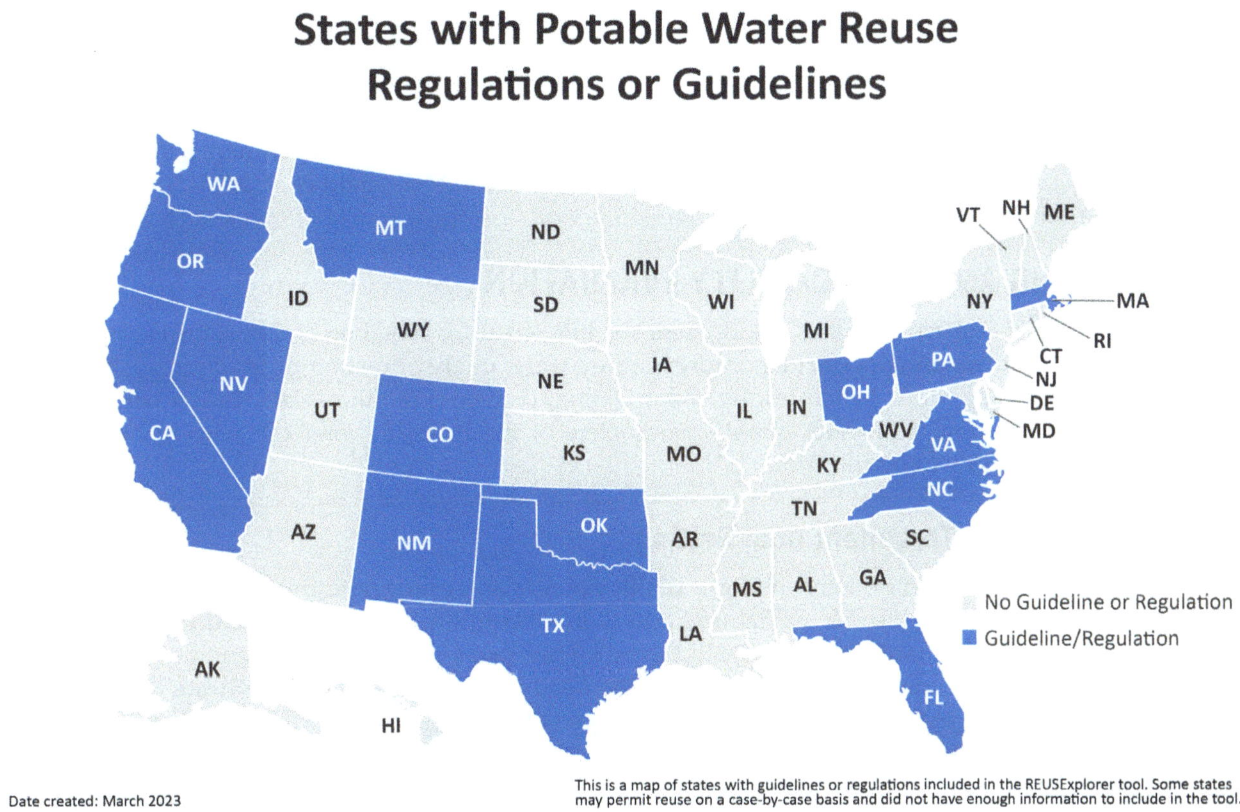

Source: Holmer (2018)

DPR—direct potable reuse, IPR—indirect potable reuse

Figure 5-2 National reuse frameworks/regulations as of 2018

Table 5-3 Recommended or required log removal values for pathogens in reuse applications

Reference	Viruses	*Giardia*	*Cryptosporidium*	Total Coliforms	Level of Treatment
Texas Commission on Environmental Quality*	8	6	5.5	None	Treated wastewater (secondary or tertiary) effluent to finished drinking water
California State Water Resources Control Board Division of Drinking Water†	12	10	10	None	Raw wastewater to groundwater replenishment (IPR)
Colorado Department of Public Health and Environment*	8	6	5.5	None	Treated wastewater (secondary or tertiary) effluent to finished drinking water

IPR—indirect potable reuse
*Recommended
†Required

Pathogen log removal credits for biofiltration may change in the future as characterization of pathogen removal continues.

These regulations typically include guidance for organic carbon and other constituents that may be removed through biofiltration. States also have different regulatory requirements for organic carbon removal. California's potable reuse requirements have a TOC limit of 0.5 milligrams per liter (mg/L) (CADDW 2018). California's treatment approach requires RO to be used. Other states, especially inland utilities, have taken the regulatory approach to rely on non–RO-based treatment trains. In either regulatory situation, biofiltration can play an important role in reaching regulatory requirements by reducing biodegradable organics. When combined with ozone, biofiltration has been identified as the best available technology to treat reuse water for aesthetic constituents, including T&O compounds, organics, and color (Brown et al. 2016, McDonald et al. 2015).

5.2 PLANNING AND APPLICABILITY EVALUATION

This section describes the process for investigating whether biofiltration is appropriate and evaluating performance for attaining water quality or water production goals through a paper study, bench testing, pilot testing, or demonstration-scale testing before implementation. The water quality parameters of the influent water can help in determining feasibility of biofiltration.

5.2.1 Treatment Goal Setting

The first step in determining if biofiltration is an appropriate treatment technology for a system is to identify contaminant(s) of interest in the source water and the desired level of contaminant removal. Evaluation of whether biofiltration has been known to remove specific contaminants of interest to the desired removal level will indicate if biofiltration is the appropriate unit process. The typical treatment goals for biofiltration in reuse are provided in Table 5-1.

5.2.2 Applicability Evaluation

Several studies have investigated the application potential of ozone-biofiltration in potable reuse. Water Research Foundation (WRF) Project #4832 is currently evaluating the removal of compounds of emerging concern (CECs) through ozone-biofiltration–based treatment of wastewater effluent to develop ozone-biofiltration design and operational guidance. Additionally, WRF #4777 investigated ozone-biofiltration for DPR application at Gwinnett County, Ga. The study demonstrated that ozone-biofiltration could achieve equal or higher water quality than current drinking water supplies when blended at 15 percent. There were also significant cost savings associated with the use of ozone-biofiltration in comparison with RO technology because the costs of mechanical evaporation for brine disposal and operating power for RO were more than twice the cost of the ozone-biofiltration process (Hooper et al. 2020, Noibi et al. 2020, Funk et al. 2018).

The major operations and maintenance (O&M) costs include "labor, power, chemicals such as liquid oxygen used for ozone generation, laboratory and monitoring costs, equipment maintenance and repair, residuals management, and other minor costs" (USEPA 2017). Biofiltration is often considered an economical option compared with RO because of the lack of brine stream that needs additional disposal and the reduced energy demand. Noibi et al. (2020) compared capital and operating costs for three advanced treatment trains for DPR and found that the costs of FAT were more than twice of those for ozone-biofiltration for facilities without access to a marine discharge or deep well injection. The cost of DPR using FAT with mechanical evaporation was estimated at $4,830/mil gal, with an estimated total capital cost of $621.6 million and annual O&M cost of $31.8 million. In comparison, the cost of DPR use of ozone-biofiltration was estimated at $1,900/mil gal, with total capital and annual O&M costs of approximately $239.9 million and $7.8 million, respectively (Noibi et al. 2020). Table 5-4 from the 2017 EPA Potable Reuse Compendium further compares ozone-biofiltration to FAT with RO concentrate disposal at various cost/effect categories for a 20-mgd plant (USEPA 2017).

Table 5-4 Cost of alternative treatment trains for a 20-mgd facility (USEPA 2017)

			FAT With RO Concentrate Disposal		
Cost/Effect		Ozone-Biofiltration*	Ocean Outfall	Mechanical Evaporation	Evaporation Ponds
Capital cost		$91 million	$120 million	$172 million	$303 million
Annual O&M cost		$4.2 million	$5.9 million	$10.9 million	$6.3 million
Annual environmental costs		$400,000	$1.6 million	$6.3 million	$2.2 million
Total TBL NPV		$173 million	$267 million	$533 million	$512 million
Cost of water (including environmental costs)	$/AF	$386 million	$596 million	$1.190 billion	$1.143 billion
	$/1,000 gal	$1.18 million	$1.83 million	$3.65 million	$3.51 million
	$/m^3	$310,000	$480,000	$960,000	$930,000
Power consumption (MWh/year)		4,400	16,000	65,400	22,000
Chemical consumption (dry tons/year)		1,770	1,860	3,020	1,860
Air emissions (tons/year)	CO_2	2,900	13,400	44,200	17,200
	Other	11	30	150	49

AF—active filtration, BAC—biologically active carbon, FAT—full advanced treatment, GAC—granular activated carbon, mgd—million gallons per day, NPV—net present value, O&M—operations and maintenance, RO—reverse osmosis, TBL—triple bottom line (financial, environmental, and social elements), UV—ultraviolet

*Costs include the entire treatment train: flocculation/sedimentation, ozone, BAC, GAC, and UV disinfection.

5.2.3 Testing

Biofiltration testing before full-scale implementation ensures biofiltration is the appropriate treatment unit, facilitates unit optimization of design and operation, improves overall plant performance, and can help to develop accurate capital and operating cost estimates (Brown et al. 2020, WRF 4719). Testing includes multiple phases, including desktop evaluation, bench-scale testing, pilot testing, and demonstration testing, as described in Chapter 2. Biofilter design, operation, and contaminant removal can be tested to answer specific questions.

Chapter 2 summarizes the testing objectives for biofiltration and testing scales, design, and testing outcomes. These testing protocols can also be applied to reuse water. Apart from these considerations, reuse applications necessitate defining treatment goals specific to IPR and DPR. Source water quality needs to be evaluated for feasibility to ozone and biofiltration and presence of precursors for DBP formation (e.g., bromide present in source water can lead to bromate formation in ozone). Ozone design and optimization needs to be conducted to achieve maximum AOC formation in ozone and removal in biofiltration, while minimizing DBP formation in ozone. Ozone can remove trace contaminants, but additional polishing by biofiltration might be necessary for certain contaminants.

Optimization of biofiltration parameters, such as media type (GAC, anthracite), media size, EBCT, and backwashing, might need special consideration. Bench and pilot-scale testing can provide meaningful information for these optimization strategies. Pilot and demonstration-scale testing should occur over multiple seasons to capture temporal changes in water quality and can last between 6 and 12 months or longer. These seasonal and temporal changes can affect ozone operation, such as dissolved ozone gas solubility and ozone demand and decay rates. These factors can further affect biofiltration performance needing operational changes in EBCT, backwashing, filter loading rate, etc. Pilot and demonstration testing periods of six to 12 months can also ensure sufficient time for media acclimation and biomass growth. Challenge testing with contaminants of specific interest in potable reuse scenarios, such as trace chemical constituents, may be incorporated with testing to assess performance. Testing strategies that encompass multiple testing scales produce the most reliable data.

5.3 DESIGN

There are several parameters that need to be considered for designing biofiltration. In general, biofiltration performance depends on not only the design and operational settings of the biofilter but also the configuration of upstream treatment processes. Moreover, the overall performance of treatment trains incorporating biofiltration depends on upstream and downstream treatment processes. Other design parameters specific to biofiltration include hydraulic loading rate (HLR) and EBCT, media type, and filter backwashing strategy. Upstream preoxidation is essential for the implementation of biofiltration in reuse scenarios because a higher fraction of the organic carbon present is recalcitrant (nonbiodegradable) downstream of secondary biological wastewater treatment.

5.3.1 Process Configurations

The configuration of a potable reuse treatment train is dependent on the treatment goals. The treatment objectives, regulatory considerations, influent water quality, and treated water quality goals are some of the key factors for not only determining biofilter design parameters but also for their order within the treatment process train. Potable reuse treatment trains are often characterized by whether they include RO. Biofiltration, typically implemented as ozone followed by biofiltration (ozone-biofiltration), may be used in both RO and non–RO-based treatment trains.

Figure 5-3 shows example process trains in which ozone-biofiltration is used and the associated pathogen removal. Other unit processes can also be considered for use in potable reuse treatment trains. For example, ozone-biofiltration can be located upstream or downstream of membrane-based treatment strategies. This may be a well-suited configuration if influent water quality has high levels of particulates or suspended solids. The membrane flux may be substantially increased by placing ozone/biofiltration upstream of MF or UF (Wang 2014). In other cases, MF or UF removes particulates upstream of biofiltration, and ozone-biofiltration is mainly used for the removal of organic constituents. Ozone-biofiltration ahead of MF/UF has been demonstrated to provide a greater benefit than after MF/UF (NWRI 2015). In some cases, the use of ozone-biofiltration may eliminate the need for RO for advanced water treatment (NWRI 2015). The facilities presented in Figure 5-4 are showcased as examples to demonstrate different process configurations with biofilters.

Ozone or other preoxidants or AOPs are used upstream of biofiltration to enhance the removal of organic carbon through biofiltration. Although Figures 5-3 and 5-4 show examples of treatment trains using ozone as the preoxidant, UV irradiation with AOP (UV/AOP) can also be used upstream of biofiltration, as discussed later in this chapter. When

Treatment Trains	Log Removal Credit		
	Cryptosporidium	Virus	Total Coliform
WWTP > UF > O3 > BAF > UV	11	14	16
WWTP > O3 > BAF > UF > UV	11	13	16
WWTP > O3 > BAF > MF > UV	11	14	16
WWTP > O3 > BAF > MF > RO > AOP	13	15	18
Log Removal Goals	10	12	9

Source: Modified from Parker et al. (2014) and WateReuse Research Foundation project 11-02 Trussell, 2012

AOP—advanced oxidation process, BAF—biologically active filter, RO—reverse osmosis, UF—ultrafiltration, UV—ultraviolet, WWTP—wastewater treatment plant

Figure 5-3 Example ozone-biofiltration treatment trains considered for direct potable reuse

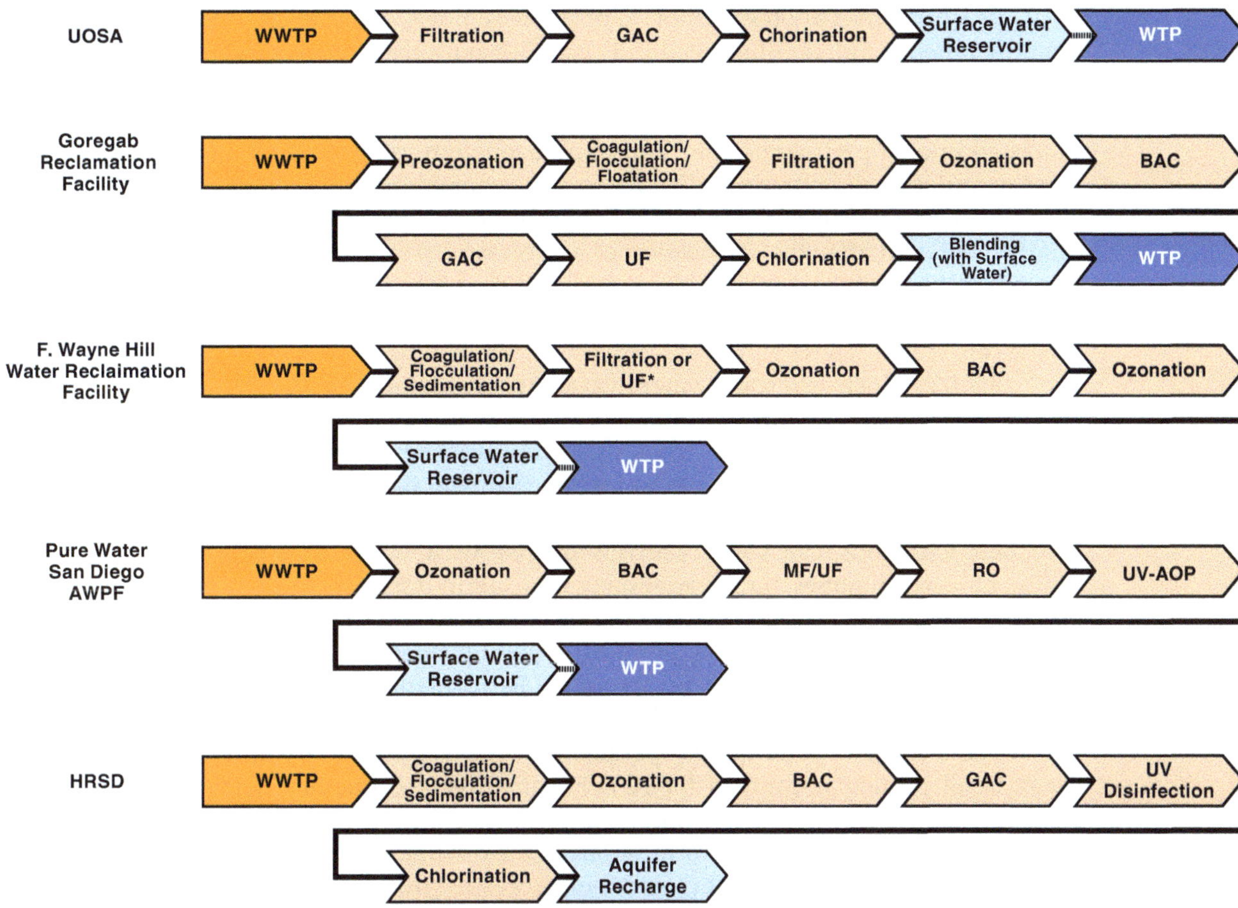

AWPF—advanced water purification facility, BAC—biologically active carbon, GAC—granular activated carbon, HRSD—Hampton Roads Sanitation District, MF—microfiltration, RO—reverse osmosis, UF—ultrafiltration, UOSA—Upper Occoquan Service Authority, UV—ultraviolet, WTP—water treatment plant, WWTP—wastewater treatment plant

Figure 5-4 Example facilities using biofiltration as a process in a potable reuse treatment train

considering biofiltration coupled with a preoxidation process, the design and operation of that oxidation process can directly influence biofiltration performance.

Ozone or UV/UV AOP may also be present downstream of biofiltration for disinfection and/or polishing, which is common to many of the treatment trains shown in Figures 5-3 and 5-4. Ozone-biofiltration is often used upstream of an RO-based treatment train, as illustrated in Figure 5-4, which shows the Pure Water San Diego advanced water purification facility (AWPF), which includes ozone-biofiltration followed by MF/UF and RO. This approach reduces membrane fouling and organic contaminants.

GAC adsorption after ozone-biofiltration can be used as a polishing step for TOC, particularly for the removal of recalcitrant organic compounds and to achieve effluent TOC goals. The ozone-biofiltration–GAC process configuration can be a particularly important strategy for meeting TOC limits when using carbon-based treatment trains. GAC adsorption can also remove other constituents such as PFAS, which are persistent in wastewater effluent (Kumar et al. 2021). As shown in Figure 5-4, the Gorengab Reclamation Facility in Windhoek, Namibia, and Hampton Roads Sanitation District (HRSD) in Virginia use GAC downstream of ozone-biofiltration. The Upper Occoquan Service Authority (Virginia) includes biologically active carbon (BAC) in their process, allowing biological activity to acclimate and transition into biofiltration in the absence of a disinfectant residual.

Coagulation is another option to promote removal of organic and particulate matter in a treatment train. As illustrated in Figure 5-4, both the F. Wayne Hill Water Resources Center (Georgia, USA) and HRSD use coagulation/flocculation/sedimentation upstream of ozone-biofiltration. The Gorengab Reclamation Facility in Windhoek, Namibia, incorporates additional pretreatment including preozonation, coagulation/flocculation/flotation, and granular media filtration (GMF) before the ozone-biofiltration process.

5.3.1.1 Design components. Design criteria should be developed for full-scale implementation after the completion of a successful pilot- or demonstration-scale biofiltration test and the selection of the location of the biofilter within the treatment train. To develop these criteria and corresponding design documents, key design parameters need to be considered. These parameters are presented in Table 5-5.

5.3.1.2 Upstream wastewater treatment considerations. Design and optimization of potable reuse treatment trains, including those with biofiltration, require an understanding of the effects from upstream wastewater treatment processes. Many different process configurations exist for the treatment of wastewater. This combined with varied targets for nutrient removal among different facilities, as well as other factors such as seasonal variations, can yield relatively diverse wastewater effluent or reuse treatment influent water qualities. Such variations can result in downstream biofilter influent water quality variations that affect hydraulic performance (e.g., unit filter run volume, head loss) and contaminant removal. For additional information, refer to WRF Project #4833, titled "Understanding Wastewater Treatment Performance on Advanced Water Treatment Processes and Finished Water Quality."

Some effluent water quality parameters linked to secondary and tertiary treatment may be of particular concern for AWPF trains using ozone-biofiltration. For example, wastewater effluent with high ammonia, nitrate, and/or nitrite concentrations can result in high ammonia or nitrate concentrations in the downstream ozone-biofiltration effluent. Depending on the downstream treatment processes, this can adversely affect regulatory compliance (e.g., nitrate), efficacy of disinfection (e.g., free chlorine), and AOP efficiency (e.g., OH radical scavenging).

Several measures can be considered to improve wastewater treatment upstream of a water reclamation facility, such as enhanced source control, improved fine screening, flow equalization (influent and return flows), tertiary filtration, and more rigorous process monitoring (NWRI 2015). Secondary treatment optimization can provide enhanced removal of nutrients, trace contaminants (Salveson et al. 2012), DBPs, and their precursors, such as nitrosamines and their precursors (Gerrity et al. 2015) and pathogen concentrations.

Studies on the effects of wastewater pretreatment (e.g., solids retention time, nitrification, denitrification, filtration) on the water reclamation facility process performance for removal of DBPs/DBP precursors and CECs are currently limited.

5.3.1.3 Upstream processes within reuse treatment train.

5.3.1.3.1 Coagulation, flocculation, and sedimentation. Coagulation is a proven process that destabilizes the charge on colloids and suspended solids, including bacteria and viruses, and removes TOC through the practice of enhanced coagulation (Kawamura 1999). Coagulation/flocculation/sedimentation can be used for potable reuse applications (e.g., HRSD's Sustainable Water Initiative for Tomorrow [SWIFT] program) to remove TOC and/or seek pathogen removal credits upstream of biofiltration (Hogard et al. 2021).

5.3.1.3.2 Preoxidation. The potential objectives of preoxidation can include oxidation of organic compounds into more assimilable forms; removal of trace organic compounds and contaminants of emerging concern, either through oxidation or a combination of oxidation and subsequent biodegradation; and pathogen inactivation. UV/AOP (with either H_2O_2 or chlorine) can also serve as a preoxidation process upstream of biofiltration,

Table 5-5 Biofilter design considerations consistent with surface water biofiltration design (see Chapter 2)

Category	Parameter
Water quality	Influent water quality
	Finished water quality goals
Process design	**Preoxidation**
	Media type
	Coagulation/flocculation pretreatment
	L/d ratio
	Loading rate and EBCT
	Underdrain design
	Chemical feeds
	Redundancy
Operations	Head loss
	Filter run time
	UFRV
	Hydraulic monitoring
	Water quality monitoring
	Startup and acclimation
Backwash	Backwash sequence (flow/loading rate, duration)
	Pressure
	Bed expansion
	Backwash water handling capacity
	Use of disinfectants
	Use of air scour

Source: Adapted from Brown et al. (2016)
EBCT—empty bed contact time, UFRV—unit filter run volume
Note: Specific design parameters that may differ in reuse applications compared with traditional drinking water treatment will be discussed further and are shown in bold.

as presented in Aurora Water's case study later in this chapter. Effectiveness of UV/AOP treatment is highly dependent on the UV transmittance and DOC of source waters (Marron et al. 2019). In addition to oxidizing organic matter, UV/AOP treatment can degrade a variety of CECs, including many pharmaceuticals, pesticides, NDMA, and 1,4-dioxane (Malley et al. 2018, Lester et al. 2016, Kruithof et al. 2007). NDMA and some pharmaceuticals can be almost completely removed using photolysis (UV light) alone, whereas other contaminants require AOP for degradation (Ngumba et al. 2020). Specific objectives are site/utility specific and may influence the oxidant selected and dosing strategy.

Ozone is a powerful oxidant and disinfectant commonly used for water/wastewater treatment and is commonly used as a preoxidant upstream of biofiltration. Research has been conducted at a variety of ozone-to-TOC ratios from 0.2 to 2.0 (Sundaram et al. 2020, Gerrity et al. 2015). The optimal ozone dose depends on several site-specific factors, including ozone demand, formation of ozone-reactive DBPs, and ability to quench excess ozone at high doses. An O_3:(TOC+nitrite) ratio of approximately 0.8, for example, has shown pathogen inactivation in some studies. This correlation is strongest with nitrite included in the ratio; however, the ratio can be effectively simplified to O_3:TOC when nitrite is not detected or is insignificant relative to the TOC concentration. Some studies have shown that this O_3/TOC ratio of approximately 0.8 can degrade TOC to more assimilable forms without exceeding bromate/NDMA targets (Sundaram et al. 2020, Funk et al.

2018, Nyfennegger et al. 2018, Salveson et al. 2019). Higher O_3/TOC ratios may result in more assimilable carbon (e.g., higher TOC removal by biofiltration) but should be tested/used with DBP formation potential in mind. Preoxidation by ozone results in high DO concentrations and can cause elevated head loss because of air binding in downstream biofilters.

It is possible to implement biofiltration for reuse purposes without the use of a preoxidant. When a preoxidant is not used, removal of remaining organics after secondary treatment is expected to be limited to the assimilable or biodegradable organic compounds (e.g., AOC or BDOC) for inert media. Biofiltration can prove to be useful in this case for the removal of specific contaminants such as manganese, T&O-causing compounds, and overall turbidity. If GAC media are used, additional removal would be observed through adsorption if contactors are reactivated on a regular basis.

5.3.1.4 HLR and EBCT. The HLR and EBCT are important parameters for sizing the biofilter volume and achieving finished water quality goals. EBCT design for biofiltration has been discussed in Chapter 2. Most drinking water utilities operate with an EBCT ranging from 5 to 15 min, depending on the media used (Dickenson et al. 2018, Brown et al. 2016). This framework has been used in water reuse applications. EBCTs toward the higher end of this range may be needed for trace chemical constituent removal because these compounds are generally more resistant to biodegradation than NOM (Brown and Lauderdale 2006, Westerhoff et al. 2005). Higher EBCTs are generally required for reuse applications, ranging from 10 to 30 min. Both capital and recurring costs of purchasing and replacing filter media along with maximizing contaminant removal should be considered when selecting an EBCT.

5.3.1.5 Media type. Media selection is another important component with respect to biofilter design. The media's capacity to support biological activity should be considered, and media size plays an important role in the available surface area for biological attachment (Zhang and Huck 1996). GAC media can provide some additional organics removal through adsorption and its ability to maintain higher concentrations of biomass (Simpson 2008). In reuse applications, it is expected that media with a larger adsorption capacity, such as GAC, would be the most beneficial because of its ability to remove organics and provide additional surface area for biomass growth. However, GAC media is generally more expensive compared with anthracite. Media size and surface area can dictate the capacity for solids removal. Common biofilter media and relative properties are presented in Table 5-6.

Factors that need to be considered when choosing between anthracite or GAC include the following:

- cost—GAC is more expensive than anthracite;
- adsorptive capacity needed—anthracite does not have adsorption capacity, GAC does have adsorption capacity;
- fluidization characteristics—anthracite is denser than GAC; and

Table 5-6 Typical filter media specifications

Media	Effective Size *mm*	UC	Density *g/mL*	Porosity *%*	Depth *in.*
Sand	0.4–0.8	1.3–1.7	2.65	40–43	10–12*
Anthracite	0.8–2.0	1.3–1.7	1.4–1.8	47–52	72†; 24–60*; 33–43‡
GAC	0.8–2.0	1.3–2.4	1.3–1.7	N/A	48†; 24–72*; 5–15‡

GAC—granular activated carbon, UC—uniformity coefficient
*Dual-media filters
†Mono-media filters
‡Tri-media filters

- resultant biological activity—both media types support effective biological activity. Some research has shown that under stressed conditions, GAC supports more robust biological activity (Emelko et al. 2006, Liu et al. 2001, Wang et al. 1995).

5.3.1.6 Backwashing strategies. Considerations for backwashing rates and durations would generally follow guidance specified in Section 2.3.2.5. Use of disinfectants in backwash water may be used to control biofouling, or, if backwash water supply is derived from a clear well, may be used for disinfection. The majority of biofilters in reuse applications use unchlorinated backwash water, particularly when implemented as part of advanced wastewater treatment. However, some drinking WTPs with planned or unplanned IPR use chlorinated or chloraminated backwash, such as Gwinnett County and Greater Cincinnati Water Works. An optimization of backwashing strategy may also be necessary in reuse applications in which preozonation could cause high or supersaturated dissolved oxygen (DO) concentrations entering the biofilter (e.g., air binding).

5.4 OPERATION

Biofilter monitoring and operational strategies during startup, shutdown, and challenging conditions should be well-understood by all operators. Additionally, troubleshooting strategies should be identified and well-known so they can be easily implemented. The proceeding sections focus on attributes specific to water reuse scenarios; refer to Chapter 2 for additional discussion of relevant strategies.

5.4.1 Monitoring

General guidance outlined in Section 2.4.1 should be referenced for most reuse applications, whether implemented as part of advanced wastewater treatment or within a drinking WTP. This section seeks to outline specific components of monitoring unique to reuse applications. As introduced in Chapter 1, monitoring parameters are described in three categories: operational parameters, water quality parameters, and microbial parameters.

5.4.1.1 Operational parameters. Effects on filter run time and head loss are well-known for many biofilters used in potable water production, particularly during warmer weather conditions. For reuse applications, particularly advanced treated water, effects on filter hydraulics may be more pronounced because of higher loading rates of electron donors (e.g., NOM, often measured as COD or TOC) and resulting higher biomass. The rate of head loss accumulation, clean bed head loss, and terminal head loss should be regularly monitored to assess whether potential excess biofouling is occurring.

Monitoring upstream chemical doses, including use of coagulants and oxidants, is important for biofilters. Carryover of flocculants after settling may affect the bioavailability of nutrients such as phosphorous. If oxidation is used upstream, oxidant residual should be monitored regularly at the filter influent to optimize residual carryover to protect the biological community.

5.4.1.2 Water quality parameters. General water quality parameters that should be monitored include turbidity, temperature, and pH. Turbidity is an important parameter for assessing pathogen removal, and thus, may be a key part of the strategy for achieving disinfection credits. Additionally, if ammonia is present at the biofilter influent, aerobic biofilters will readily nitrify ammonia to nitrate; if anaerobic conditions exist (e.g., DO < 0.5 mg/L) within the biofilter, denitrification may also occur. Nitrification consumes alkalinity and denitrification produces alkalinity. Thus, alkalinity is also an important monitoring parameter. An example treatment train with biofiltration in the advanced wastewater treatment from Gwinnett County is shown in Figure 5-5. Total nitrogen was conserved from secondary effluent through the UF membrane and GMF, and finally through preozonation/biofiltration (labeled as "FWH BAC Effluent").

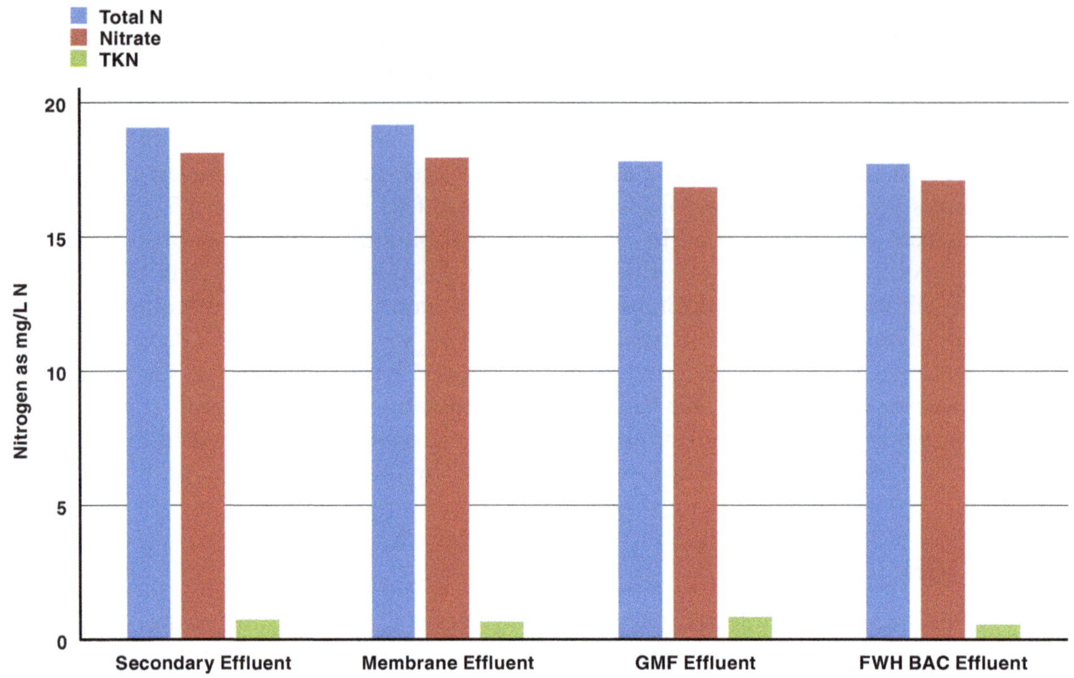

BAC—biologically active carbon, GMF—granular media filtration, TKN— total Kjeldahl nitrogen

Figure 5-5 Nitrogen species through advanced wastewater treatment (Funk et al. 2018)

Organic carbon monitoring should include TOC, DOC, BDOC, and/or AOC, as well as wastewater-specific parameters such as COD, soluble COD, BOD, and volatile fatty acids. TOC and DOC are important for characterizing organic carbon removal, whereas BDOC and AOC are used to characterize the loading of biodegradable substrate.

Other constituents typically monitored in reuse applications include trace contaminants such as PPCPs and EDCs; DBPs such as THMs, HAAs, nitrosamines, and nitrogenous DBPs; pesticides and herbicides; and contrast media (e.g., medical waste).

Although NDMA removal is well-documented in aerobic biofilters, previous research has indicated that NDMA may be formed when implemented as part of advanced treatment and upstream secondary treatment does not include nitrification. If oxygen is completely consumed in the biofilter and conditions become nitrate-reducing (e.g., denitrifying), then NDMA may be produced (Stanford et al. 2017), possibly associated with wastewater-derived soluble microbial products (Krasner et al. 2012). Thus, NDMA monitoring is required.

In Chapter 2, Table 2-9 summarizes the recommended monitoring for water quality parameters. Apart from these parameters, other parameters need to be monitored regularly for reuse applications. Table 5-7 provides a summary of these additional parameters.

5.4.1.3 Microbial parameters. Microbial activity may be assessed by measuring adenosine triphosphate (ATP) concentration on filter media, the biofilm formation rate, or DO consumption. However, DO consumption should only be used for monitoring filters without direct upstream ozonation (Hooper et al. 2019). Limited data are also published from microbial community analysis using 16S ribosomal RNA sequencing, and metagenomics have been used to identify class, family, or genus-level identification, characterize microbial community diversity, and identify organisms with specific functional capabilities. An example microbial community analysis for biofilters at Gwinnett County's F. Wayne Hill Water Resources Center and the Shoal Creek Filter Plant drinking WTP is shown in Figure 5-6. Amoeba, known to harbor opportunistic pathogens, as well as genus-level identification of opportunistic pathogens, such as Legionella, have been observed in some

Table 5-7 Recommended monitoring for general water quality parameters

Parameter	Sampling Location	Sampling Method	Sampling Frequency	Performance Indication
Alkalinity	Biofilter influent and effluent	Grab	Regular (e.g., weekly); frequent in case of any upsets in the upstream WWTP	Indicates nitrification/ denitrification in biofilter.
Nitrogen (ammonium, nitrate, nitrite)	Biofilter influent and effluent	Grab inline if supplementing ammonia in the process	Regular (e.g., weekly)	Affects biological growth and activity. Needs to comply with final effluent limitations.
Trace contaminants (PPCPs, EDCs, PCBs, PFAS)	Biofilter influent and effluent	Grab	Occasional (e.g., quarterly); more frequent for specific constituents and during startup/ acclimation	Removal rate of target compounds indicates biofilter performance.
DBPs (TTHM, HAA, bromate, NDMA)	Biofilter influent and effluent	Grab	Regular (e.g., once or twice a week)	Indicates treatment performance for removal of DBP precursors.

DBP—disinfection by-product, EDC—endocrine disrupting compound, HAA—haloacetic acid, NDMA—N-nitrosodimethylamine, PCB—polychlorinated biphenyl, PFAS—polyfluoroalkyl substances, PPCP—pharmaceutical and personal care product, TTHM—total trihalomethane, WWTP—wastewater treatment plant

biofilters in potable reuse applications (Gerrity et al. 2018). However, when DPR-treated water was compared with conventional potable water, there was no significant increase in the incidence of opportunistic pathogens or antibiotic resistance genes in simulated distribution system premise plumbing pipe-loops (Garner et al. 2019).

5.4.2 Initial Startup/Acclimation

Transient operating periods include startup, acclimation, and hibernation. It is important to know how the biofilter will perform during these periods in terms of organics and turbidity removal. The acclimation period is affected by several hydraulic and water quality parameters, illustrated in Figure 5-6, which can influence the complexity and/or length of the acclimation period. Nutrient addition could allow faster acclimation (Lauderdale et al. 2012). Additional factors affecting biofilter acclimation are depicted in Figure 5-7.

In reuse scenarios, special care may be given to encourage beneficial microbial growth and enhance treatment. For example, the acclimation time for biological NDMA removal was shown to be greater than the time needed for removal of readily biodegradable organics (Sundaram et al. 2020). Table 5-8 summarizes startup and acclimation considerations that may be coupled with mitigation strategies to improve treatment.

5.4.3 Steady-State Operation

Steady-state condition is typically defined as biofilters operating under normal conditions postacclimation. During a steady-state operating period, there are many daily considerations to note, and the utility should have the ability to troubleshoot any common challenges that could arise. For example, Figure 5-8 shows a DOC profile from filter startup to shutdown, demonstrating that longer filter runs improve biofilter performance.

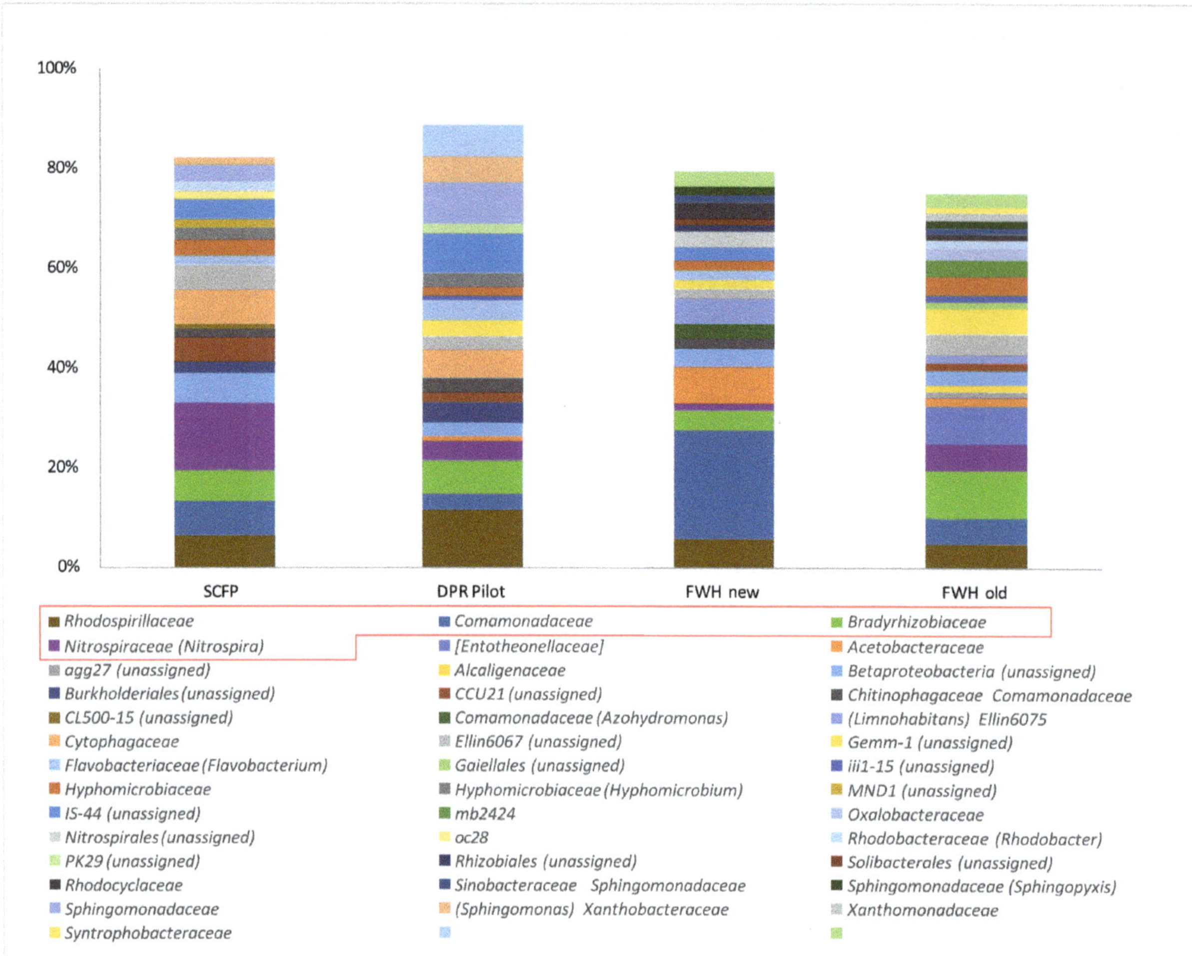

DPR—direct potable reuse, FWH – F Wayne Hill, GAC—granular activated carbon, SCFP—Shoal Creek Filter Plant, WRC—Water Resources Center

Figure 5-6 Example microbial community analysis report (Funk et al. 2018)

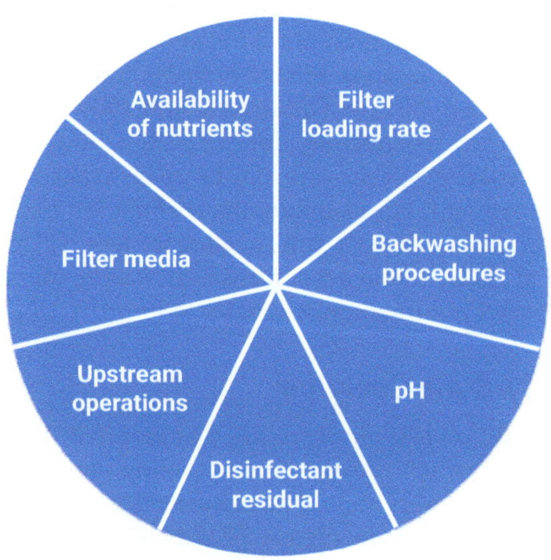

Source: Brown et al. (2016)

Figure 5-7 Parameters affecting biofilter acclimation

Table 5-8 Startup and acclimation considerations for reuse applications

Startup/Acclimation Considerations	Potential Mitigation Strategies
Greater organics/PPCPs loading	Consider extending EBCTs to allow for greater removal, optimize nutrient loading to enhance biological growth, and investigate seeding of organisms (Riley et al. 2018, Halle et al. 2015, Lauderdale et al. 2012).
Greater nutrient substrate loading	Consider enhancements to optimize nitrifier growth, such as increasing EBCT, pH maintenance between 7.5 and 8.5 SU, temperature monitoring, and DO control to maintain DO around 3–4 mg/L if not in excess from ozone addition (Rattier et al. 2014, Jones and Hood 1980, Stenstrom and Poduska 1980).
High ozone dose and associated DO concentrations	Investigate methods for quenching DO and reducing high DO concentrations that may be inhibitory to organisms (quenching agents such as peroxide) (Vaidya et al. 2020).

EBCT—empty bed contact time, DO—dissolved oxygen, SU—standard unit, PPCP—pharmaceutical and personal care product

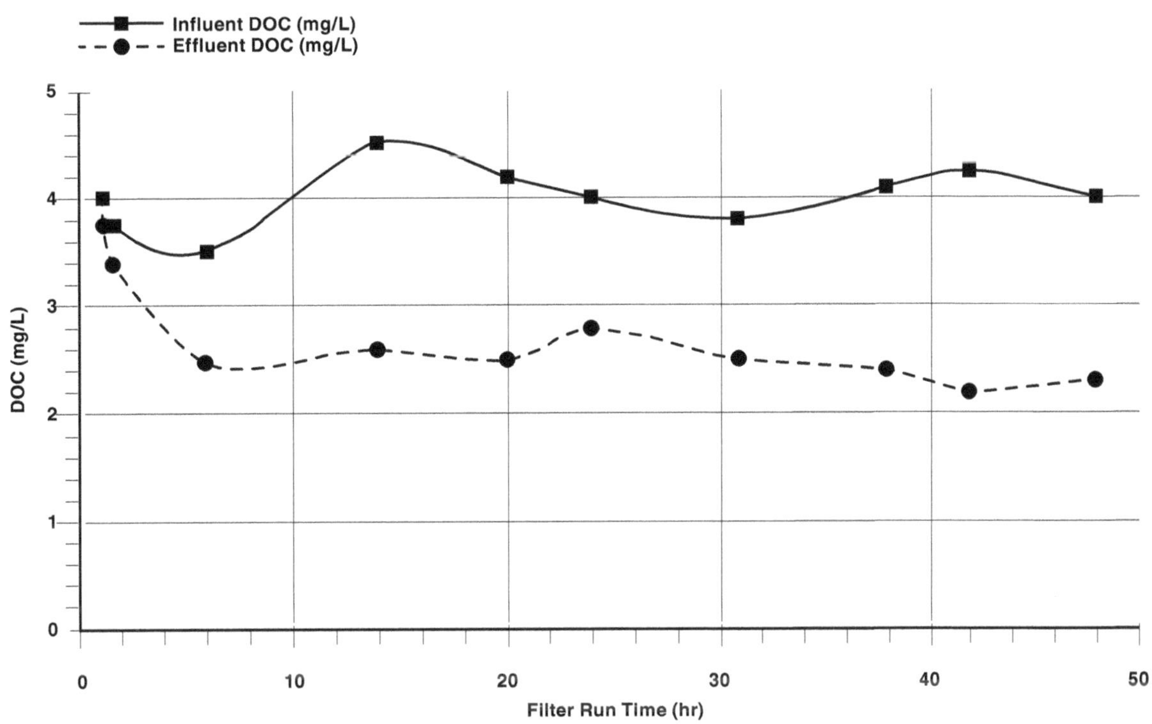

DOC—dissolved organic carbon

Figure 5-8 DOC profile from filter startup to shutdown

5.4.4 Shutdown and Restart

Biofilter shutdown is sometimes required for seasonal maintenance, low flow conditions, or handling a process upset. However, once operation is suspended, biofilter conditions can change significantly, consequently extending the reacclimation period during restart. During a shutdown event, biofilter media can be left dry or wet/submerged. Both conditions bring unique challenges, described in Figure 5-8. Thus, biofilters should be thoroughly backwashed before restart to remove any inactive biomass and other sloughed materials that may have accumulated. Additionally, when biofilters are placed off-line and out of service for an extended period (e.g., more than a few days), daily air scouring can be used to maintain aerobic conditions (Hooper et al. 2019, Brown et al. 2016; Figure 5-9).

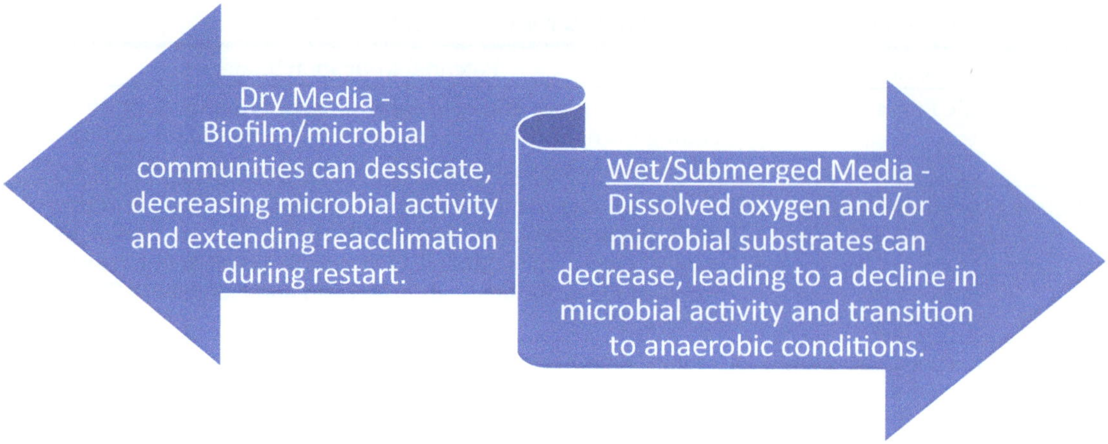

DO—dissolved oxygen

Figure 5-9 Media idling strategies

Table 5-9 Example biofilter optimization strategies and associated level of effort for implementation

Strategy	Relative Effort	Explanation
Nutrient augmentation	Minimal	Requires chemical metering pump for phosphate and/or ammonia dosed with a low throughout. Capital and chemical costs are relatively low (Lauderdale et al. 2018).
Backwash rate and duration adjustment	Minimal	Requires study to evaluate backwashing efficacy and basic media characterization studies such as media particle size distribution, floc retention analysis, and media loss analysis.
pH adjustment	Minimal to moderate	Requires chemical metering pump and dose control logic to maintain pH.
Backwashing disinfectant type and dose adjustment	Moderate	Modification of disinfectant type may require chemical dosing system for quenching agent or use of alternative oxidant. Chemical metering pumps and chemical costs would be required.
Preoxidation type and dose adjustment	Moderate to complex	Requirements for pathogen LRVs or removal of other water quality parameters may dictate selection of oxidant type and dose.
EBCT Adjustment	Moderate to complex	Longer EBCTs may allow for increased contaminant removal. This may be achieved by placing additional filters on-line. If additional filtration units are not readily available, costs for expansion could be significant.
Media type changeout	Complex	Requires change out of media and potentially upgrades to filter basins and underdrains.

EBCT—empty bed contact time, LRV—log removal value

5.4.5 Optimization

General guidance outlined in Section 2.4.5 should be referenced for most reuse applications, whether implemented as part of advanced wastewater treatment or within a drinking WTP. This section seeks to outline specific components of optimization unique to reuse applications.

Table 5-9 summarizes optimization strategies that have been implemented for increasing bioactivity and associated degradation of target contaminants. Higher COD present in the biofilter influent in reuse applications may result in greater accumulation of biomass and hydraulic effects compared with biofiltration used for potable water production. Nutrient augmentation using either phosphate or ammonia has the potential in some facilities to mitigate excursions with excess extracellular polymeric substance production causing increased filter head loss accumulation rates and decreased filter run times in drinking water applications (Lauderdale et al. 2012), and thus, could also be considered for reuse applications. If low concentrations of phosphate are added for optimization (e.g., <0.5 mg-P/L), effects on the downstream treatment processes and corrosion control should be considered. pH adjustment has also been used as a control measure to maintain bioavailable ortho-phosphate as well as to improve contaminant removal. Thus, pH optimization may be beneficial depending on influent water quality and treatment goals.

5.4.6 Troubleshooting

Common challenges that may arise during biofilter operation and mitigation measures are presented in Section 2.4.6.

One issue with reuse applications implementing ozone-biofiltration is supersaturated DO in the biofilter influent due to upstream ozone. Water supersaturated with DO can off-gas and, in some cases, result in elevated head loss because of air binding in downstream biofilters. For an ozone dose of 5 mg/L, the oxygen dose applied is about 45 mg/L. Hooper et al. (2019) reported DO concentrations up to 32 mg/L in biofilter influent water after ozonation. Air binding can be mitigated by "bumping" the filter, but caution should be used with this method to ensure that subsequent backwash adequately cleans the filter bed. Because volatilization of DO occurs at the top of the bed or within the filter bed, DO monitoring of biofilters is not recommended at facilities that practice preozonation. If significant shifts in oxygen concentrations occur because of operational changes (e.g., higher ozone dose or lower ozone demand), accumulation of reactive oxygen species such as peroxide and superoxide may inhibit bioactivity and thus response should be monitored (Imlay 2013, Korshunov and Imlay 2006).

5.5 CASE STUDIES

The following sections provide a detailed description of facilities implementing potable reuse and included cases associated with IPR at a drinking water treatment facility (Aurora Water) and IPR for surface water augmentation (City of San Diego) and IPR for aquifer recharge (HRSD).

These facilities are listed in Table 5-10, in which location, IPR category, and biological treatment goals are outlined.

5.5.1 Indirect Reuse at a Drinking Water Treatment Facility

Peter D. Binney Water Purification Facility (BWPF), Aurora, Colo.
Author: Kevin Linder, Aurora Water

In 2010, the City of Aurora, Colo., commissioned one of the most unique drinking water delivery projects in operation. The Prairie Waters Project (PWP) was an innovative approach to use a water right the City has in place allowing a portion of the supply to be used to extinction. The first step in the PWP is extracting water out of the South Platte River downstream of the Denver Metro area, an urban watershed, with multiple large wastewater plant discharge points adding to the source. The first step in this nonmembrane IPR advanced treatment system is riverbank filtration (RBF). The RBF effluent then goes through an aquifer recharge and recovery (ARR) step, and then the water is conveyed 34 miles through three pump stations to BWPF.

Table 5-10 IPR case studies

Facility, Location	IPR Category	Selected Biological Treatment Goals
BWPF, Aurora, Colo.	IPR at a drinking water treatment facility	• Turbidity <0.10 NTU • Reduce T&O compounds • TOC removal • Stabilize AOC downstream of advanced oxidation
Pure Water San Diego, San Diego, Calif.	IPR via surface water augmentation	• Act as an additional pathogen barrier: 1-log *Cryptosporidium*, 6-log *Giardia*, and 6-log virus • Provide additional barrier for organic compounds that are poorly removed by RO • Reduce chemical requirements and operational costs for membrane treatment • Improve RO concentrate quality
HRSD, Suffolk, Va.	IPR via aquifer recharge	• Turbidity <0.10 NTU • TOC removal • Manganese removal • NDMA removal

AOC—assimilable organic carbon, BWPF—Peter D. Binney Water Purification Facility, HRSD—Hampton Roads Sanitation District, IPR—indirect potable reuse, NDMA—N-nitrosodimethylamine, RO—reverse osmosis, T&O—taste and odor, TOC—total organic carbon

5.5.1.1 Facility description. This facility has two treatment trains, as shown in Figure 5-10. The advanced South Platte (SP) treatment train takes the water being delivered from the RBF/ARR at North Campus and treats it through an advanced treatment regimen consisting of precipitative softening, a UV/hydrogen peroxide advanced oxidation step followed by biologically active filtration (BAF) with carbon, and a GAC adsorption final barrier before being blended with the other source in the chlorine contact basin. The other train at BWPF, the Aurora Reservoir (AR) train, is a conventional treatment train treating a high-quality mountain source using ferric chloride and cationic polymer, potassium permanganate, and prechlorine with BAF processes before going onto the blended disinfection step with free chlorine primary disinfection and chloramine secondary disinfection going into the distribution system.

5.5.1.2 Biological treatment goals. BWPF has the following biofiltration goals that were implemented at the plant startup in 2010:

- combined filter effluent (CFE) turbidity <0.10 NTU 100 percent of the time for both treatment trains,
- individual filter effluent (IFE) turbidity <0.10 NTU 100 percent of the time for both treatment trains,
- reduce algal metabolite (geosmin and 2-Methylisoborneol (MIB)) levels from Aurora Reservoir on the AR train,
- reduce T&O compounds in both treatment trains,
- TOC removal in both treatment trains, and
- stabilize AOC downstream of advanced oxidation in SP train.

5.5.1.3 Operations and performance. At this facility, biofiltration has resulted in higher CFE and IFE turbidities and significantly shorter filter runs than traditional multimedia filters. BWPF investigated the overall effects adding prechlorine in the coagulant addition step of both treatment trains to evaluate the effect on settling and filtration of particulate versus effect on biological performance of the filter and disinfection unit process. Bench-scale work demonstrated improved particle destabilization by reducing negative surface charge of the particle being coagulated, resulting in improved coagulation of particulate, including algal cells, translating into increased effectiveness of settling and biofiltration turbidity removal. Operations staff tenaciously analyzed the decay rate of chlorine

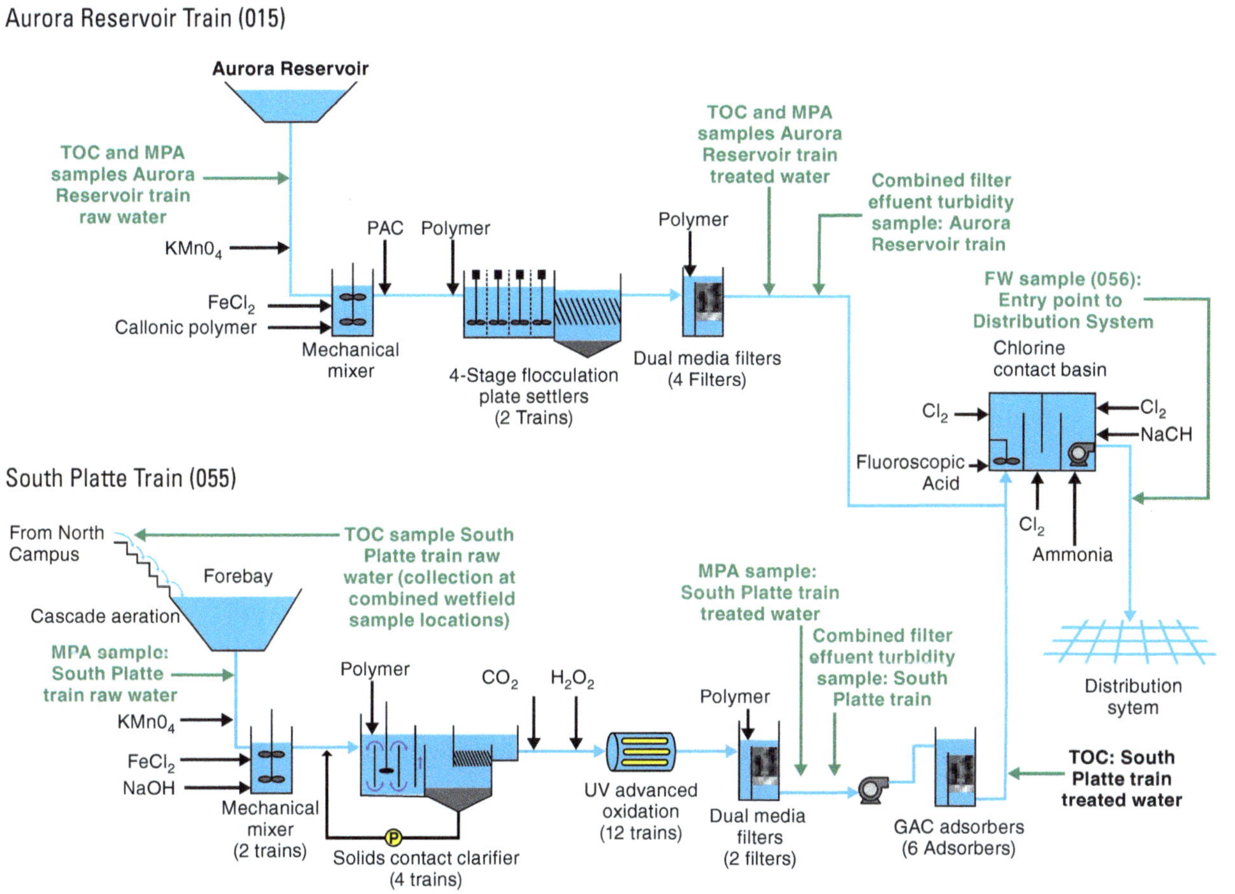

FW—fresh water, GAC—granular activated carbon, MPA—microscopic particulate analysis, PAC—powdered activated carbon, TOC—total organic carbon, UV—ultraviolet

Figure 5-10 Binney Water Purification Facility treatment schematic

residual to ensure no free chlorine made it to the biological filter influent. Biological performance was evaluated using ATP levels and TOC removals. Both of those metrics were not significantly affected at the bench scale by the addition of prechlorine before BAF. The bench scale also demonstrated no significant increase in disinfection by-products resulting from the addition of prechlorine in the coagulation step.

With successful bench-scale trial results, full-scale implementation was initiated. The AR train-initiated prechorine first. After several weeks of positive results on the AR train that confirmed the bench-scale findings, prechlorination was implemented on the advanced SP treatment train. The performance of the BAF on both trains was improved with no observable increase in DBPs or detrimental effect on the efficacy of contaminant removal through the BAF.

5.5.1.4 Lessons learned.

- It is critical to define chlorine demand frequently and adjust prechorine dose accordingly, always erring on the side of underfeeding to ensure no free chlorine makes it to the BAF.
- Most surface waters contain some level of ammonia, which will react with the free chlorine used in the prechlorination step. BWPF staff has found total chlorine concentrations of <0.20 mg/L have been safe for the biota at BWPF.
- This process is plant specific, so bench- and/or pilot-scale testing is recommended before full-scale implementation.

5.5.2 IPR via Surface Water Augmentation—or Advanced Wastewater Treatment

Pure Water San Diego, San Diego, Calif.

Authors: Dr. Daniel Gerrity, Southern Nevada Water Authority, and Dr. Aleksey Pisarenko, Trussell Technologies, Inc.

In 2011, the City of San Diego, Calif., commissioned an advanced water purification demonstration project, "Pure Water San Diego," as part of a larger plan to implement IPR via surface water augmentation. At that time, IPR via groundwater replenishment had been an established practice in California for decades, but Pure Water San Diego aimed to become the first project in California to augment a drinking water reservoir with recycled water. Pure Water San Diego is now expected to encompass a series of AWPFs augmenting multiple reservoirs throughout the County of San Diego. The initial phase of Pure Water San Diego involved a combination of large-scale demonstration, extensive public outreach, and comprehensive engineering and scientific assessments with oversight by expert panels, regulators, and other stakeholders. By 2018, this process facilitated adoption of California's Surface Water Augmentation Regulations and the start of Phase 1 construction of the 30-mgd Pure Water Facility (San Diego 2020).

5.5.2.1 Facility description. The demonstration-scale AWPF received tertiary effluent from the adjacent North City Water Reclamation Plant. At the time of testing, the full-scale water reclamation plant had a capacity of 30 mgd and received primarily domestic wastewater. The wastewater was treated with activated sludge with biological nutrient removal, secondary clarification, anthracite media filtration, side-stream partial

Table 5-11 Aurora Water Binney WTP operational parameters

Parameter	SP Train	AR Train
Plant design capacity (mgd)	16.67	33.3
Average production (mgd)	6	12
Maximum production (mgd)	10	33
Source water type	Urban watershed	Lake
Plant process scheme	Softening, UV/AOP, BAF, GAC	Conventional—BAF
Primary coagulant	NaOH	Ferric chloride
Primary coagulant dosage (mg/L)	180–220	25–50
Polymer type	Cationic	Cationic
Polymer dosage (mg/L)	1–4	1–3
Settling basin residence time (h)	NA	NA
Filter aid polymer type	NA	NA
Filter aid dosage (mg/L)	NA	NA
Number of filters	2–6*	4–8*
Primary disinfectant	Chlorine	Chlorine
Primary disinfectant dosage	NA	NA
Point of disinfection	CCB (postfiltration)	CCB (postfiltration)
Secondary disinfectant type	Chloramine	Chloramine
Secondary disinfectant dosage	NA	NA
Point of disinfection	CCB effluent	CCB effluent

AOP—advanced oxidation process, BAF—biologically active filtration, CCB—chlorine contact basin, mgd—million gallons per day, SP—South Platte, UV—ultraviolet, WTP—water treatment plant

*Valving in the filter influent channel allows some BAF to be used on either train depending on the production rate of the trains.

Table 5-12 Aurora Water Binney WTP filter design components

Parameter	SP Treatment Train	AR Treatment Train
Media type	GAC	GAC
Media specification	1.4 mm/72"	1.4 mm/72"
Type of GAC	Bituminous	Bituminous
Virgin or reactivated	Virgin	Virgin
EBCT (min)	5.9	5.3
Total filter media depth (in.)	72	72
Filter area per filter	780	780
HLR (gpm/ft^2)	9.9	9.9
Type of underdrain system	IMS	IMS
BW water type	Unchlorinated	Unchlorinated
BW rate	High rate: 22 gal/min/ft^2	High rate: 22 gal/min/ft^2
BW sequence	2-min air scour, 3-min low-rate BW, 3-min high-rate BW.	2-min air scour, 3-min low-rate BW, 3-min high-rate BW.
Bed expansion during BW	20–25%	20–25%
BW frequency	Average of two times per week at 7–8 gpm/ft^2	Average of one time per week at 7–8 gpm/ft^2
Average head loss at BW Initiation	14 ft	14 ft
Air scour use	Yes	Yes
Surface wash use	No	No
Media extraction system	Water induction	Water induction

AR—Aurora Reservoir, BW—backwash, EBCT—empty bed contact time, HLR—hydraulic loading rate, GAC—granular activated carbon, IMS—integrated membrane system, SP—South Platte, WTP—water treatment plant

demineralization by electrodialysis reversal, and disinfection with chlorination. Historically, the reclaimed water was distributed to irrigation, landscaping, and industrial customers through more than 79 miles of distribution pipelines across northern San Diego.

The demonstration-scale AWPF, formerly known as the Pure Water Demonstration Facility, received and treated 1 mgd of the full-scale tertiary effluent (i.e., treated wastewater before disinfection) with ozone, BAC with an EBCT of 15 min, MF or UF, RO, and UV/AOP (Figure 5-11). Ozone was dosed at an O_3/TOC ratio of approximately 1.2 to achieve a target ozone CT of 3.8 mg-min/L and 1-log *Cryptosporidium* inactivation (based on a temperature of 20.5°C)

5.5.2.2 Facility biological treatment goals. At a minimum, California's Surface Water Augmentation Regulations require "FAT," consisting of RO and an AOP. MF or UF is generally included for RO pretreatment and to satisfy certain pathogen log removal value requirements. For the AOP, UV-based treatment is generally used because of its efficacy against a wide range of target pathogens and certain organic compounds, including NDMA. Although not required in a regulatory context, ozone-BAF was integrated into Pure Water San Diego's demonstration-scale AWPF to achieve several objectives:

- Act as an additional pathogen barrier: 1-log *Cryptosporidium*, 6-log *Giardia,* and 6-log virus;
- Provide additional barrier for organic compounds that are poorly removed by RO;
- Reduce chemical requirements and operational costs for membrane treatment; and
- Improve RO concentrate quality.

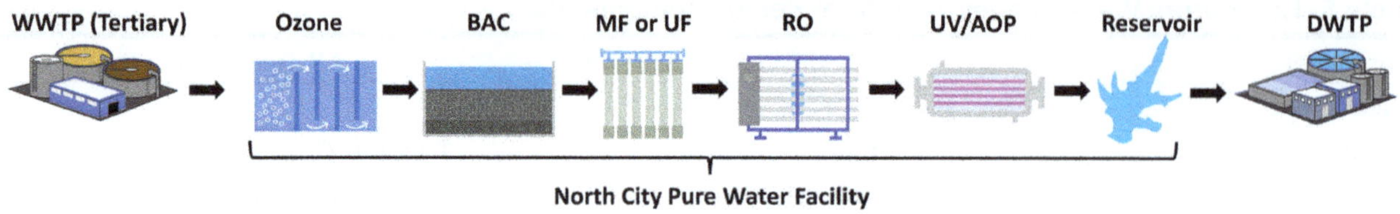

AOP—advanced oxidation process, AWPF—advanced water purification facility, BAC—biologically active carbon, DWTP—drinking water treatment plant, MF—microfiltration, RO—reverse osmosis, UF—ultrafiltration, UV—ultraviolet

Figure 5-11 Pure Water San Diego: Phase 1 treatment schematic

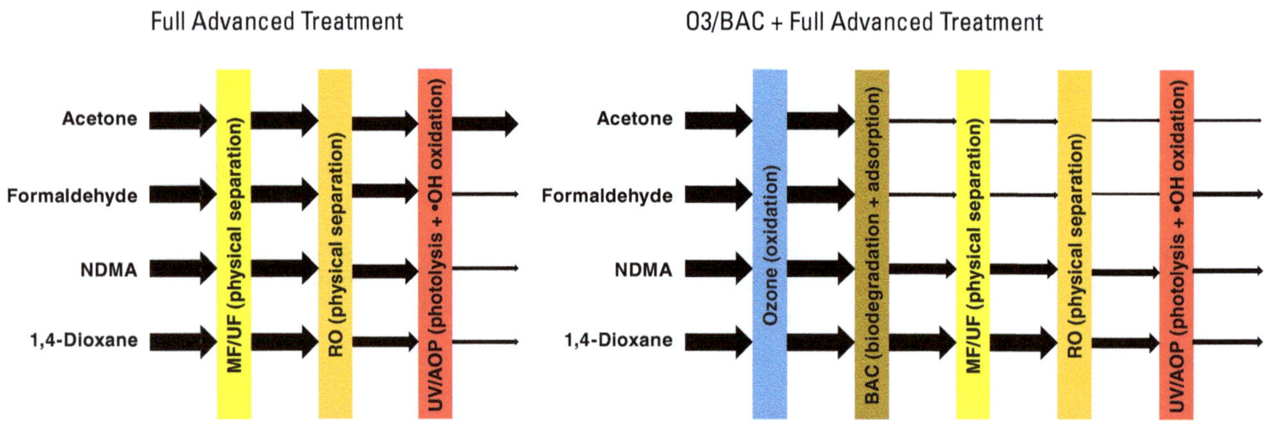

Source: Adapted from Tackaert et al. (2019)

AOP—advanced oxidation process, AWPF—advanced water purification facility, BAC—biologically active carbon, MF—microfiltration, NDMA—N-nitrosodimethylamine, RO—reverse osmosis, UF—ultrafiltration, UV—ultraviolet

Figure 5-12 Observed removals of spiked acetone, formaldehyde, NDMA, and 1,4-dioxane at the Pure Water San Diego AWPF

5.5.2.3 Operations and performance.

5.5.2.3.1 Added robustness. In addition to inorganic constituents, RO and UV/AOP serve as robust barriers against a wide variety of organic compounds, but they are not absolute barriers. For example, the West Basin Municipal Water District observed formaldehyde in its product water after implementing ozonation upstream of its membranes (i.e., in the absence of biofiltration), and the Orange County Water District observed a one-time spike of acetone in its RO feed and permeate. Tackaert et al. (2019) demonstrated that by incorporating ozone-BAC into Pure Water San Diego's AWPF, spikes in biodegradable organic compounds that are known to persist through RO, such as formaldehyde and acetone, could be attenuated by more than 65 to 90 percent (Figure 5-12). NDMA may also form during ozonation (Gerrity et al. 2015), and RO is only a partial barrier for this low molecular weight organic compound, with a typical removal of <25 percent (Tackaert et al. 2019). Although limited formation of NDMA was observed in the case of San Diego, biofiltration may be a critical barrier for NDMA attenuation in other ozone-based systems to avoid overloading the RO and UV/AOP processes (Bacaro et al. 2019). Finally, ozone provides added robustness against a wide range of trace organic compounds (Lee et al. 2012), including regulated compounds such as 1,4-dioxane (Tackaert et al. 2019).

5.5.2.3.2 Cost savings due to reduced membrane pretreatment requirements. Chloramination has traditionally been used to control membrane biofouling. Beyond the operational costs associated with the addition of chlorine and ammonia, reactions between chloramines and organic precursors are known to form DBPs, such as NDMA. Potable reuse systems in California must then rely on RO and UV/AOP to remove NDMA to below the 10-ng/L notification level. Studies have shown that NDMA can even reform in the product water in the presence of residual chloramines.

Pure Water San Diego evaluated the feasibility of using ozone-BAC for mitigation of membrane fouling and to reduce chemical use. Ozonation has the potential to significantly transform bulk organic matter, thereby reducing organic fouling, and biofiltration is effective at eliminating biodegradable DOC, thereby eliminating a bacterial food source and reducing biological growth on membranes. At the demonstration-scale AWPF, testing demonstrated that, with upstream ozonation and biofiltration, chloramines could be maintained at <1 mg/L in the RO feed, resulting in chemical cost savings of up to $0.5 million annually while still adequately controlling membrane fouling. Although not verified during the study, it is expected that lower chloramine exposure would also reduce oxidative damage to the RO membranes and lengthen the life of the membrane modules. Finally, lower residual chloramine concentrations would also lead to chemical and energy cost savings in relation to the UV/AOP. Chloramines absorb UV light and act as hydroxyl radical scavengers, which can significantly decrease the efficiency of UV photolysis and advanced oxidation.

5.5.2.3.3 Improved RO feed, permeate, and concentrate quality. As an ancillary benefit, implementation of ozone-BAC also improved the quality of the RO feed, concentrate, and permeate in the Pure Water San Diego AWPF (Tackaert et al. 2019). Ozone-BAC achieved a 40 to 50 percent reduction in TOC, which resulted in an approximately 25 percent reduction in the RO permeate TOC and a reduction of 40 percent in the RO concentrate TOC. The corresponding transformation of bulk organic matter is shown in Figure 5-13. Because ozonation is effective in attenuating a broad range of trace organic compounds and is an effective barrier against many target pathogens, both the RO permeate and concentrate would be expected to contain lower concentrations of trace organic compounds, as well.

5.5.2.4 Lessons learned. The City of San Diego historically relied on imported water to satisfy its water demand, but stakeholders and policymakers realized the water supply portfolio would need to be diversified to increase the region's drought resilience. Moreover, the City was facing the prospect of spending almost two billion dollars to upgrade the Point Loma wastewater treatment plant (WWTP) to improve the quality of

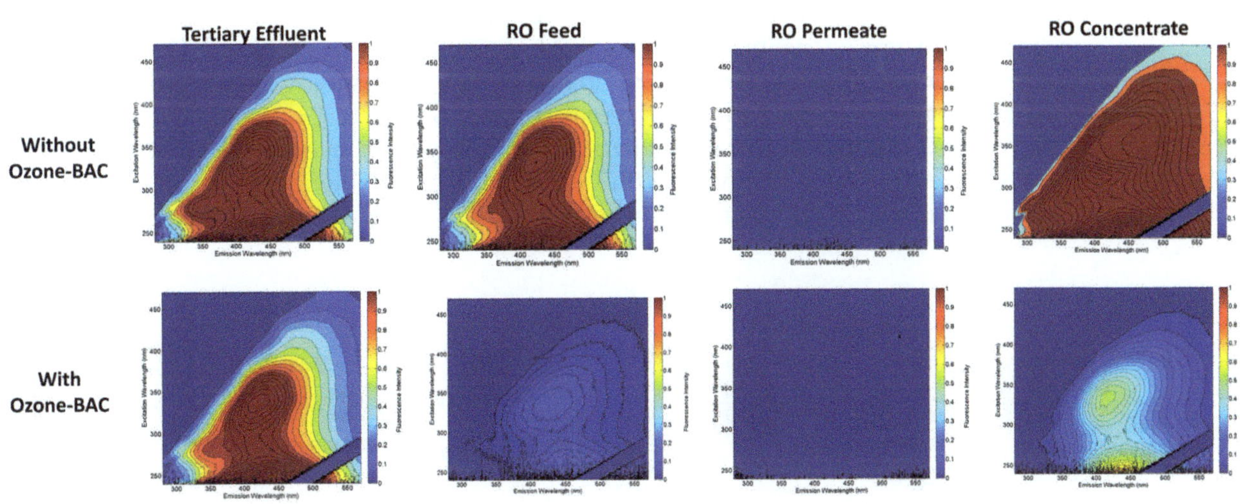

BAC—biologically active carbon, RO—reverse osmosis

Figure 5-13 Fluorescence excitation emission matrixes (top) without ozone-BAC and (bottom) with ozone-BAC

wastewater discharged to the ocean. Instead, potable reuse was identified as a means of producing a safe, reliable, sustainable, local water supply and reducing ocean discharge by up to 50 percent, thereby offering pragmatic legitimacy for Pure Water San Diego. Moral legitimacy was established principally through successful demonstration of the AWPF, oversight by expert panels, and an education and outreach program in which the public could tour the demonstration facility. Finally, cognitive legitimacy was established through parallel planning and/or implementation of numerous potable reuse projects throughout California. A coordinated research program, both inside California and at the national level, was also critically important in identifying and closing research gaps, some of which were specifically addressed by integrating ozone-biofiltration into Pure Water San Diego's AWPF. Ultimately, Pure Water San Diego is expected to provide up to a third of the City's water supply by 2035 (San Diego, 2018).

5.5.3 Advanced Treatment of Wastewater Effluent for Aquifer Recharge

HRSD, Suffolk, Va.

Authors: Ramola Vaidya, Virginia Tech; Dr. Charles Bott, HRSD; and Germano Salazar-Benites, HRSD

The rising concern of land subsidence, declining aquifer levels, nutrients discharged to the Chesapeake Bay, and seawater intrusion in Eastern Virginia motivated HRSD to implement the SWIFT. This program aims to recharge the Potomac aquifer with the effluent from an AWPF. The source water to the advanced treatment facility will be the secondary effluent from a WWTP. The initial planning and design of this program began in 2015 followed by a pilot plant operation from 2016 to 2018 to optimize the treatment technology. A demonstration facility (1 mgd) that has been recharging the aquifer was built in 2018 to demonstrate that the treatment process is effective at a meaningful scale and is producing water compatible with the groundwater. The first full-design efforts have begun with an objective of up to 100 mgd of recharge water by 2030.

5.5.3.1 Facility description. The 1-mgd treatment process is briefly described in Figure 5-14. Secondary effluent from HRSD's Nansemond treatment plant undergoes coagulation, flocculation, sedimentation followed by side-stream ozonation, BAF with carbon, GAC adsorption, and UV disinfection. The final effluent is pH-adjusted and disinfected using combined chlorine or chloramines to prevent biofouling in the wellhead. Because bromide concentrations were high in the secondary effluent, preformed monochloramine was added before ozone injection to control bromate formation by ozone. This monochloramine residual postozone was dechlorinated before BAF using sodium bisulfite to avoid any effect on the microbial growth in the BAF.

5.5.3.2 Biological filtration goals. The flow postozone was split into four biofilters that were operated to have an EBCT of 12 min when all biofilters were in service. The goal of biofiltration was to remove the organic carbon that was broken down by ozonation

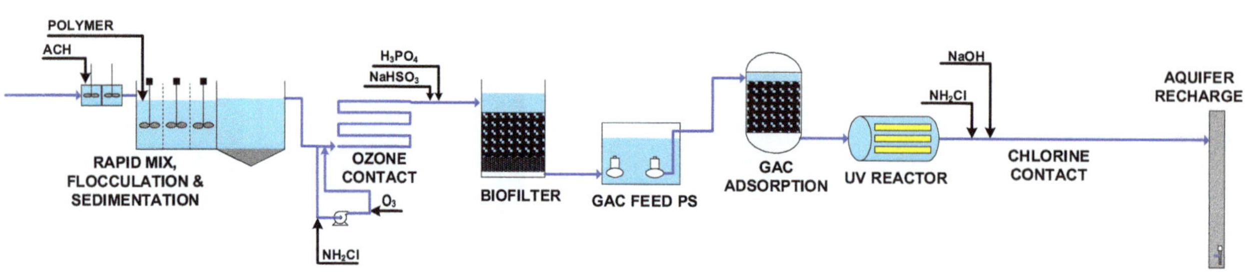

ACH—aluminum chlorohydrate, GAC—granular activated carbon, SWIFT—Sustainable Water Initiative for Tomorrow, UV—ultraviolet

Figure 5-14 SWIFT treatment process schematic

and provide filtered water turbidity consistent with drinking water requirements. This was achieved by a combination of biodegradation and filtration. The goal was to achieve turbidity below 0.1 NTU consistently.

5.5.3.3 Operation and performance.

Turbidity removal: The goal was to achieve turbidity below 0.1 NTU consistently, and the filters were backwashed if the turbidity increased above 0.1 NTU. Turbidity and TOC removal in BAF were both affected by the upstream flocculation/sedimentation and ozone operation, and any upsets in these upstream processes resulted in the increase in BAF effluent turbidity.

Organic carbon removal: The activated carbon media was not exhausted before startup, and it took a period of three months for the media to be completely exhausted and for the TOC to reach a steady-state removal. BAF removed 20 to 30 percent of TOC from the settled water during normal operation.

Manganese removal: BAF also removed manganese. The removal in manganese increased slowly with media acclimation and was completely removed after two to three months. This duration required for the complete removal of manganese coincided with the time period needed to achieve complete nitrification in BAF.

NDMA removal: NDMA was formed as a DBP during ozonation. This NDMA was removed by biodegradation in the BAFs. It was critical to adjust the sodium bisulfite dose used for dechlorinating monochloramine residual postozonation because excess sodium bisulfite affected the biofilter performance and appears to have negatively affected NDMA removal by BAF.

5.5.3.4 Lessons learned.

High DO from ozone resulted in rapid head loss accumulation: The ozone dose was controlled to have a minimum of three log virus removal and 1.5-log *Giardia* removal. This resulted in ozone doses of 5–6 mg/L and DO concentrations greater than 40 mg/L in the ozone effluent. There was no place for this high DO to equilibrate with the atmosphere postozone, resulting in oxygen bubbles getting trapped in the BAF media. This gas binding within the media caused artificial head loss buildup in the BAF. This caused the biofilters to backwash frequently, resulting in shorter run times, as shown in Figure 5-15, before May 22, 2018. A filter "bump" protocol was implemented starting May 22, 2018, which involved switching off the influent and effluent flow to release the accumulated gas and improve filter run times, as seen in Figure 5-15. The bumps were initiated once the BAF head loss increased above 6 ft, resulting in bumps every 8–12 hours. For a full-scale SWIFT plant, it is important to design the appropriate tankage and equipment between the ozone contactor and BAF to allow the DO to drop below concentrations needed to avoid the "bump" protocol and prevent artificial head loss buildup.

Measuring turbidity during backwash decreased water usage: BAFs were backwashed according to either head loss accumulation, turbidity breakthrough, or run time. The backwash settings included air scour, combined air and water wash, low-water wash, high-water wash, and a final low-water wash. The high-rate wash was initially set at 23 gpm/ft^2 for 10 min. To evaluate the efficiency of backwashing and media cleaning, turbidity samples were collected during backwashing, and Figure 5-16 shows the turbidity values collected every minute during the high-rate backwash. These values showed that the media was cleaned by 5 min, and the duration of the high-rate wash could be reduced. Thus, the high-rate wash was decreased to 5 min, resulting in a shorter time required for backwashing and for the BAF to be out of service. Lowering the high-rate wash to 5 min decreased the water used for backwashing from 16,000 gallons to 8,000 gallons. Thus, this change also saved the amount of water used for backwashing.

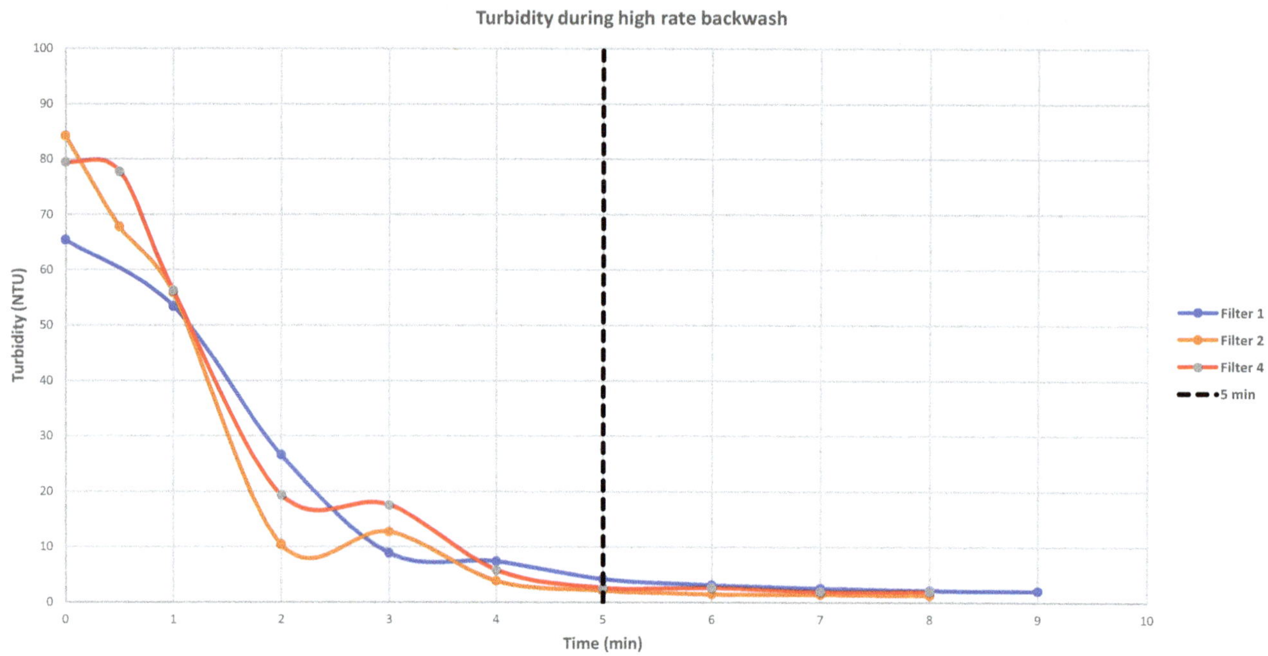

Figure 5-15 Effect of implementation of "bump" procedure

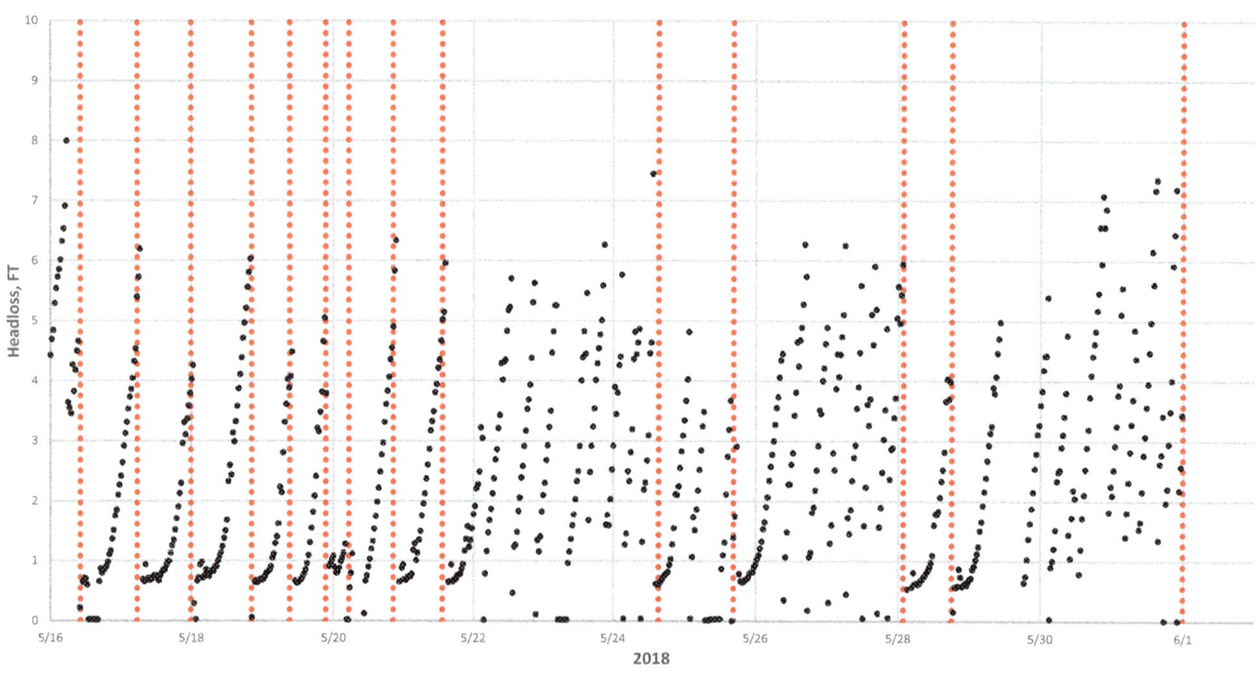

Figure 5-16 Backwash waste turbidity samples during high-rate backwash for optimizing biologically active filter backwash

5.6 REFERENCES

Bacaro, F., E. Dickenson, Trenholm, R., and Gerrity, D. 2019. "N-Nitrosodimethylamine (NDMA) Formation and Mitigation in Potable Reuse Treatment Trains Employing Ozone and Biofiltration." *Environmental Science: Water Research and Technology*. 5:713-725. https://doi.org/10.1039/C8EW00926K

Brandhuber, P., S. Clark, W. Knocke, and J. Tobiason. 2013. "Guidance for the Treatment of Manganese." Project #4373. Denver: WRF (Water Research Foundation).

Brown, J. and C.V. Lauderdale. 2006. "Efficient, Simultaneous Destruction of Multiple Drinking Water Contaminants Using Biological Filtration." *Florida Water Resources Journal*.

Brown, J., G. Upadhyaya, J. Carter, T. Brown, and C. Lauderdale. 2016. "North American Biofiltration Knowledge Base." Project #4459. Denver: WRF.

Brown, J., G. Upadhyaya, J. Nyfennegger, G. Pope, S. Bassett, A. Evans, J. Carter, V. Nystrom, S. Black, C. Alito, C. Lauderdale, J. Hooper, B. Finnegan, L. Strom, L. Williams Stephens, E. Palmer, E. Dickenson, S. Riley, W. Wert, L. Weinrich, and P. Keenan. 2020. "Biofiltration Guidance Manual for Drinking Water Facilities." Project #4719. Denver: WRF.

CADDW (California State Water Resources Control Board's Division of Drinking Water). 2018. "Regulations Related to Recycled Water." https://www.waterboards.ca.gov/drinking_water/certlic/drinkingwater/documents/lawbook/RWregulations_20181001.pdf. Accessed Oct. 16, 2024.

California Department of Public Health. 2013. "NDMA and Other Nitrosamines - Drinking Water Issues." http://www.waterboards.ca.gov/drinking_water/certlic/drinkingwater/NDMA.shtml (accessed Aug. 15, 2022).

Crittenden, J.C., R.R. Trussell, D.W. Hand, K.J. Howe, and G. Tchobanoglous. 2012 (3rd ed.). *MWH's Water Treatment: Principles and Design*. New York John Wiley & Sons.

Dickenson, E., E. Wert, E. Verdugo, K. Greenstein, J. Brown, G. Upadhyaya, P.G. Pope, O. Schneider, R. Marfil-Vega, and R.S. Summers. 2018. "Simultaneous Removal of Multiple Chemical Contaminants Using Biofiltration." Project #4559. Denver: WRF.

Emelko, M.B., P.M. Huck, B.M. Coffey, and E.F. Smith. 2006. "Effects of Media, Backwash, and Temperature on Full-Scale Biological Filtration." *Journal AWWA*. 98(12):61-73.

Funk, D., J. Hooper, M. Noibi, J. Goldman, R. Oliva, C. Schulz, K. Bell, D. Castaneda, C.H. Huang, and E. Machek. 2018. "Ozone Biofiltration Direct Potable Reuse Testing at Gwinnett County." Project #4777. Denver: WRF.

Garner, E., M. Inyang, E. Garvey, J. Parks, C. Glover, A. Grimaldi, E. Dickenson, J. Sutherland, A. Salveson, M. Edwards, and A. Pruden. 2019. "Impact of Blending for Direct Potable Reuse on Premise Plumbing Microbial Ecology and Regrowth of Opportunistic Pathogens and Antibiotic Resistant Bacteria" *Water Research*, 151: 75-86 https://doi.org/10.1016/j.watres.2018.12.003

Gerrity, D., E. Owens-Bennett, T. Venezia, B. Stanford, M. Plumlee, J. Debroux, and R.S. Trussell. 2014. "Applicability of Ozone and Biological Activated Carbon for Potable Reuse." *Ozone: Science & Engineering*. 36(2):123-137. https://doi.org/10.1080/01919512.2013.866886.

Gerrity, D., A. Pisarenko, E. Marti, R. Trenholm, F. Gerringer, J. Reungoat, and E. Dickenson. 2015. "Nitrosamines in Pilot-Scale and Full-Scale Wastewater Treatment Plans With Ozonation." *Water Research*. 72:251-261. https://doi.org/10.1016/j.watres.2014.06.025.

Gerrity, D., M. Arnold, E. Dickenson, D. Moser, J. Sackett, and E. Wert. 2018. "Microbial Community Characterization of Ozone-Biofiltration Systems in Drinking Water and Potable Reuse Applications." *Water Research*. 135:207-219. https://doi.org/10.1016/j.watres.2018.02.023.

Gifford, M., A. Selvy, and D. Gerrity. 2018. "Optimizing Ozone-Biofiltration Systems for Organic Carbon Removal in Potable Reuse Applications." *Ozone: Science & Engineering*. 40(6):427-440.

Halle, C., P.M. Huck, and S. Peldszus. 2015. "Emerging Contaminant Removal by Biofiltration: Temperature, Concentration, and EBCT Impacts." *Journal AWWA*. 107(7):E364-E379.

Hogard, S., G. Salazar-Benites, R. Pearce, T. Nading, L. Schimmoller, C. Wilson, J. Heisig-Mitchell, and C. Bott. 2021. "Demonstration-Scale Evaluation of Ozone–Biofiltration–Granular Activated Carbon Advanced Water Treatment for Managed Aquifer Recharge." *Water Environment Research*. 93(8):1157-1172.

Holmer, M. 2018. "Reuse Regulation Roundup: The Standards for Reuse." *Texas Desalination Association*.

Hooper, J., D. Funk, K. Bell, M. Noibi, K. Vickstrom, C. Schulz, E. Machek, and C. Huang. 2020. "Pilot Testing of Direct and Indirect Potable Water Reuse Using Multi-stage Ozone-Biofiltration Without Reverse Osmosis." *Water Research*. 169:1151178. https://doi.org/10.1016/j.watres.2019.115178.

Hooper, J., K. Vickstrom, P. Evans, C. Alito, S. Black, and C. Lauderdale. 2019. "Guidance Manual for Monitoring Biological Filtration of Drinking Water." Project 4620. Denver: WRF.

Imlay, J.A. 2013. "The Molecular Mechanisms and Physiological Consequences of Oxidative Stress: Lessons From a Model Bacterium." *Nature Reviews Microbiology*. 11(7):443-454.

Jones, R. and M. Hood. 1980. "Effects of Temperature, pH, Salinity, and Inorganic Nitrogen on the Rate of Ammonium Oxidation by Nitrifiers Isolated From Wetland Environments." *Microbial Ecology*. 6(4):339-347. https://doi.org/10.1007/BF02010496.

Kawamura, S. 1999. "Design and Operation of High-Rate Filters." *Journal AWWA*. 91(12):77-90.

Kohl, P.M. and S.J. Medlar. 2006. "Occurrence of Manganese in Drinking Water and Manganese Control." Project #2863. Denver: WRF.

Kolpin, D.W., E.T. Furlong, M.T. Meyer, E.M. Thurman, S.D. Zaugg, L.B. Barber, and H.T. Buxton. 2002. "Pharmaceuticals, Hormones, and Other Organic Wastewater Contaminants in US Streams, 1999– 2000: A National Reconnaissance." *Environmental Science & Technology*. 36(6):1202-1211.

Korshunov, S. and J.A. Imlay. 2006. "Detection and Quantification of Superoxide Formed Within the Periplasm of *Escherichia coli*." *Journal of Bacteriology*. 188(17):6326-6334.

Krasner, S., C.F.T. Lee, S. Liang, W. Mitch, U.V. Gunten, and P. Westerhoff. 2012. "Impact of a Nitrified Biofilter on NDMA Formation." Proc. AWWA Annual Conf. and Exposition, Dallas, Tex.

Kruithof, J., P. Kamp, and B. Martijn. 2007. "UV/H2O2 Treatment: A Practical Solution for Organic Contaminant Control and Primary Disinfection." *Ozone: Science & Engineering*. 29(4):273-280. doi: 10.1080/01919510701459311.

Kumar, P., L. Rodriguez-Gonzalez, A. Salveson, D. Ammerman, and E. Steinle-Darling. 2021. "Per- and Polyfluoroalkyl Substance Removal in Carbon-Based Advanced Treatment for Potable Reuse." *AWWA Water Science*. 3(4):e1244. https://doi.org/10.1002/aws2.1244

Kumar, A. and I. Xagoraraki. 2010. "Pharmaceuticals, Personal Care Products and Endocrine-Disrupting Chemicals in US Surface and Finished Drinking Waters: A Proposed Ranking System." *The Science of the Total Environment*. 408(23):5972-5989.

Lauderdale, C., C. Alito, J. Hooper, K. Dowdell, and E. Wert. 2018. "Optimizing Biofiltration for Various Source Water Quality Conditions." Denver: WRF.

Lauderdale, C., P. Chadik, M.J. Kirisits, and J. Brown. 2012. "Engineered Biofiltration: Enhanced Biofilter Performance Through Nutrient and Peroxide Addition." *Journal AWWA*. 104(5):E298-E309.

Lee, C.O., K.J. Howe, and B.M. Thomson. 2012. "Ozone and Biofiltration as an Alternative to Reverse Osmosis for Removing PPCPs and Micropollutants From Treated Wastewater." *Water Research*. 46(4):1005-1014.

Lester, Y., D. Aga, N. Love, R. Singh, I. Morrissy, and K. Linden. 2016. "Integrative Advanced Oxidation and Biofiltration for Treating Pharmaceuticals in Wastewater." *Water Environment Research*. 88(11):1985-1993. doi: 10.2175/106143016X14504669767454.

Liu, X., P.M. Huck, and R.M. Slawson. 2001. "Factors Affecting Drinking Water Biofiltration." *Journal AWWA*. 93(12):90-101. https://doi.org/10.1002/j.1551-8833.2001.tb09358.x

Lundgren, B.V., A. Grimvall, and R. Sävenhed. 1988. "Formation and Removal of Off-flavour Compounds During Ozonation and Filtration Through Biologically Active Sand Filters." *Water Science & Technology*. 20(8-9):245-253.

Malley, J., A. Bowen, M. White, Z. Angelini, and A. Packham. 2018. "UV photolysis, AOPs and/or Biofiltration for Micropollutants in Water Reuse." *IUVA News*. 20(3):17-23.

Marron, E., W. Mitch, U. Gunton, and D. Sedlak. 2019. "A Tale of Two Treatments: The Multiple Barrier Approach to Removing Chemical Contaminants During Potable Water Reuse." *Accounts of Chemical Research.* 52(3):615-622. doi.org/10.1021/acs.accounts.8b00612.

Marti, E.J., A.N. Pisarenko, J.R. Peller, and E.R.V. Dickenson. 2015. "N-nitrosodimethylamine (NDMA) Formation From the Ozonation of Model Compounds." *Water Research.* 72:262-270. https://doi.org/10.1016/j.watres.2014.08.047.

McDonald, E., M. Nellor, S. Trussell, J. Drewes, S. Duranceau, D. Lawler, S. Snyder, and G. Tchobanoglous. 2015. "Final Report: Direct Potable Reuse Resource Document." Austin, Tex.: Texas Water Development Board.

Miege, C., J.M. Choubert, L. Ribeiro, M. Eusèbe, and M. Coquery. 2009. "Fate of Pharmaceuticals and Personal Care Products in Wastewater Treatment Plants–Conception of a Database and First Results." *Environmental Pollution.* 157(5):1721-1726.

Ngumba, E., A. Gachanja, and T. Tuhkanen. 2020. "Removal of Selected Antibiotics and Antiretroviral Drugs During Post-treatment of Municipal Wastewater With UV, UV/Chlorine, and UV/Hydrogen Peroxide." *Water and Environment Journal.* 34(4):697-703.

Noibi, M., J. Hooper, D. Funk, and K. Bell. 2020. "Direct Potable Reuse Using Full Advanced Treatment Compared With Non-Reverse Osmosis Based Ozone and Biofiltration Treatment at Gwinnett County: A Cost Comparison." *AWWA Water Science.* 2(6):e1210.

NWRI (National Water Research Institute). 2015. "Framework for Direct Potable Reuse." WRRF 14-20. Alexandria, Va.: WRRF (WateReuse Research Foundation).

Nyfennegger, J., J. Brown, C. Lauderdale, N. Ross, G. Upadhyaya, S. Burns, and C. Owen. 2018. "New Data Tracking Tool Monitors Biofilter Performance." *Opflow.* 44(3):22-25. https://doi.org/10.5991/OPF.2018.44.0028.

Pernet-Coudrier, B., L. Clouzot, G. Varrault, M.-H. Tusseau-Vuillemin, A. Verger, and J.-M. Mouchel. 2008. "Dissolved Organic Matter From Treated Effluent of a Major Wastewater Treatment Plant: Characterization and Influence on Copper Toxicity." *Chemosphere.* 73(4):593-599.

Rattier, M., J. Reungoat, and W. Keller. 2014. "Removal of Micropollutants During Tertiary Wastewater Treatment by Biofiltration: Role of Nitrifiers and Removal Mechanisms." *Water Research.* 54:89-99. https://doi.org/10.1016/j.watres.2014.01.030.

Riley, S., D. Ahoor, and T. Cath. 2018. "Enhanced Biofiltration of O&G Produced Water Comparing Granular Activated Carbon and Nutrients." *Science of the Total Environment.* 640-641:419-428. https://doi.org/10.1016/j.scitotenv.2018.05.228.

Schimmoller, L. and M.J. Kealy. 2014. "Fit for Purpose Water: The Cost of Overtreating Reclaimed Water." Alexandria, Va.: WRRF and Bureau of Reclamation.

Simpson, D.R. 2008. "Biofilm Processes in Biologically Active Carbon Water Purification." *Water Research.* 42: 2839-2848.

Sirivedhin, T. and K.A. Gray. 2005. "2. Comparison of the Disinfection By-product Formation Potentials Between a Wastewater Effluent and Surface Waters." *Water Research*. 39:1025-1036.

Stanford, B., E. Dickenson, E. Wert, and M. Inyang. 2017. "Controlling Trace Organic Compounds Using Alternative, Non-FAT Technology for Potable Water Reuse." Project #1704. Denver: WRF.

Stenstrom, M. and R. Poduska. 1980. "The Effect of Dissolved Oxygen Concentration on Nitrification." *Water Research*. 14:643-649. https://doi.org/10.1016/0043-1354(80)90122-0.

Sun, Y., B. Angelotti, M. Brooks, B. Dowbiggin, P.J. Evans, B. Devins, and Z.-W. Wang. 2018. "A Pilot-Scale Investigation of Disinfection By-product Precursors and Trace Organic Removal Mechanisms in Ozone-Biologically Active Carbon Treatment for Potable Reuse." *Chemosphere*. 210:539-549. https://doi.org/10.1016/j.chemosphere.2018.06.162

Sundaram, V., K. Pagilla, T. Guarin, L. Li, R. Marfil-Vega, and Z. Bukhari. 2020. "Extended Field Investigations of Ozone-Biofiltration Advanced Water Treatment for Potable Reuse." *Water Research*. 172:115513.

Tackaert, R.A., A.N. Pisarenko, E.C. Chen, A. Kolakovsky, B.M. Pecson, J.E. Drewes, R. Rhodes Trussell, R. Shane Trussell. 2019. "Demonstrating Process Robustness of Potable Reuse Trains During Challenge Testing With Elevated Levels of Acetone, Formaldehyde, NDMA, and 1,4-Dioxane." *Journal of Water Supply: Research and Technology–AQUA*. 68(5):313-324.

Trussell, R.R., A. Salveson, S.A. Snyder, R.S. Trussell, D. Gerrity, and B.M. Pecson. 2013. "Potable Reuse: State of the Science Report and Equivalency Criteria for Treatment Trains." Alexandria, Va.: WRRF.

USEPA. 2017. "Potable Reuse Compendium." https://www.epa.gov/sites/production/files/2018-01/documents/potablereusecompendium_1.pdf (accessed XXXX).

USEPA. 2012. "Guidelines for Water Reuse." September 2012. https://www.epa.gov/sites/production/files/2019-08/documents/2012-guidelines-water-reuse.pdf (accessed XXXX).

USEPA. 2002. "Guidelines for Water Reuse." 2002. https://www.epa.gov/sites/default/files/2019-08/documents/2004-guidelines-water-reuse.pdf (accessed XXXX).

Vaidya, R., C.A. Wilson, G. Salazar-Benites, G., A. Pruden, and C. Bott. 2021. "Factors Affecting Removal of NDMA in an Ozone-Biofiltration Process for Water Reuse." *Chemosphere*. 264:128333.

Vaidya, R., C.A. Wilson, G. Salazar-Benites, A. Pruden, and C. Bott. 2020. "Implementing Ozone-BAC-GAC in Potable Reuse for Removal of Emerging Contaminants." *AWWA Water Science*. 2(5):e1203.

Wang, J.Z., R.S. Summers, and R.J. Miltner. 1995. "Biofiltration Performance: Part 1, Relationship to Biomass." *Journal AWWA*. 87(12):55-63.

Wang, Y. 2014. "Assessment of Ozonation and Biofiltration as a Membrane Pre-treatment at a Full-Scale Drinking Water Treatment Plant." University of Waterloo thesis. Waterloo, O.N. Wang_Yulang.pdf (uwaterloo.ca)

Westerhoff, P., R.S. Summers, Z. Chowdhury, and S. Kommineni. 2005. "Ozone-Enhanced Biofiltration for Geosmin and MIB Removal." Project #2775. Denver: WRF.

Zearley, T.L. and R.S. Summers. 2012. "Removal of Trace Organic Micropollutants by Drinking Water Biological Filters." *Environmental Science & Technology*. 46(17):9412-9419.

Zhang, H., L. Gu, B. Liu, H. Gan, K. Zhang, H. Jin, and X. Yu. 2016. "An Improved Biofilter to Control the Dissolved Organic Nitrogen Concentration During Drinking Water Treatment." *Environmental Science and Pollution Research International*. 23(18):18137-18144.

Zhang, S. and P.M. Huck. 1996. "Removal of AOC in Biological Water Treatment Processes: A Kinetic Modeling Approach." *Water Research*. 30:1195-1207.

Glossary

2-methylisoborneol (MIB) ($C_{11}H_{20}O$) A musty, camphor-smelling chemical produced by certain cyanobacteria and actinomycetes. This odorous compound can be perceived at low nanogram-per-liter concentrations. Removal at a treatment plant requires either ozone (O_3) oxidation or activated carbon adsorption.

adenosine triphosphate (ATP) (C10H16N5O13P3) A compound found in all living cells in which energy is stored in high-energy phosphate bonds. Its components are purine adenine, ribose, and three phosphoric acid groups.

adsorption (1) The attraction and adhesion of molecules of a gas, liquid, or dissolved substance to a surface. Adsorption is generally a passive and reversible process. Granular or powdered activated carbon is often used as an adsorption medium. (2) The interaction of an analyte with the surface of a matrix. This interaction can form the basis for the extraction of analytes from water or the chromatographic separation of compounds.

advanced oxidation process (AOP) A process in which the oxidative capacity of a parent compound is modified to make oxidation–reduction reactions more rapid or complete. For example, hydrogen peroxide (H_2O_2) may be added to ozone (O_3), or hydrogen peroxide may be photolyzed by ultraviolet light, to generate hydroxyl radicals, which are strong, nonspecific oxidizing agents. Other AOPs include the ozone–biodegradation process (O_3/bio), ozone–ultraviolet light process (O_3/UV), pulsed ultraviolet light–hydrogen peroxide process (pulsed UV/H_2O_2), and sonolyis.

aerobic (1) IN REFERENCE TO MICROORGANISMS, PERTAINING TO THE requirement for oxygen for growth. (2) In reference to environmental conditions, pertaining to the presence of oxygen.

aerobic treatment A biologic treatment technique used to oxidize and remove from water-soluble or fine materials, usually organic in nature.

air scour (1) The practice of admitting air through the underdrain system of a filter to help complete cleaning of media during filter backwash. (2) A technique using high-volume, high-velocity, low-pressure air in water to remove tuberculation and other internal corrosion particles, improving water quality and restoring the flow-carrying capacity of a water pipeline, with minimal disruption to service.

aliphatic compound An organic compound with carbon atoms arranged in a branched or unbranched open chain rather than a ring. If an aliphatic compound contains double bonds, it is said to be unsaturated; if not, it is saturated.

anaerobic biological treatment Biological treatment process that occurs in the absence of free dissolved oxygen and oxidized compounds.

anoxic biological treatment Biological treatment process that occurs in the absence of free dissolved oxygen, in which oxidized compounds such as nitrate and nitrite are used to drive metabolic reactions; denitrification is an example of an anoxic process.

anoxic condition A state in which free dissolved oxygen is absent but in which oxygen is available as a terminal electron acceptor through compounds such as nitrate (NO_3^-) or sulfate (SO_4^{2-}).

anthracite coal A particulate form of coal that is used in granular media filters to remove particles from water. Anthracite coal is typically used in dual-media filters in combination with sand. See also anthracite coal–sand filter; dual-media filter.

anthropogenic Pertaining to or involving the impact of humans.

assimilable organic carbon (AOC) The fraction of organic carbon that can be used by specific microorganisms and converted to cell weight. AOC also represents a potential for biological regrowth in distribution systems. Ozone (O_3) can convert organic matter in water to assimilable organic carbon, whereas biological filtration can reduce the AOC level. See also biological filtration.

atrazine (2-chloro-4-ethylamino-6-isopropyl amino1,3,5-triazine) (C8H14N5Cl) A herbicide and plant growth regulator used primarily on corn and soybeans. It is slightly soluble in water, and its presence in drinking water is regulated by the US Environmental Protection Agency at a maximum contaminant level of 0.003 milligrams per liter.

autotrophic Pertaining to organisms capable of synthesizing organic nutrients directly from simple inorganic substances such as carbon dioxide (CO_2), nitrogen (N), and nitrate ion (NO_3^-).

backwash (1) The process of cleansing filter media of particles that have been removed during the filtration or adsorption process. Backwashing involves reversing the flow through a filter to dislodge the particles. Backwash water either is treated and returned to the plant influent or is disposed. (2) The passing of a fluid (feedwater, treated water, or other fluid) or air through an ion exchange column or some types of microporous membranes for cleaning purposes to remove particles.

bench-scale study An experimental study to evaluate the performance of unit processes performed in laboratory surroundings. For example, a jar test to evaluate the coagulation, flocculation, and sedimentation processes is conducted on a batch of water, and the apparatus is of a size that will fit on a laboratory "bench."

bench-scale testing An experimental study to evaluate the performance of unit processes performed in laboratory surroundings. For example, a jar test to evaluate the coagulation, flocculation, and sedimentation processes is conducted on a batch of water, and the apparatus is of a size that will fit on a laboratory "bench."

best available technology (BAT) (1) Under Section 1412(b)(4)(D) of the Safe Drinking Water Act, the best technology, treatment techniques, or other means that the US Environmental Protection Agency administrator finds available to water systems—after examination for efficacy under field conditions, not solely under laboratory conditions, and taking cost into consideration—for meeting National Primary Drinking Water Standards. For the purpose of setting maximum contaminant levels for synthetic organic chemicals, any best available technology must be at least as effective as granular activated carbon. Water systems are not required to use the best available technology set under Section 1412 to meet a regulation; they may use any technology acceptable to the state primacy agency. (2) Under Section 1415(9)(1)(A) of the Safe Drinking Water Act, the best technology, treatment techniques, or other means that the US Environmental Protection Agency administrator finds available (taking costs into consideration). To be eligible for a variance, a water system must agree to apply the best available technology designated under Section 1415 before applying for a variance. Best available technology designations by the US Environmental Protection Agency under Sections 1412 and 1415 are separate and may or may not be the same.

biocatalyst A substance, such as an enzyme or hormone, of which a fractional percentage notably affects the rate of a chemical reaction, without the substance itself being consumed or undergoing a chemical change. Most catalysts accelerate reactions, but a few retard them (the latter substances are called negative catalysts or inhibitors).

biodegradable organic carbon (BDOC or BOC) That portion of the organic carbon in water that can be mineralized by heterotrophic microorganisms. Ozone (O_3) can convert organic matter in water to biodegradable organic carbon, whereas biological filtration can reduce the biodegradable organic carbon level.

biodegradable organic matter (BDOM or BOM) That portion of the organic matter in water that can be degraded by microorganisms. Biodegradable organic carbon is a subset of biodegradable organic matter.

biodegradation The breakdown of organic matter by microorganisms.

biofilm A layer of microorganisms held together by a matrix of organic polymers and found at the interface between water and a solid substrate in an aquatic environment. Biofilms in natural environments are heterogeneous and contain a range of microenvironments conducive to development of a diverse bacterial population. A biofilm can be detrimental to a process (say, by fouling a water treatment membrane) or beneficial (as in a biological treatment system). Biofilms are common in pipes used to distribute finished water to consumers and can adversely affect the quality of the water.

biofouling A phenomenon in which the performance of a unit process is compromised by biological growth. For example, biofouling can occur in a membrane process when microorganisms grow on the membrane surface and result in a premature and excessive loss of flow rate, or flux, through the membrane.

biologically active filter (BAF) Granular media (e.g., activated carbon, anthracite coal, or sand) that has developed a biofilm capable of degrading organic matter, ammonia (NH_3), or both.

biologically active filtration (BAF) The removal of biodegradable organic matter by passing water over a fixed medium covered with a biofilm.

biostability A condition and quality of water in a distribution system in which the growth potential of bacteria is at or near an equilibrium.

biota Living organisms, including bacteria, plants, and animals, in a given ecosystem.

biotransformation The metabolism of a chemical. Metabolism can either detoxify or actually mediate the toxicity of a chemical. Metabolic reactions in mammalian systems are frequently placed into two categories, referred to as phase I and II reactions. Phase I reactions generally result in the production of reactive intermediates that are often the mediators of toxic responses. Phase II reactions generally involve conjugation reactions with moieties (chemical groups) that increase the water solubility of the chemical and thus make it easier to eliminate from the body. These latter reactions most frequently lead to reductions in toxicity, but definite exceptions to this rule exist.

Brownian motion The random movement of particles suspended in a fluid medium resulting from molecular bombardment. Brownian motion accounts for the diffusion mechanism in particle removal by filtration and in perikinetic flocculation.

bubble diffuser Diffusers are a pollution control technology. They produce a plethora of very small air bubbles which rise slowly from the floor of a wastewater treatment plant or sewage treatment plant aeration tank and provide substantial and efficient mass transfer of oxygen to the water. The oxygen, combined with the food source, sewage, allows the bacteria to produce enzymes which help break down the waste so that it can settle in the secondary clarifiers or be filtered by membranes. (link here)

by-products A group of disinfection by-products containing organic nitrogen. Haloacetonitriles and N-nitrosodimethylamine are two typical examples.

cascade aerator An aerating device built in the form of either steps or an inclined plane, on which is placed staggered projections arranged to break up the water and bring it into contact with air.

chemically enhanced backwashing In membrane treatment, the process of adding chemicals to the permeate during backwash to enhance the effectiveness of the backwash for inactivating biological growth and removing foulants.

chemical oxygen demand (COD) An operationally defined surrogate parameter in water analysis that is a measure of the constituents susceptible to oxidation by strong chemical oxidants.

chloramination The process of disinfecting water with chloramines.

chloramines Disinfectants produced from the mixing of chlorine (Cl_2) and ammonia (NH_3). The general formula is NH_xCl_y, where x can be 0, 1, or 2; y can be 1, 2, or 3. Typically, monochloramine (NH_2Cl) and a small percentage of dichloramine ($NHCl_2$) are formed, depending on the pH and the chlorine-to-ammonia ratio that reacts. Under certain circumstances, nitrogen trichloride (trichloramine, $NHCl_3$) can be formed. In the presence of the bromide ion, bromamines can be produced. In the presence of organic nitrogen, organic chloramines can also form; however, organic chloramines are not considered to be disinfectants.

chlorine contact tank (CCT) A detention basin provided primarily to secure the diffusion of chlorine through the liquid. It is also called a chlorination chamber.

clean bed A bed of granular filter media that has been cleaned or backwashed. Immediately following a thorough backwash, the particles that formerly adhered to the surface of the media are removed and the interstitial spaces between media grains are clear and free of particulate matter.

clean bed head loss The head loss that occurs after a filter has been backwashed and restored to a clean condition. Clean bed head loss is measured at the start of a filter run. It is also known as starting head loss.

clean-in-place (CIP) system Facilities used for in situ cleaning.

Clean Water Act (CWA) The US law, codified generally as 33 US Code 1251–1387, that establishes a regulatory and enforcement program administered by the US Environmental Protection Agency to control pollutant discharges into US waters.

clearwell A tank or vessel used for storing treated water. Typical examples of storage needs include (1) finished water storage to prevent the need to vary the rate of filtration with variations in distribution system demand and (2) backwash water for filters. Clearwells are located on site at a water treatment plant. A clearwell is also called a filtered-water reservoir.

coagulation The process of destabilizing charges on particles in water by adding chemicals (coagulants). Natural particles in water have negative charges that repel other material and thereby keep it in suspension. In coagulation, positively charged chemicals are added to neutralize or destabilize these charges and allow the particles to accumulate and be removed by physical processes such as sedimentation or filtration. Commonly used coagulants include aluminum and iron salts and cationic polymers.

colloid A small, discrete solid particle in water that is suspended (not dissolved) and will not settle by gravity because of molecular bombardment

co-metabolism Process by which microorganisms transform substrates into organic products but do not obtain energy, carbon, or nutrients from the conversion. This process offers a biological alternative for the removal of pollutants from contaminated aquifers because the microorganisms transform the pollutant substrate (e.g., trichloroethylene) into intermediate degradation products but fail to multiply.

compounds of emerging concern (CECs) include nanoparticles, pharmaceuticals, personal care products, estrogen-like compounds, flame retardants, detergents, and some industrial chemicals with potential significant effects on human health and aquatic life. (US Environmental Protection Agency, link here)

contaminant Any physical, chemical, biological, or radiological substance or matter in water.

continuous-flow stirred-tank reactor (CFSTR) A reactor in which the contents are completely mixed on a continuous basis. In theory, flow introduced to the continuous-flow stirred-tank reactor is immediately mixed to a uniform concentration upon entering the reactor, and therefore fractions of the flow will be exiting as soon as they enter. Steady-state conditions prevail, and no accumulation of constituents occurs in the reactor with time. Continuous-flow stirred-tank reactors are a fundamental component of reactor theory, although few true applications exist in water treatment. A rapid mix basin for adding coagulant chemicals, if mechanically mixed at high intensity, is one example of a treatment process that approximates this behavior.

coprecipitation The simultaneous precipitation of a normally soluble component with a macrocomponent from the same solution by the formation of mixed crystals, by adsorption, occlusion, or mechanical entrapment.

corrosion The gradual deterioration or destruction of a substance (usually a metal) or its properties as a result of a reaction with the substance's surroundings.

cyanotoxin Complex organic compounds produced by many types of cyanobacteria such as *Anabaena, Aphanizomenon,* and *Microcystis*. Toxins produced by *Anabaena* and *Aphanizaomenon* are neurotoxins (affecting the nervous system), and those produced by *Microcystis* are hepatoxins (affecting the liver).

cylindrospermopsin The toxin cylindrospermopsin is produced by the cyanobacteria *Cylindrospermopsis raciborskii, Aphanizomenon ovalisporum,* and *Umezakia natans*. Most common in tropical and arid zone water bodies, it has also been detected in North and South America, Europe, and New Zealand. The main mechanism of this toxin is believed to be inhibition of protein synthesis. Ingesting contaminated water or fish can cause hepatitis-like symptoms, kidney damage, and intestinal illness. The common presence of cylindrospermopsin in drinking water, particularly in tropical regions, poses a public health problem. Treatment techniques, including chlorination, ozonation, and ultraviolet radiation, have the ability to degrade cylindrospermopsin, but the formation of by-products may be of concern if they remain toxic.

demineralization The removal of dissolved minerals and inorganic constituents from water to meet water quality standards, improve aesthetic quality, or reduce corrosivity. Commonly used treatment techniques for demineralization include chemical coagulation or lime softening for heavy metals and activated alumina. See also activated, alumina, coagulation, deionization, desalting, ion exchange, and lime softening.

direct potable-water reuse (DPWR) The use of community wastewaters treated to a sufficient degree that they are acceptable for drinking and direct discharge into a potable-water distribution system.

This practice has not been adopted by or approved for any community in the United States.

Disinfectants and Disinfection By-Products Rule (DBPR) A national primary drinking water regulation promulgated by the US Environmental Protection Agency to regulate drinking water disinfectants and by-products of disinfection. Stage 1 became effective in 2001 and set maximum contaminant levels (MCLs) for total trihalomethanes (TTHMs) and haloacetic acid (HAA5) and established required levels of disinfectants in the distribution system. Stage 2 was published in 2006. It required that water utilities monitor to determine the locations of highest occurrence of TTHMs and HAA5 in their systems and changed how compliance with the MCLs was calculated.

disinfection by-product (DBP) A chemical by-product of the disinfection process. Disinfection by-products are formed by the reaction of the disinfectant, natural organic matter, and the bromide ion (Br^-). Some disinfection by-products are formed through halogen (e.g., chlorine or bromine) substitution reactions; i.e., halogen-substituted by-products are produced. Other disinfection by-products are oxidation by-products of natural organic matter (e.g., aldehydes—RCHO). Concentrations are typically in the microgram-per-liter or nanogram-per-liter range.

dissolved inorganic carbon (DIC) The fraction of inorganic carbon (the carbonate [CO_3^{2-}], bicarbonate [HCO_3^-], and dissolved carbon dioxide [CO_2]) in water that passes through a 0.45-micrometer pore-diameter filter.

dissolved organic carbon (DOC) That portion of the organic carbon in water that passes through a 0.45-micrometer pore-diameter filter. For most drinking water sources, the dissolved organic carbon fraction represents a very high percentage of the total organic carbon pool. It is composed of individual compounds as well as nonspecific humic material, although humic substances account for a large portion of dissolved organic matter in natural waters. Typically, the dissolved organic carbon level provides some indication of the amount of disinfection by-product precursors in a water source. After filtration, dissolved organic carbon is determined in the same manner as total organic carbon. Organic carbon concentrations should be reported as dissolved organic carbon only if the sample has been filtered through a 0.45-micrometer pore-diameter filter before analysis.

dissolved oxygen (DO) The concentration of oxygen in aqueous solution, which is often expressed in units of milligrams per liter. It is usually determined by one of two methods: a dissolved oxygen probe or Winkler titration.

electrodialysis reversal (EDR) An electrodialysis process in which the electrical polarity of the electrodes is reversed on a set time cycle, thereby reversing the direction of ions in the system for membrane scaling and fouling control.

empty bed contact time (EBCT) A standard convention or measure of the time during which a water to be treated is in contact with the treatment medium. The empty bed contact time is calculated by dividing the empty volume in a contactor that will be occupied by the treatment medium by the flow rate.

$$EBCT = V/Q$$

Where (in any consistent set of units):

V = *the volume of the vessel*
Q = *the flow rate*

Because the treatment medium, such as granular activated carbon, will occupy some volume, the empty bed contact time overestimates the actual time that the flow resides in the contactor.

endocrine disrupting compound (EDC) An exogenous chemical substance or mixture that alters the function(s) of the endocrine system and thereby causes adverse effects to an organism, its progeny, or (sub)populations.

endocrine disruptor An exogenous chemical substance or mixture that alters the function(s) of the endocrine system and thereby causes adverse effects to an organism, its progeny, or (sub) populations.

extracellular enzymes A catalyst that is discharged and acts on the outer part of the cell. These enzymes are also termed exoenzymes. They assume an essential role during decay or decomposition. Moreover, they corrupt complex particles like cellulose and hemicellulose into smaller particles. (link here)

extracellular polymeric substance (EPS) A substance released extracellularly by microbes, usually bacteria. These substances are primarily polysaccharides and exist as highly hydrated gels and fibers surrounding, encapsulating, or connecting a consortium of microbial species in an aquatic environment (i.e., a biofilm).

Filter Backwash Recycling Rule (FBRR) A US Environmental Protection Agency rule requiring systems that recycle to return specific recycle flows through all processes of the water system's existing conventional or direct filtration system or at an alternate location approved by the primacy agency.

fixed bed (1) The filter or ion exchange medium retained in a vessel. (2) A media bed that is "contained," that is, filled to the top or to the restraining barrier with filter media and is not capable of being expanded during backwashing.

fixed-bed bioreactor (FXB or FBBR) Advanced biological wastewater treatment technologies used for efficiently treating wastewaters with high to extremely high organic contamination levels. (link here)

flocculant A water-soluble organic polyelectrolyte that is used alone or in conjunction with inorganic coagulants, such as aluminum or iron salts, to agglomerate solids present in water to form large, dense floc particles that settle rapidly.

flocculation The water treatment process following coagulation that uses gentle stirring to bring suspended particles together so they will form larger, more separable clumps called floc.

flow equalization The use of storage tanks to control a changing flow of water and make it nearly uniform with time.

fluidized-bed bioreactor (FBR) A submerged attached growth process that can be used for either aerobic or anaerobic applications where wastewater is fed upward to a bed of sand or activated carbon; the fluid velocity causes the bed to expand and form a biofilm that removes soluble organic matter.

fouling (1) The accumulation of undesirable foreign matter in a filter or ion exchange media bed, causing clogging of pores or coating of surfaces and inhibiting or limiting the proper operation of the bed and the treatment system. (2) The reduction in adsorption capacity of granular activated carbon as a result of other material occupying the adsorption surface. For example, granular activated carbon can be fouled by naturally occurring organic matter in drinking water, thereby limiting the absorption of specific microcontaminants. (3) In a membrane water treatment process, the deposition of material such as colloidal matter, microorganisms, and metal oxides on the membrane surface or in its pores, causing a change in membrane performance (e.g., flux decline). (4) A gelatinous, slimy accumulation resulting from the activity of organisms in water. This type of fouling may be found on concrete, masonry, and metal surfaces (tuberculation).

full advanced treatment (FAT) The treatment of an oxidized wastewater using a reverse osmosis and an oxidation treatment process.

geosmin ($C_{12}H_{22}O$) The common name for *trans*-1,10-dimethyl-*trans*-9-decalol, an earthy-smelling chemical produced by certain cyanobacteria and actinomycetes. This odorous compound can be perceived at low nanogram-per-liter concentrations. Geosmin removal at a treatment plant requires either ozone (O_3) oxidation or activated carbon adsorption.

granular activated carbon (GAC) A form of particulate carbon manufactured with increased surface area per unit mass to enhance the adsorption of soluble contaminants. Granular activated carbon is used in fixed-bed contactors in water treatment and is removed and regenerated (reactivated) when the adsorption capacity is exhausted. In some applications, granular activated carbon can be used to support a biological population for stabilizing biodegradable organic material.

gravity filter A rapid granular filter of the open type for which the operating level is placed near the hydraulic grade line of the influent and through which the water flows by gravity.

haloacetic acid (HAA) (CX_3COOH, where X = Cl, Br, H in various combinations) A class of disinfection by-products formed primarily during the chlorination of water containing natural organic matter. When bromide (Br^-) is present, a total of nine chlorine-, bromine-and-chlorine–, or bromine-substituted species may be formed. Trihalomethanes and haloacetic acids (HAAs) are the two most prevalent classes of by-products formed during chlorination and are subject to regulation under the Disinfectants and Disinfection By-Products Rule.

haloacetonitrile (HAN) ($CX_3{\equiv}N$, where X = Cl, Br, H in various combinations) A class of disinfection by-products formed primarily during the chlorination of water containing natural organic matter. Some amino acids (e.g., from algae) have been found to be precursors to haloacetonitriles (HANs). When bromide (Br^-) is present, a total of four chlorine-chlorine-and bromine-, or bromine-substituted species may be formed. Monitoring for haloacetonitriles was required in the Information Collection Rule.

halonitromethane (HNM) A class of disinfection by-products formed in chlorinated water. There are nine species containing chlorine and/or bromine (two monohalogenated, three dihalogenated, and four trihalogenated species). The most commonly measured one is chloropicrin.

head loss A reduction of water pressure (head) in a hydraulic or plumbing system. Head loss is a measure of (1) the resistance of a medium bed (or other water treatment system), a plumbing system, or both to the flow of the water through it, or (2) the amount of energy used by water in moving from one location to another. In water treatment technology, head loss is basically the same as pressure drop.

helminth An intestinal parasite.

heterotrophic plate count (HPC) A bacterial enumeration procedure used to estimate bacterial density in an environmental sample, generally water. Other names for the procedure include total plate count, standard plate count (SPC), plate count, and aerobic plate count.

high-density polyethylene (HDPE) A polymer with many industrial applications, including use as a pipe material. Polyethylene ($(H_2C{:}CH_2)x$) pipe can be used in low-pressure applications for transporting potable water and other liquids.

human machine interface (HMI) A means by which a human operator interacts with a supervisory control and data acquisition (SCADA) system. The human–machine interface of a SCADA system consists of color cathode ray tubes, mapboards, projection video, alarm annunciators, and alarm loggers.

hydraulic flocculation Flocculation attained by the flow of water through a baffled basin or channel. No mechanical stirring is used in hydraulic flocculation.

hydraulic loading The amount of water applied to a given treatment process, usually expressed as volume per unit time or volume per unit time per unit surface area.

hydraulic loading rate (HLR) The amount of water applied to a given treatment process, usually expressed as volume per unit time or volume per unit time per unit surface area.

incidental biofiltration Operation of carbon filters without an oxidant residual.

indirect potable-water reuse (IPWR) The use of water from streams that have upstream discharges of wastewater.

inorganic compound A substance that is not derived from hydrocarbons. Examples include sodium chloride ($NaCl$), ferric hydroxide ($Fe(OH)_3$), and aluminum sulfate ($Al_2(SO_4)_3$).

interception (1) The process by which precipitation is caught and held by foliage, twigs, and branches of trees, shrubs, and other vegetation and then lost by evaporation, never reaching the surface of the ground. (2) The amount of precipitation intercepted.

Interim Enhanced Surface Water Treatment Rule (IESWTR) A rule promulgated Dec. 16, 1998, by the US Environmental Protection Agency (63 FR 69478–69521) that set requirements to better protect consumers against *Cryptosporidium* and other pathogens. The IESWTR applies to all public water systems that use surface water or groundwater under direct influence of surface water, and serve 10,000 people or more. See also Long Term 1 Enhanced Surface Water Treatment Rule; Long Term 2 Enhanced Surface Water Treatment Rule.

ion exchange (IX or IE) A reversible chemical process in which ions from an insoluble permanent solid medium (the ion exchanger—usually a resin) are exchanged for ions in a solution or fluid mixture surrounding the insoluble medium. The superficial physical structure of the solid is not affected. The direction of the exchange depends on the selective attraction of the ion exchanger resin for the certain ions present and the concentrations of the ions in the solution. Both cation and anion exchange are used in water conditioning. Cation exchange is commonly used for water softening.

ion exchange softening A treatment used to remove calcium and magnesium from water by ion exchange resins.

log removal A shorthand term for log10 removal, used in reference to the Surface Water Treatment Rule and the physical–chemical treatment of water to remove, kill, or inactivate pathogenic organisms such as *Giardia lamblia* and viruses. A 1-log removal equals a 90 percent reduction in density of the target organism; a 2-log removal equals a 99 percent reduction; a 3-log removal equals a 99.9 percent reduction; and so on. See also common logarithm.

Long Term 1 Enhanced Surface Water Treatment Rule (LT1ESWTR) A rule promulgated Jan. 14, 2002, by US Environmental Protection Agency (67 FR 1812–1844) that requires small public water systems using surface water to meet strengthened filtration requirements. The LT1ESWTR requires systems to calculate levels of microbial inactivation to ensure that microbial protection is not jeopardized if systems make changes to comply with requirements of the Stage 1 Disinfectants and Disinfection By-products Rule (Stage 1 DBPR). This rule, which addresses subpart H systems serving fewer than 10,000 people, is similar to the Interim Enhanced Surface Water Treatment Rule (IESWTR) established for larger systems.

Long Term 2 Enhanced Surface Water Treatment Rules (LT2ESWTR) A rule promulgated Jan. 5, 2006, by the US Environmental Protection Agency (71 FR 653–702) to reduce the possibility of disease incidence associated with *Cryptosporidium* and other pathogenic microorganisms in drinking water. The LT2ESWTR supplements existing regulations by targeting additional *Cryptosporidium* treatment requirements to higher-risk systems. This regulation also contains provisions to mitigate risks from uncovered finished water storage facilities and to ensure that systems maintain microbial protection as they take steps to reduce the formation of disinfection by-products. The LT2ESWTR applies to all systems that use surface water or groundwater under the direct influence of surface water.

maximum contaminant level (MCL) A value defined under the Safe Drinking Water Act Section 1401(3) as the maximum permissible level (concentration) of a contaminant in water delivered to any user of a public water system. Maximum contaminant levels are the legally enforced standards in the United States.

membrane biofilm reactor (MBfR) A biological wastewater treatment device or system that uses a membrane for solids–liquid separation and where the biomass is attached to the membrane surface rather than suspended in the process fluid. It is also called a membrane-immobilized biofilm system. This treatment technique has also been studied for the removal of perchlorate and nitrate from drinking water.

membrane filtration (MF) Filtration using a microfiltration or ultrafiltration membrane.

method detection limit (MDL) As defined by the US Environmental Protection Agency in 40 CFR Part 136 Appendix B for use in regulatory programs, the minimum concentration of a substance that can be measured and reported with 99 percent confidence that the analyte concentration is greater than zero. In practice, it is a value calculated from the absolute precision of replicate determinations. Many regulatory programs make use of the method detection limit. The concept and application of the method detection limit have been vigorously debated in recent years. The method detection limit is often mistakenly called the minimum detection limit.

microbiota Collectively, the microscopic animals and plants of a region or habitat.

microcystin Most frequently occurring toxin in surface water supplies produced by *Microcystis cyanobacteria*.

microcystin-LR One of more than 50 microcystins, toxins produced by some cyanobacteria (blue-green algae). Microcystin-LR is a hepatotoxin produced by the cyanobacterium Microcystis aeruginosa and is one of the most common of the cyanobacterial toxins found in water. It has been responsible for poisoning of both humans and animals. The proposed guideline value for this toxin in water in Canada is 0.0015 milligrams per liter; no proposed guideline value exists in the United States at this writing.

microorganism A microscopic organism, either plant or animal, invisible to the naked eye. Examples are algae, bacteria, fungi, protozoa, and viruses.

monomedia filtration A method of filtration in which particles are removed with a single media type. Monomedia filters are often deeper than conventional filters, up to 6 feet (1.8 meters), and use large diameter media (from 1.2 to more than 1.8 millimeters in effective size) to reduce head loss. The deeper monomedia can provide additional storage for collected particles. Often used with ozone, monomedia configurations use anthracite or granular activated carbon as a result of their larger particle size compared to sand. See also effective size; granular media.

moving bed A unique application of granular activated carbon in which a single activated carbon column can approach the efficiency of several columns in series. This is accomplished by the removal of spent activated carbon from one end of the granular activated carbon bed and the addition of fresh granular activated carbon at the other end with little or no interruption in the process.

National Pollutant Discharge Elimination System (NPDES) permit The regulatory agency document issued by either a federal or state agency that is designed to control all discharges of pollutants from point sources into US waterways. These permits regulate discharges into navigable waters from all point sources of pollution, including industries, municipal wastewater treatment plants, sanitary landfills, large agricultural feed lots, and return irrigation flows.

National Sanitation Foundation (NSF) International A noncommercial, not-for-profit agency located in Ann Arbor, Mich., and devoted to research, education, and service. Services provided include development of consensus standards, voluntary product testing and certification with policies and practices that protect the integrity of registered trademarks, education and training, and research and demonstration, all relating to public health and the environmental sciences. NSF International was formerly called the National Sanitation Foundation.

natural organic matter (NOM) A heterogeneous mixture of organic matter that occurs ubiquitously in both surface water and groundwater, although its magnitude and character differ from source to source. Natural organic matter contributes to the color of a water, and it functions as disinfection by-product precursors in the presence of such disinfectants as chlorine (HOCl). Humic substances (e.g., fulvic acid) represent a significant fraction of natural organic matter in surface water sources.

nitrification The process of formation of nitrate (NO_3^-) from reduced inorganic nitrogen compounds. Nitrification in the environment is carried out primarily by autotrophic bacteria and some chemoorganotrophic bacteria.

nitrifying bacteria Involved in the process of nitrification. The *Nitrosomonas* genus of bacteria oxidize ammonia (NH_3) to nitrite (NO_2^-). The *Nitrobacter* genus of bacteria oxidize nitrite to nitrate (NO_3^-). Both groups are chemolithotrophic organisms, found together in soils, that obtain energy for carbon dioxide (CO_2) fixation by aerobic oxidation of nitrite and nitrate. Some chemoorganotrophic bacteria—such as isolates of the genera *Vibrio, Bacillus, Mycobacterium,* and *Streptomyces*—can also oxidize ammonia to nitrite or nitrate.

N-nitrosodimethylamine [NDMA] A synthetic organic chemical that is a known animal carcinogen and is classified as a probable human carcinogen. It has been used in rocket fuels, as a solvent, and as a rubber accelerator. In addition, it may be formed during drinking water treatment by the reaction of secondary amines (R1–NH–R2), nitrite (NO_2^-), and acid.

oligotrophic Pertaining to water bodies that are poor in such nutrients as phosphorus, nitrogen, and calcium but have abundant oxygen at all depths. Such waters support little growth of plankton.

operations and maintenance (O&M) The ongoing process of carrying out activities necessary to fulfill the mission of an organization and to keep a system in such condition as to be able to achieve those objectives. Operations represent organized procedures for enabling a system to perform its intended function;

maintenance represents organized procedures for keeping the system (equipment, plants, facilities) in such condition that it is able to continue performing its intended function.

organic carbon compound A carbon-containing compound that is derived from living organisms, e.g., a hydrocarbon, alcohol, ether, aldehyde, ketone, carboxylic with organic substances. Others are synthetic organic chemicals.

oxidation reduction potential (ORP) The potential required to transfer electrons from the oxidant to the reductant, used as a qualitative measure of the state of oxidation in treatment systems. It is a dimensionless value. The more positive the value, the more oxidizing the solution; more negative values represent more reducing conditions. Aerobic waters in equilibrium with the atmosphere have a pE value of approximately +13. Anaerobic waters may have a pE of –4. Because electron activity varies with the pH of water, referring to graphs of pE as a function of pH is often useful. In this manner, the range of thermodynamically stable species can be illustrated.

oxygen demand (OD) The quantity of oxygen used in the oxidation of substances in a specified time, at a specified temperature, and under specified conditions.

ozonation The process of applying ozone (O_3) to a liquid for disinfection.

ozone-biofiltration An established method for treatment water containing humates and other undesirable substances. Under the influence of ozone as a strong oxidation agent, the organic carbon compounds are split and the concentrations of assimilable organic materials increase. (link here)

per- and polyfluoroalkyl substances (PFAS) Per- and polyfluoroalkyl substances (PFAS) are a group of man-made chemicals that includes PFOA, PFOS, GenX, and many other chemicals. Examples of where PFAS can be found include cleaners, textiles, leather, paper and paints, firefighting foams, and wire insulation.

pharmaceuticals and personal care products (PPCPs) A very broad, diverse collection of thousands of chemical substances, including prescription and over-the-counter therapeutic drugs, fragrances, cosmetics, sunscreen agents, diagnostic agents, nutraceuticals, biopharmaceuticals, and many others. They are potential health and environmental risks that generally enter the environment in treated wastewater effluent or terrestrial runoff.

pilot-scale testing Evaluation, on a scale larger than laboratory scale but smaller than full scale, of the amenability of water to treatment by particular operations or processes.

plasticizer An organic compound added to a high-molecular-weight polymer to facilitate processing, increase the flexibility, and increase toughness of the final product by internal modification of the polymer molecule. Among the more important plasticizers are nonvolatile organic liquids and low-melting solids (e.g., phthalate, adipate).

polisher A treatment stage placed at the end of other treatment to bring the water to a more highly conditioned and more perfect state, e.g., a mixed bed of ion exchange media installed as the final treatment step in the deionization process to remove last traces of undesirable ions.

polishing filter A filter installed for use after the primary water treatment stage to remove any traces of undesirable matter or to "polish" the water. See also polisher.

polymerase chain reaction (PCR) An assay or protocol that allows exponential amplification (cloning) of a specific deoxyribonucleic acid sequence from the original sequence (template).

potable reuse The reuse of reclaimed water for potable purposes.

powdered activated carbon (PAC) Activated carbon composed of fine particles and providing a large surface area for adsorption. Powdered activated carbon is typically added as a slurry on an intermittent or continuous basis to remove taste- and odor-causing compounds or trace organic contaminants and is not reused.

preoxidation Application of an oxidant before a water treatment step. Better usage is to specify where the oxidant is added, e.g., source water oxidation or prefiltration oxidation.

proportional–integral–derivative (PID) control Based on the deviation between the actual output and desired output according to the proportion of deviation, the integral of deviation, and the derivative of deviation.

quantitative (real-time) polymerase chain reaction (qRTPCR) A more sensitive technique that detects low-copy-number RNA and can be applied to detect, e.g., enteric viruses and viable microorganisms in water and to measure specific gene expression. (link here)

radical scavenger Any substance that is added or naturally present in a system or mixture and that consumes or inactivates free radicals. For example, during advanced oxidation processes, hydroxyl radicals can be efficiently trapped by the bicarbonate ion (HCO_3^-), which is a common component of natural waters. In some laboratory experiments, tert-butanol ($(CH_3)_3COH$) has been used as a hydroxyl radical scavenger.

recalcitrant Constituents that are not biodegradable or are only biodegradable with difficulty.

recarbonation The introduction of carbon dioxide (CO_2) into the water, after precipitative softening using excess lime for magnesium removal, to lower the pH of the water.

redox reaction A reaction in which electron transfers occur. Oxidation and reduction always occur simultaneously, and the substance that gains electrons (and is thus reduced) is termed the oxidant.

reverse osmosis (RO) A pressure-driven membrane separation process that removes ions, salts, and other dissolved solids and nonvolatile organics. The separation capability of the process is controlled by the diffusion rate of solutes through a membrane barrier and by sieving; it is dependent on the membrane type. In potable-water treatment, reverse osmosis is typically used for desalting, specific ion removal, and natural and synthetic organics removal.

riverbank filtration (RBF) A process of collecting water in an infiltration gallery located within a bank along a river to allow the river water to pass through the soil in the riverbank. Riverbank filtration provides particle removal, as well as partial removal of organic compounds.

Safe Drinking Water Act (SDWA) Public Law 93-523, enacted Dec. 16, 1974, establishing Title XIV of the US Public Health Service Act, codified generally as 42 US Code 300f-300j-11. It required the US Environmental Protection Agency to set national primary (health-related) drinking water regulations that were the first to apply to all public water systems, as defined by the act, in the United States.

saxitoxin A neurotoxin produced by planktonic algae (dinoflagellates primarily), including *Alexandrium tamarense*, *Gymnodinium catenatum*, and *Pyrodinium bahamense*. Filter-feeding mollusks ingest the algal species, thereby concentrating the toxins that they produce. When people consume these contaminated shellfish, they quickly succumb to the effects of the toxin, a syndrome known as paralytic shellfish poisoning (PSP). Symptoms of PSP are predominantly neurological and include tingling, burning, numbness, drowsiness, incoherent speech, and respiratory paralysis. Saxitoxin is reported to be one of the most toxic nonprotein substances known. The LD50 (lethal dose–50 percent) in mice is approximately 8 micrograms per kilogram, which suggests that a single dose of 0.2 milligrams would be lethal for the average-weight human.

secondary maximum contaminant level (SMCL) Nonmandatory water quality standards for 15 contaminants. These "secondary maximum contaminant levels" (SMCLs) are nonforceable. They are established as guidelines to assist public water systems in managing their drinking water for aesthetic considerations, such as taste, color, and odor. These contaminants are not considered to present a risk to human health at the SMCL.

sedimentation A treatment process using gravity to remove suspended particles.

soil-aquifer treatment The treatment achieved as reclaimed water passes through the soil vadose zone to an aquifer, and subsequently as it flows through the aquifer.

solids retention time (SRT) The average time the activated-sludge solids are in a treatment system.

spraying aerator An aerator consisting of a pressure nozzle through which water is propelled into the air in a fine spray. A spray aerator is also called a nozzle aerator.

standard operating procedure (SOP) A set of instructions or guidelines directing personnel to take certain actions under certain conditions.

steady-state analysis A pipe network analysis that assumes that consumptions, supplies, storage-tank levels, and certain other variables are unchanging for one specific point in time.

stoichiometry The mathematical and theoretical study of how elements combine in predetermined quantities to form compounds.

subfluidization flow rate The rate of a fluid passing up through a suspension such that the particles in the suspension remain in contact with one another.

substrate (1) A substance used by organisms for growth. (2) A substance acted upon by an enzyme. (3) The basic—and usually most abundant—component of a medium, or a sample of that component.

supervisory control and data acquisition (SCADA) A computer-monitored alarm, response, control, and data acquisition system used by drinking water facilities to monitor their operations.

Surface Water Treatment Rule (SWTR) The common name for the US Environmental Protection Agency regulation promulgated June 29, 1989, that set maximum contaminant level goals for *Giardia lamblia*, viruses, and *Legionella*, as well as National Primary Drinking Water Regulations for public water systems using surface water sources or groundwater under the direct influence of surface water. These regulations include (1) criteria under which filtration (including coagulation and sedimentation, as appropriate) are required and procedures by which the states are to determine which systems must install filtration, and (2) disinfection requirements.

taste and odor (T&O)

taste

- (1) To ascertain the flavor of a substance by taking a portion into the mouth; to perceive or recognize by the sense of taste. (2) The act of tasting. Taste is one of the special senses that perceives and distinguishes the sweet, sour, bitter, or salty quality of a dissolved substance. It is mediated by taste buds on the tongue. Strictly speaking, taste refers to the four basic tastes just discussed, whereas flavor refers to the blend of taste and smell sensations evoked by a substance in the mouth.

odor

- A quality of something that stimulates the olfactory organ (i.e., its smell). Many compounds have a characteristic odor that serves as an effective means of identification.

trace organic compound (TOrC) includes pharmaceuticals, personal care products, surfactants, pesticides, flame retardants, and others. Also known as Contaminants of Emerging Concern (CEC). (link here)

transmembrane pressure (TMP) (1) For a pressure-driven membrane process, the net driving force across the membrane. (2) For microfiltration and ultrafiltration with negligible osmotic pressure differential across the membrane, the hydraulic pressure differential from the feed side to permeate side.

trichloroethylene (TCE) ($CHCl=CCl_2$) A volatile organic compound that has been used as a metal degreaser, as a solvent, in dry cleaning, and as a fumigant. Trichloroethylene (also called trichloroethene) is regulated in drinking water by the US Environmental Protection Agency at a maximum contaminant level of 0.005 milligrams per liter.

trihalomethane (THM) (CHX_3) Any of numerous organic compounds named as derivatives of methane (CH_4) in which three halogen atoms (chlorine, bromine, iodine, singly or in combination) are substituted for three of the hydrogen atoms. Trihalomethanes (THMs) are formed during the disinfection of water with free chlorine. Because of their carcinogenic potential and other possible health effects, the presence of these compounds in drinking water is regulated by the US Environmental Protection Agency at a maximum contaminant level of 0.080 milligrams per liter for the sum of the four regulated THMs, referred to as total trihalomethanes (TTHMs).

turbidity (1) A condition in water caused by the presence of suspended matter, resulting in the scattering and absorption of light. (2) Any suspended solids that impart a visible haze or cloudiness to water and can be removed by treatment. (3) An analytical quantity, usually reported in nephelometric turbidity units, determined by measurements of light scattering. The turbidity in finished water is regulated by the US Environmental Protection Agency.

ultrafiltration (UF) A pressure-driven membrane process that separates submicron particles (down to a 0.01-micrometer size or less) and dissolved solutes (down to a molecular weight cutoff of approximately 1,000 daltons) from a feed stream by using a sieving mechanism that is dependent on the pore size rating of the membrane.

ultraviolet advanced oxidation processes (UV/AOP) A process used to oxidize target compounds in water by irradiating them with ultraviolet light, either with or without the addition of hydrogen peroxide (H_2O_2). The UV light causes double bonds in organic compounds to vibrate, eventually liberating electrons and changing the molecular structure. When H_2O_2 is used, UV light irradiation results in the formation of hydroxyl radicals (•OH), which are very strong oxidizing agents.

ultraviolet transmittance (UVT) A measure of the fraction of incident light transmitted through the water column. UVT is the ratio of the light entering the water to that exiting the water and is usually reported for a path length of 1 centimeter. (If an alternate path length is used, it should be specified.) UVT is often represented as a percentage and is related to the UV absorbance by the following equation: %UVT = 100 × 10-*A* (where *A* is the ultraviolet absorbance).

As the ultraviolet absorbance increases, UVT decreases.

unit filter run volume [UFRV] The volume of water that is processed through a filter in a single filter run between backwashes. The purpose of determining the unit filter run volume is to compare the throughput of a filter under different filtration rates. Although filter run times will decrease as filtration rate increases, the unit filter run volume may stay the same or increase because of the higher filtration rate. The unit filter run volume is expressed as volume/area/run (e.g., gallons/square feet/run).

volatile organic compound [VOC] A class of organic compounds that includes gases and volatile liquids. Many volatile organic compounds are used as solvents. Several of these compounds are regulated by the US Environmental Protection Agency.

Water Pollution Control Federation (WPCF) The former name of the Water Environment Federation.

Water Research Foundation (WRF) Formerly known as the AWWA Research Foundation (AwwaRF), a member-supported international nonprofit organization that sponsors research to enable water utilities, public health agencies, and other professionals to provide safe and affordable drinking water to the public. Its website is www.waterresearchfoundation.org.

Abbreviation & Acronym List

ANSI	American National Standard Institute
AOC	assimilable organic carbon
AOP	advanced oxidation process
ATP	adenosine triphosphate
AWWA	American Water Works Association
BAF	biologically active filters
BAT	best available technology
BDOC	biodegradable organic carbon
BDWTC	Biological Drinking Water Treatment Committee
BOM	biodegradable organic matter
CAT	Biofilter Conversion Assessment Tool
CCT	chlorine contact tank
CEB	chemically enhanced backwashing
CEC	compounds of emerging concern
CFSTR	continuous-flow stirred-tank reactor
COD	chemical oxygen demand
comammox	complete ammonia oxidation
CWA	Clean Water Act
DBP	disinfection by-product
> **THM**	trihalomethane
> **HAA**	haloacetic acid
DBPR	Disinfectants and Disinfection Byproducts Rule
DIC	dissolved inorganic carbon
DMS	dimethyl sulfide
DO	dissolved oxygen
DOC	dissolved organic carbon
DPR/ DPWR	direct potable water reuse
EDC	endocrine disrupting compound
EBCT	empty bed contact time
EDR	electrodialysis reversal
EfOM	effluent organic matter
EPA	Environmental Protection Agency
EPS	extracellular polymeric substance
FAT	full advanced treatment
FBR	fluidized-bed bioreactor
FBRR	Filter Backwash Recycling Rule
FXB	fixed-bed bioreactor
GAC	granular activated carbon
HDPE	high-density polyethylene
HLAR	head loss accumulation rate
HLR	hydraulic loading rate
HMI	Human Machine Interface
HPC	heterotrophic plate count
IESWTR	Interim Enhanced Surface Water Treatment Rule
IPWR	indirect potable reuse
IX	ion exchange

LT1ESWTR	Long Term 1 Enhanced Surface Water Treatment Rule
LT2ESWTR	Long Term 2 Enhanced Surface Water Treatment Rule
MBDEN	moving-bed denitrification
MBfR	membrane biofilm reactor
MCL	maximum contaminant level
MDL	method detection limit
MF	membrane filtration
MIB	2-methylisoborneol
MNE	MicroNiche Engineering
NDMA	*N*-Nitroso-dimethylamine
NOM	natural organic matter
NPDES	National Pollutant Discharge Elimination System
NSF	National Sanitation Foundation
O&M	operations and maintenance
OD	oxygen demand
ORP	oxidation-reduction potential
PCR	polymerase chain reaction
PFAS	per- and polyfluoroalkyl substance
PID	proportional-integral-derivative
qRTPCR	quantitative (real-time) polymerase chain reaction
RO	reverse osmosis
RBF	riverbank filtration
SDWA	Safe Drinking Water Act
SRT	solids retention time
SOP	standard operating procedure
SWTR	Surface Water Treatment Rule
T&O	taste and odor
TOrC	trace organic compound
TMP	transmembrane pressure
TCE	trichloroethylene
UF	ultrafiltration
UV/AOP	ultraviolet advanced oxidation processes
UFRV	unit filter run volumes
USEPA	US Environmental Protection Agency
UVT	ultraviolet transmittance
VOC	volatile organic compounds
WPCF	Water Pollution Control Federation
WRF	Water Research Foundation

Index

Note: *f* indicates figure; *t* indicates table

E

F

G

H

I

K

L

M

U

W

AWWA Manuals

M1, *Principles of Water Rates, Fees, and Charges*, #30001
M2, *Instrumentation and Control*, #30002
M3, *Safety Management for Water Utilities*, #30003
M4, *Water Fluoridation Principles and Practices*, #30004
M5, *Water Utility Management*, #30005
M6, *Water Meters—Selection, Installation, Testing, and Maintenance*, #30006
M7, *Problem Organisms in Water: Identification and Treatment*, #30007
M9, *Concrete Pressure Pipe*, #30009
M11, *Steel Pipe—A Guide for Design and Installation*, #30011
M12, *Simplified Procedures for Water Examination*, #30012
M14, *Backflow Prevention and Cross-Connection Control: Recommended Practices*, #30014
M17, *Fire Hydrants: Installation, Field Testing, and Maintenance*, #30017
M19, *Emergency Planning for Water and Wastewater Utilities*, #30019
M20, *Water Chlorination/Chloramination Practices and Principles*, #30020
M21, *Groundwater*, #30021
M22, *Sizing Water Service Lines and Meters*, #30022
M23, *PVC Pipe—Design and Installation*, #30023
M24, *Planning for the Distribution of Reclaimed Water*, #30024
M25, *Flexible-Membrane Covers and Linings for Potable-Water Reservoirs*, #30025
M27, *External Corrosion Control for Infrastructure Sustainability*, #30027
M28, *Rehabilitation of Water Mains*, #30028
M29, *Water Utility Capital Financing*, #30029
M30, *Precoat Filtration*, #30030
M31, *Distribution System Requirements for Fire Protection*, #30031
M32, *Computer Modeling of Water Distribution Systems*, #30032
M33, *Flowmeters in Water Supply*, #30033
M36, *Water Audits and Loss Control Programs*, #30036
M37, *Operational Control of Coagulation and Filtration Processes*, #30037
M38, *Electrodialysis and Electrodialysis Reversal*, #30038
M41, *Ductile-Iron Pipe and Fittings*, #30041
M42, *Steel Water-Storage Tanks*, #30042
M44, *Distribution Valves: Selection, Installation, Field Testing, and Maintenance*, #30044
M45, *Fiberglass Pipe Design*, #30045
M46, *Reverse Osmosis and Nanofiltration*, #30046
M47, *Capital Project Delivery*, #30047
M48, *Waterborne Pathogens*, #30048
M49, *Quarter-Turn Valves: Head Loss, Torque, and Cavitation Analysis*, #30049
M50, *Water Resources Planning*, #30050
M51, *Air Valves: Air-Release, Air/Vacuum and Combination*, #30051
M52, *Water Conservation Programs—A Planning Manual*, #30052
M53, *Microfiltration and Ultrafiltration Membranes for Drinking Water*, #30053
M54, *Developing Rates for Small Systems*, #30054
M55, *PE Pipe—Design and Installation*, #30055
M56, *Nitrification Prevention and Control in Drinking Water*, #30056
M57, *Algae: Source to Treatment*, #30057
M58, *Internal Corrosion Control in Water Distribution Systems*, #30058
M60, *Drought Preparedness and Response*, #30060
M61, *Desalination of Seawater*, #30061
M62, *Membrane Applications for Water Reuse*, #30062
M63, *Aquifer Storage and Recovery*, #30063
M65, *On-Site Generation of Hypochlorite*, #30065

M66, *Cylinder and Vane Actuators and Controls—Design and Installation,* #30066

M68, *Water Quality in Distribution Systems,* #30068

M69, *Inland Desalination and Concentrate Management,* #30069

M77, *Condition Assessment of Water Mains,* #30077

www.ingramcontent.com/pod-product-compliance
Ingram Content Group UK Ltd.
Pitfield, Milton Keynes, MK11 3LW, UK
UKHW052011041026
15590UKWH00008B/2275

9 781647 171957